Advances in Metabolic Profiling of Biological Samples

Advances in Metabolic Profiling of Biological Samples

Editor

Joana Pinto

MDPI • Basel • Beijing • Wuhan • Barcelona • Belgrade • Manchester • Tokyo • Cluj • Tianjin

Editor
Joana Pinto
Associate Laboratory i4HB,
UCIBIO-REQUIMTE,
Department of Biological
Sciences, Laboratory of
Toxicology
Faculty of Pharmacy of the
University of Porto
Porto
Portugal

Editorial Office
MDPI
St. Alban-Anlage 66
4052 Basel, Switzerland

This is a reprint of articles from the Special Issue published online in the open access journal *Metabolites* (ISSN 2218-1989) (available at: www.mdpi.com/journal/metabolites/special_issues/Metabolic_Profiling).

For citation purposes, cite each article independently as indicated on the article page online and as indicated below:

LastName, A.A.; LastName, B.B.; LastName, C.C. Article Title. *Journal Name* **Year**, *Volume Number*, Page Range.

ISBN 978-3-0365-7423-3 (Hbk)
ISBN 978-3-0365-7422-6 (PDF)

Contents

About the Editor

Joana Pinto

Joana Pinto is a Researcher at the Associate Laboratory Institute for Health and Bioeconomy (i4HB) and Applied Molecular Biosciences Unit (UCIBIO/REQUIMTE), Faculty of Pharmacy of the University of Porto, Portugal. With a BSc in Biochemistry (2008), an MSc in Biomolecular Methods (2010), and a PhD in Biochemistry (2015) from the University of Aveiro, she has extensive expertise in metabolic profiling using advanced analytical techniques such as NMR spectroscopy, GC-MS, and LC-MS, as well as bioinformatics analysis and interpretation of metabolomic datasets. Her research has focused on various areas, including biomarker discovery for prenatal disorders (gestational diabetes, fetal malformations and chromosomal disorders) and several cancers (glioblastoma, pancreatic, renal, prostate, bladder), investigating the impact of environmental chemicals on the metabolic profile of primary fish hepatocytes, and understanding the complexity of cork/wine interactions in collaboration with the cork industry. Her current research lines focus on applying metabolomics to 1) gain a better understanding of the mechanisms of toxicity of xenobiotics, and 2) personalised medicine in urological cancers aiming to predict metabolic responses to therapeutic options. She has led and participated in numerous research projects with competitive funding, including one as Co-Principal Investigator which explored the potential of GC-MS-based metabolomics for the non-invasive detection of urological cancers. Her research has been recognised with six awards and distinctions in prestigious national and international conferences.

Preface to "Advances in Metabolic Profiling of Biological Samples"

Metabolomics has been a powerful approach for studying low-molecular-weight metabolites and their interactions within a biological system in a wide range of research fields (e.g., clinical and biomedical research, toxicology, microbiology, nutritional, environmental). The biological samples analysed include blood serum/plasma, urine, tissues, cells, saliva, cerebrospinal fluid, and faeces. Due to the chemical diversity and concentration range of all metabolites present in biological samples, there are still several challenges from sample collection to metabolite annotation that need to be addressed. This Special Issue is dedicated to reviews and original articles covering the current methodological and technological advancements in the pre-analytical handling of biological samples, sample preparation protocols, analytical approaches for untargeted and targeted metabolic profiling, and metabolite annotation tools. As guest editor, I would like to thank all the authors for their remarkable studies, the peer reviewers, and the *Metabolites* Editorial Office for their support and contributions.

Joana Pinto
Editor

metabolites

Editorial

Advances in Metabolic Profiling of Biological Samples

Joana Pinto [1,2]

[1] Associate Laboratory i4HB, Institute for Health and Bioeconomy, Department of Biological Sciences, Laboratory of Toxicology, Faculty of Pharmacy, University of Porto, 4050-313 Porto, Portugal; jipinto@ff.up.pt
[2] UCIBIO–REQUIMTE, Department of Biological Sciences, Laboratory of Toxicology, Faculty of Pharmacy, University of Porto, 4050-313 Porto, Portugal

Citation: Pinto, J. Advances in Metabolic Profiling of Biological Samples. *Metabolites* **2023**, *13*, 534. https://doi.org/10.3390/metabo13040534

Received: 24 March 2023
Accepted: 6 April 2023
Published: 9 April 2023

Metabolomics constitutes a promising approach to clinical diagnostics, but its practical implementation in clinical settings is hindered by the requirement for rapid and efficient analytical methods. This Special Issue provides valuable insights into the development of novel sample preparation protocols and analytical methods for rapid metabolite analysis in biofluids and tissues. Specifically, a range of articles is presented [1–3], each addressing different aspects of this challenge. Campanella et al. [2] proposed a fast and straightforward method for the analysis of saliva by attenuated total reflectance Fourier-transformed infrared spectroscopy (ATR–FTIR) and Raman spectroscopy for large-scale preclinical studies. The effects of saliva collection and processing were investigated by vibrational spectroscopy and liquid chromatography (LC). This study proposed a novel method for saliva collection via the deposition of multiple spots onto low-cost polypropylene sheets, revealing reliable and reproducible ATR–FTIR spectra. Bordanaba-Florit et al. [1] developed a rapid (6 min runtime per sample) and sensitive method for the quantification of steroid hormone compounds (androgens, estrogens, progestogens, and corticoids) by liquid chromatography–high-resolution mass spectrometry (LC–HRMS). The performance of this methodology was tested in several rat tissues (adrenal glands, testis, prostate, liver, and brain), human urine, and urinary extracellular vesicles. Riccio et al. [3] developed a rapid (10 min) analytical method for the analysis of volatile organic compounds (VOCs) in the urine headspace using gas chromatography coupled with ion mobility spectrometry (GC–IMS). The method provided high sensitivity, yielded linearity at the ppb levels, and enabled the identification of 23 molecules (e.g., ketones, aldehydes, alcohols, and sulfur compounds) in a cohort of 115 urine samples. VOCs are molecules released as products of metabolic pathways in the human body, and alterations in their levels have been related to cancer [4,5].

Some authors have explored the ability to extract meaningful information from small sample volumes, which is especially important in clinical settings where sample availability can be limited. He et al. [6] developed a sample preparation method for simultaneously extracting non-polar and polar metabolites from limited amounts of mouse muscle tissues (5–50 mg dry weight). Overall, 109 lipids (e.g., oxylipins, fatty acids, and lysophospholipids) and 62 polar targeted metabolites (e.g., amino acids, sugars, and nucleotides) were successfully detected via ultra-performance liquid chromatography–tandem mass spectrometry (UPLC–MS/MS) and capillary electrophoresis–mass spectrometry (CE–MS). The influence of the sample isolation speed on metabolite stability revealed that rapid (<15 min) muscle tissue collection is crucial, particularly for more oxidative muscles. These findings will be critical for metabolomic mechanistic studies of sarcopenia (the age-related loss of muscle mass and function).

In addition to rapid analysis, the comprehensive characterization of biological samples is also essential. To this end, Bekhti et al. [7] propose superior analytical methods for the more holistic characterization of meconium, including the integration of two complementary LC–HRMS platforms. Overall, up to 229 polar and non-polar metabolites were successfully identified in human meconium at a high confidence level, belonging to amino acids, carbohydrates, nucleosides, and nucleotides, among other chemical classes. This methodology was applied to investigate the progressive evolution of the meconium metabolic profile in healthy newborns during the first three days of life.

Other articles have focused on the detection of specific metabolite classes [8–11], such as bile acids, sphingoid bases, fatty acids, and advanced glycation end products (AGEs), which have been implicated in many diseases. Zhang et al. [8] developed and validated a high-throughput method for the comprehensive analysis of bile acids in human and rodent fecal samples. Overall, 58 bile acids were quantified by ultra-high-performance liquid chromatography coupled with mass spectrometry (UPLC–MS). This analytical method was applied to identify and quantify BAs in end-stage renal disease patients. The profiling of bile acids is particularly important for gut-microbiome-related health studies since a complex interplay between bile acids and gut microbiota has been reported [12]. Morano et al. [9] compared different pre-analytical procedures, derivatization protocols, and chromatographic conditions for the quantification of sphingolipids in human plasma via liquid chromatography–tandem mass spectrometry (LC–MS/MS). The authors identified several critical steps for obtaining accurate results, such as single-phase extraction followed by an alkaline methanolysis, the choice of appropriate columns in order to efficiently separate complex sphingolipids and sphingoid bases, and the effectiveness of the derivatization procedure for solely non-phosphorylated species. Comprehensive two-dimensional gas chromatography coupled with mass spectrometry ($GC \times GC - MS$) can be a powerful tool to investigate saturated and unsaturated fatty acids in lipidomics studies, as shown by Bhatt et al. [10]. The authors optimized the derivatization and extraction of multiple classes of fatty acid methyl esters (saturated, monounsaturated, and polyunsaturated) in plasma without requiring in-depth MS/MS investigations. This method successfully distinguished boar-tainted and untainted pigs based on their serum fatty acid compositions. In another study, Yan et al. [11] established an untargeted HILIC–MS method for the comprehensive analysis of AGEs in biological samples (plasma, feces, and urine). The authors tested different columns and mobile phases in Maillard model systems. The proposed method revealed good reproducibility and AGE coverage in the presence of other endogenous metabolites from biological matrices. Elevated levels of AGEs have been associated with several pathologies, including cardiovascular disease, diabetes mellitus, cancer, and Alzheimer's disease [13].

The metabolite annotation process is a pivotal step in metabolomics and frequently represents a bottleneck in the discovery of biologically relevant metabolites (e.g., biomarkers). In this context, Renai et al. [14] undertook a novel investigation of the pertinence of Feature-Based Molecular Networking (FBMN) in combination with two novel, nutritionally relevant mass spectral libraries (~300 reference molecules) to increase the accuracy of metabolite annotation and explore the postprandial kinetics of the metabolites present in biological samples. This approach enabled the annotation of 67 berry-related and human-endogenous metabolites in the urine from individuals taking *Vaccinium* supplements, revealing similar or higher performance when compared with other annotation workflows. In addition, this tool linked several metabolite classes with phase II (early postprandial) and phase I (late postprandial) metabolism.

Finally, Viegas et al. [15] integrated measurements of glucose 6-phosphate, pentose phosphate pathway, and de novo lipogenesis fluxes in mice to provide a holistic assessment of hepatic glucose and fructose metabolism. The authors showed that a combination of deuterated water and [U-^{13}C]hexose sugar can quantify these fluxes in mice under natural feeding conditions through the ^{2}H- and ^{13}C-based nuclear magnetic resonance (NMR) analysis of liver glycogen and triglyceride. The results demonstrated that glucose 6-phosphate accounted for 40–60% of lipogenic acetyl-CoA and that 10–12% was oxidized by the pentose phosphate pathway. The NADPH produced from flux in the pentose phosphate pathway accounted for a minority (~30%) of the total de novo lipogenesis requirements. This study provides critical information for understanding hepatic sugar metabolism under pathological conditions.

Overall, this Special Issue provides pivotal methodological and technological advancements for the rapid and comprehensive analysis of metabolites in clinical settings. These advances are crucial for the translation of metabolomics into routine clinical practice and have the potential to revolutionize disease diagnosis and personalized medicine. As a Guest Editor, I would like to thank all the authors for their remarkable studies, the peer reviewers, and the *Metabolites* Editorial Office for their support and contributions.

Conflicts of Interest: The author declares no conflict of interest.

References

1. Bordanaba-Florit, G.; Liempd, S.; van Cabrera, D.; Royo, F.; Falcón-Pérez, J.M. Simultaneous Quantification of Steroid Hormones Using hrLC-MS in Endocrine Tissues of Male Rats and Human Samples. *Metabolites* **2022**, *12*, 714. [CrossRef] [PubMed]
2. Campanella, B.; Legnaioli, S.; Onor, M.; Benedetti, E.; Bramanti, E. The Role of the Preanalytical Step for Human Saliva Analysis via Vibrational Spectroscopy. *Metabolites* **2023**, *13*, 393. [CrossRef] [PubMed]
3. Riccio, G.; Baroni, S.; Urbani, A.; Greco, V. Mapping of Urinary Volatile Organic Compounds by a Rapid Analytical Method Using Gas Chromatography Coupled to Ion Mobility Spectrometry (GC–IMS). *Metabolites* **2022**, *12*, 1072. [CrossRef] [PubMed]
4. Pinto, J.; Amaro, F.; Lima, A.R.; Carvalho-Maia, C.; Jerónimo, C.; Henrique, R.; de Bastos, M.L.; Carvalho, M.; de Pinho, P.G. Urinary Volatilomics Unveils a Candidate Biomarker Panel for Noninvasive Detection of Clear Cell Renal Cell Carcinoma. *J. Proteome Res.* **2021**, *20*, 3068–3077. [CrossRef] [PubMed]
5. Lima, A.R.; Pinto, J.; Azevedo, A.I.; Barros-Silva, D.; Jerónimo, C.; Henrique, R.; de Lourdes Bastos, M.; Guedes de Pinho, P.; Carvalho, M. Identification of a biomarker panel for improvement of prostate cancer diagnosis by volatile metabolic profiling of urine. *Br. J. Cancer* **2019**, *121*, 857–868. [CrossRef] [PubMed]
6. He, Y.; van Mever, M.; Yang, W.; Huang, L.; Ramautar, R.; Rijksen, Y.; Vermeij, W.P.; Hoeijmakers, J.H.J.; Harms, A.C.; Lindenburg, P.W.; et al. A Sample Preparation Method for the Simultaneous Profiling of Signaling Lipids and Polar Metabolites in Small Quantities of Muscle Tissues from a Mouse Model for Sarcopenia. *Metabolites* **2022**, *12*, 742. [CrossRef]
7. Bekhti, N.; Castelli, F.; Paris, A.; Guillon, B.; Junot, C.; Moiron, C.; Fenaille, F.; Adel-Patient, K. The Human Meconium Metabolome and Its Evolution during the First Days of Life. *Metabolites* **2022**, *12*, 414. [CrossRef] [PubMed]
8. Zhang, X.; Liu, X.; Yang, J.; Ren, F.; Li, Y. Quantitative Profiling of Bile Acids in Feces of Humans and Rodents by Ultra-High-Performance Liquid Chromatography–Quadrupole Time-of-Flight Mass Spectrometry. *Metabolites* **2022**, *12*, 633. [CrossRef]
9. Morano, C.; Zulueta, A.; Caretti, A.; Roda, G.; Paroni, R.; Dei Cas, M. An Update on Sphingolipidomics: Is Something Still Missing? Some Considerations on the Analysis of Complex Sphingolipids and Free-Sphingoid Bases in Plasma and Red Blood Cells. *Metabolites* **2022**, *12*, 450. [CrossRef] [PubMed]
10. Bhatt, K.; Dejong, T.; Dubois, L.M.; Markey, A.; Gengler, N.; Wavreille, J.; Stefanuto, P.-H.; Focant, J.-F. Lipid Serum Profiling of Boar-Tainted and Untainted Pigs Using GC×GC–TOFMS: An Exploratory Study. *Metabolites* **2022**, *12*, 1111. [CrossRef] [PubMed]
11. Yan, Y.; Hemmler, D.; Schmitt-Kopplin, P. HILIC-MS for Untargeted Profiling of the Free Glycation Product Diversity. *Metabolites* **2022**, *12*, 1179. [CrossRef] [PubMed]
12. Cai, J.; Sun, L.; Gonzalez, F.J. Gut microbiota-derived bile acids in intestinal immunity, inflammation, and tumorigenesis. *Cell Host Microbe* **2022**, *30*, 289–300. [CrossRef] [PubMed]
13. Perrone, A.; Giovino, A.; Benny, J.; Martinelli, F. Advanced Glycation End Products (AGEs): Biochemistry, Signaling, Analytical Methods, and Epigenetic Effects. *Oxidative Med. Cell. Longev.* **2020**, *2020*, 3818196. [CrossRef]
14. Renai, L.; Ulaszewska, M.; Mattivi, F.; Bartoletti, R.; Del Bubba, M.; van der Hooft, J.J.J. Combining Feature-Based Molecular Networking and Contextual Mass Spectral Libraries to Decipher Nutrimetabolomics Profiles. *Metabolites* **2022**, *12*, 1005. [CrossRef]
15. Viegas, I.; Di Nunzio, G.; Belew, G.D.; Torres, A.N.; Silva, J.G.; Perpétuo, L.; Barosa, C.; Tavares, L.C.; Jones, J.G. Integration of Liver Glycogen and Triglyceride NMR Isotopomer Analyses Provides a Comprehensive Coverage of Hepatic Glucose and Fructose Metabolism. *Metabolites* **2022**, *12*, 1142. [CrossRef] [PubMed]

metabolites

Article

The Role of the Preanalytical Step for Human Saliva Analysis via Vibrational Spectroscopy

Beatrice Campanella [1], Stefano Legnaioli [1], Massimo Onor [1], Edoardo Benedetti [2] and Emilia Bramanti [1,*]

[1] Institute of Chemistry of Organometallic Compounds (ICCOM), Consiglio Nazionale delle Ricerche(CNR), 56124 Pisa, Italy
[2] Hematology Unit of Azienda Ospedaliero Universitaria Pisana (AOUP), 56100 Pisa, Italy
* Correspondence: bramanti@pi.iccom.cnr.it; Tel.: +39-050-315-2293

Abstract: Saliva is an easily sampled matrix containing a variety of biochemical information, which can be correlated with the individual health status. The fast, straightforward analysis of saliva by vibrational (ATR-FTIR and Raman) spectroscopy is a good premise for large-scale preclinical studies to aid translation into clinics. In this work, the effects of saliva collection (spitting/swab) and processing (two different deproteinization procedures) were explored by principal component analysis (PCA) of ATR-FTIR and Raman data and by investigating the effects on the main saliva metabolites by reversed-phase chromatography (RPC-HPLC-DAD). Our results show that, depending on the bioanalytical information needed, special care must be taken when saliva is collected with swabs because the polymeric material significantly interacts with some saliva components. Moreover, the analysis of saliva before and after deproteinization by FTIR and Raman spectroscopy allows to obtain complementary biological information.

Keywords: saliva; ATR-FTIR; sample processing; Raman

Citation: Campanella, B.; Legnaioli, S.; Onor, M.; Benedetti, E.; Bramanti, E. The Role of the Preanalytical Step for Human Saliva Analysis via Vibrational Spectroscopy. *Metabolites* **2023**, *13*, 393. https://doi.org/10.3390/metabo13030393

Academic Editor: Joana Pinto

Received: 10 February 2023
Revised: 27 February 2023
Accepted: 6 March 2023
Published: 8 March 2023

1. Introduction

Saliva is a matrix rich of biochemical information. The term "*salivaomics*" was introduced in 2008 to indicate the complexity and the importance of knowing the various "omic" constituents of saliva (https://iadr.abstractarchives.com/abstract/2008Dallas-100600/salivaomics-knowledge-base-skb, accessed on 27 February 2023). It is quite clear that whole-mouth saliva contains a variety of high- (proteins and nucleic acids) or low-molecular-weight compounds (salts, organic and inorganic acids, sugars, and nitrogenous bases.) and that its analysis might disclose clinically relevant information regarding the oral and systemic health status [1,2] (and references therein). Saliva collection is noninvasive and straightforward; it has high patient compliance, and it can be easily repeated [3,4]. For this reason, many biological and bioanalytical techniques (chromatographic and spectroscopic) have been developed in the last 15 years to investigate *salivaomics* through targeted and untargeted methods [5,6].

Attenuated total reflectance-Fourier transformed infrared spectroscopy (ATR-FTIR) is a nondestructive/microdestructive, fast, and cost-effective spectroscopic approach that requires in principle minimal sample handling to collect information from biological samples, tissues, cells, or biofluids

Several reviews report on the application of mid-infrared (IR) as a promising tool in human saliva [2,7–13]. The analysis of saliva as a diagnostic specimen by ATR-FTIR in tandem with chemometric analysis has experienced a rapid growth over the last decade, and even more in the last 2–3 years. In 1996, a new quantitative method based on transmittance FTIR was developed to evaluate thiocyanate concentrations in 5 μL of dried human saliva [14] using the band at 2058 cm^{-1}. More than 10 years later, Khaustova et al. developed an ATR-FTIR method to rapidly assess the biochemical properties of the saliva (total protein concentration, glucose, secretory immunoglobulin A, urea, amylase, cortisol, and inorganic phosphate) [15].

Recently, FTIR has been applied to study saliva from diabetic patients [16–21] and patients with oral pathologies [22,23] and to identify cancer biomarkers [4,24,25] and COVID-19-related biomarkers [26–29]. Recently, ATR-FTIR spectra in tandem with chemometric have been employed to analyze the spectral changes in semen, saliva, and urine in violent crimes during dry out, allowing to estimate their time since deposition [30].

Raman spectroscopy can yield complementary information to IR spectroscopy as the two techniques rely on different processes and selection rules. The inherently weak Raman signals of biological molecules, often overwhelmed by sample fluorescence, are counterbalanced by the fact that Raman spectra are mostly unaffected by water bands and exhibit sharper signals compared to IR. The application of Raman to saliva analysis was recently reviewed by Hardy et al. [31].

Although sample preparation in vibrational spectroscopy is minimal, several methodological features are critical to obtain reproducible, comparable spectra of saliva [8,13,32]. Thus, it is crucial to standardize the preanalytical step, including both saliva sampling [2,33] and sample preparation, to obtain time- and cost-effective procedures and to minimize sample handling and possible contaminations.

Saliva composition depends on the collection method, as well as on the nature and duration of salivation stimulation, subject hydration status, collection timing, etc. [2,33]. In many studies, vibrational spectra are acquired on dried samples, adopting a drying time variable between 3 min (directly drying the saliva sample onto the plate of the ATR device) and 24 h (after drying onto various supports for ATR and scattering analysis).

Table 1 summarizes the main works published in which FTIR spectroscopy was employed for the analysis of saliva, focusing on the brief descriptions of the preanalytical steps.

Table 1. Sample preparation for FTIR analysis.

Application	Preanalytical Step	Ref.
COVID-19 Positive patients vs. controls	3 µL saliva (sampling not specified) on the ATR crystal and dried at RT for 15 min.	[28]
Screening Test for COVID-19	WS (sampling not specified) deposited onto a transflection substrate, dried (10 min), and analyzed by ATR.	[34]
COVID-19 Positive patients vs. controls	5 µL saliva (sampling not specified) on aluminum foil and air-dried at RT overnight.	[27]
Diabetic patients vs. controls	50 µL of unstimulated WS by expectoration dried under vacuum on BaF_2 windows.	[35]
Diabetic patients vs. controls	3 µL saliva by spitting and dried at RT for 15 min on the ATR crystal.	[16,21,28]
Correlation FTIR spectra/surface tension; FTIR spectra/age and gender	50 µL WS (collected by spitting) on zinc selenide, dried at 37 °C for 60 min, and analyzed by ATR.	[36,37]
Burning mouth syndrome (BMS) vs. controls	30 µL WS (collected by spitting) on platinum, dried at 40 °C, and analyzed in diffuse reflectance mode.	[23]
Salivary gland tumor vs. controls	20 µL WS (collected by spitting) on zinc selenide, dried at RT, and analyzed by ATR.	[25]
Correlation FTIR spectra/biochemical composition	50 µL WS (collected by spitting) on zinc selenide, dried at 37 °C for 60 min, and analyzed by ATR.	[38]
Effects of saliva sample preparation	10 µL WS or saliva collected by spitting methods or cotton swab, dried as is or after centrifugation on germanium crystal or saliva concentrate after 4 h at 60 °C; analyzed by ATR.	[4]
Periodontitis vs. controls	50 µL WS collected by spitting, dried onto BaF_2, and analyzed by transmittance FTIR.	[39]
Diabetes and periodontitis vs. controls	WS collected by spitting, dried, and analyzed onto ATR crystal.	[20]
Salivary profile of athletes	WS collected by spitting; 1.5 mg of dried saliva analyzed onto FTIR-ATR crystal.	[40]

Table 1. *Cont.*

Application	Preanalytical Step	Ref.
Psoriasis status	30 µL WS collected by spitting deposited on a circular aluminum reflective surface, dried in hot air flow, and analyzed by transflectance FTIR.	[41]
Maximal Progressive Test in Athletes	100 µL WS collected by spitting, dried, and analyzed by ATR.	[42]
Folic Acid Deficient Pregnant Women vs. controls	WS collected by spitting deposited on TlBr crystal, dried, and analyzed by transmittance FTIR.	[43]
Physiological stress vs. controls	2 µL saliva (sampled by Salivette, Sarstedt) deposited on the ATR crystal, and dried at RT for 15 min.	[44]
Breast cancer patients vs. controls	Lyophilized saliva (sampled by Salivette, Sarstedt), dried, and analyzed by ATR crystal.	[24]
Oral Submucous Fibrosis (OSMF) vs. controls	3–5 µL saliva collected by Salivette, Sarstedt, dried, and analyzed by ATR.	[22]
Detection of SARS-CoV-2 Infection	WS collected by pharyngeal cotton swabs directly analyzed onto ATR crystal.	[26]

RT = room temperature; WS = whole saliva.

Basically, in all the works, the spectra were recorded on dried samples, i.e., after the removal of water. Water bands may indeed affect the sensitivity and reproducibility in the detection of several sample components, especially for IR.

In the last few years, our research group has extensively studied the salivary metabolites by liquid and gas chromatography approaches [45–50]. The analysis of saliva by ATR-FTIR and Raman provides complementary, fast, and holistic information on the sample, which includes low-molecular-weight (MW) metabolites and (macromolecules proteins, carbohydrates, and lipids), both having a high diagnostic value for local and systemic disorders.

The aim of this work is to investigate the effect of saliva sampling (spitting method or sampling with commercial polymeric swab) on the vibrational spectra (ATR-FTIR and Raman) acquired before and after deproteinization with two methods (protein precipitation with ethanol or using 3 kDa cut-off centrifugation units). The spitting method may indeed simplify the sampling, meeting patient compliance (especially for children) and reducing costs and risks. Saliva contains about 0.1–1.5 mg/mL protein [51], and the saliva deproteinization may simplify the spectral information, allowing the analyst to focus on the analytical window of interest. In all cases, information remains complex, and the coupling with chemometrics is crucial to extract information from the vibrational spectra. An easy "printing" of sample dried spots (SDSs) prepared on polypropylene (PP) sheets onto ATR crystal is described for the fast, interference-free acquisition of FTIR spectra.

Our work implements the information recently reported by Paschotto et al. [4], who investigated ATR-FTIR absorption of saliva sampled with different collection methods (spitting method vs. soaking) and processing protocols (dried unprocessed, dried supernatant after centrifugation, and dried concentrate), confirming the need of standardized collection–processing protocols based on the biochemical component analysis. Paschotto et al. investigated the effects of sampling using cotton swabs, and they applied centrifugation conditions at low *g* values, probably removing cells and bacteria. They did not investigate the deproteinization effect, nor were both FTIR and Raman spectroscopy used. In our work, the concentrations of the main metabolites in saliva after the various sample handling procedures were also determined by RP-HPLC-DAD [49] to focus on the possible artefacts of saliva sampling and sample handling.

2. Experimental Design

2.1. Chemicals

Sulfuric acid for HPLC analysis was employed (V800287 VETEC ≥ 85% Sigma-Aldrich, Milan, Italy). Methanol for RP-HPLC was purchased from Carlo Erba (Rodano, Italy). Preparation and dilution of samples and solutions were performed gravimetrically using ultrapure MilliQ water (18.2 MΩ cm^{-1} at 25 °C, Millipore, Bedford, MA, USA). Standard solutions for HPLC (TraceCERT®, 1000 mg/L in water) were purchased from Sigma-Aldrich, Milan, Italy. Analyte stock and working solutions were prepared as previously reported [49].

2.2. Experimental Design: Saliva Sample Collection and Processing

Whole, nonstimulated saliva samples were collected from 10 nominally healthy volunteers. The study was performed in accordance with the Declaration of Helsinki. Written informed consent was obtained from all volunteers who agreed to provide saliva samples. A fasting period of at least 8 h was required, and volunteers did not brush or rinse the oral cavity with mouthwash before sampling. Exclusion criteria included the existence of any oral disease or a systemic pathology, alcohol consumption, smokers, or systemic medication usage. The pattern of samples analyzed was the following: The volunteers were asked to spit into sterile polypropylene tubes (about 2 mL for each subject). Saliva samples were pooled, homogenized in vortex, and stored in a freezer at −20 °C. For the analysis, pooled saliva was thawed at room temperature and subdivided into two processing groups: one half ("salivette" in this work) was loaded onto Salivette® swabs (2 mL/swab) for 5 min as physiological time for the adsorption of the whole saliva, centrifuged at 4500× *g* for 10 min at 4 °C (Eppendorf™ 5804R Centrifuge), and pooled again. Second half was used as is (unprocessed saliva, "saliva" in this work). This procedure was chosen to perform the methodological comparison exactly on the same sample, avoiding changes in saliva composition due to presence of the swab.

Both saliva and salivette samples were fractionated in three parts: (i) a part was analyzed as is (named saliva and salivette); (ii) a part was deproteinized by ultracentrifugation (30 min) using Microcon® Centrifugal Filters with cut-off 3 kDa (Merk, Milan, Italy) (named saliva_CO and salivette_CO); (iii) a total of 100 µL of saliva or salivette was mixed with 900 µL ethanol (EtOH) (10-fold dilution), cooled at −20 °C for 2 h, and centrifuged at 14,000 rpm (10,000× *g*) for 30 min in a refrigerated centrifuge (named saliva_EtOH and salivette_EtOH). The solution remaining in the upper part of 3 kDa cut-off filtering units was also analyzed by ATR-FTIR to characterize the HMW compounds ("HMWsaliva_CO" and "HMWsalivette_CO").

2.3. ATR-FTIR Analysis

Five drops (50 µL each) of sample were deposited onto a polypropylene (PP) sheet by a micropipette (Eppendorf Research Plus pipette, Eppendorf AG) and air-dried at room temperature overnight. Spectra were recorded in ATR mode on sample dried spots (SDSs) using a Frontiers FTIR spectrometer (Perkin Elmer, Milan, Italy), equipped with a diamond-attenuated total reflectance (ATR) sampling accessory. The flat sample press tip (2 mm diameter) was employed to "stamp" the sample from the SDSs (Figure 1). After this, the PP sheet was removed. The microamount "printed" on the ATR diamond window was enough to obtain reliable and reproducible spectra. Using this method, at least 3 spectra can be recorded from 3 different areas of one single SDS. Spectra were recorded in 4000–600 cm^{-1} spectral range with a 4 cm^{-1} resolution, with 32 scans for the background and the sample. For each analysis, the diamond sampling window and the sample press tip were cleaned with 70% ethanol v/v. Mid-infrared (MIR) spectra were acquired on 3–5 different SDSs. Saliva_EtOH and salivette_EtOH sample spectra were acquired after the deposition of 3 µL of the samples directly onto the ATR crystal as ethanol evaporates in less than 15 s. HMWsaliva_CO and HMWsalivette_CO samples were analyzed by wiping (w) the tip wetted with the sample onto ATR crystal (samples dried in less than 15 s) or by "printing" (p) from SDSs.

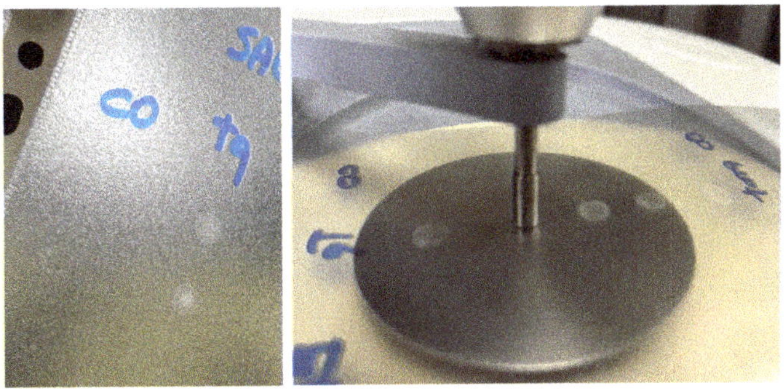

Figure 1. Saliva sample dried spot (SDS) from 50 μL deposition onto PP sheet and "printing" on ATR-FTIR crystal.

2.4. Raman Analysis

Five drops (10 μL each) of sample were deposited onto a glass slide covered with an aluminum foil and air-dried at room temperature overnight. Spectra were recorded with a Renishaw inVia confocal micro-Raman system, coupled with an optical Leica DLML microscope equipped with an NPLAN objective 50×. The laser sources were a diode laser with a wavelength of 785 nm and an He–Ne laser with a wavelength of 633 nm. The spectrometer consisted of a single-grating monochromator (1200 or 1800 lines mm^{-1} according to the selected laser wavelength), coupled with a CCD detector, a RenCam 578 × 400 pixels (22 μm × 22 μm) cooled by a Peltier element. The spectral calibration of the instrument was performed on the 520.5 cm^{-1} band of a pure silicon crystal. Spectra were acquired with 633 nm laser source at 5.5 mW and with 785 nm laser source at 40 mW, 5 accumulations of 10 s each.

2.5. RP-HPLC-DAD Analysis

Saliva, salivette, saliva_CO, and salivette_CO samples were 5-fold diluted in 5 mM sulfuric acid, filtered using a 0.20 μm RC Mini-Uniprep (Agilent Technologies, Milan, Italy) filter, injected in the HPLC system (V_{inj} = 5 μL), and analyzed as previously reported [49]. Saliva_EtOH and salivette_EtOH were directly injected in the HPLC system (V_{inj} = 5 μL).

2.6. Data Processing

Principal component analysis (PCA) was carried out on the mean-centered column-wise spectra to investigate possible clustering of samples. ATR spectra were standardized by using standard normal variate (SNV) to minimize unwanted contributions (e.g., global intensity effects or baseline shifts). Raman spectra were treated to remove cosmic rays, and then Savitzky–Golay (zero-order derivative, third-degree polynomial order, and a window size equal to 9 data points) and Asymmetric Least Squares algorithms were applied for smoothing and baseline correction, respectively.

The analysis was performed with the open-source Chemometric Agile Tool (CAT) program (http://www.gruppochemiometria.it/index.php/software/19-download-the-r-based-chemometric-software, accessed on 27 February 2023) and by a tailored in-house R-script (R version 3.6.3 (R Development Core Team 2012) and R-Studio, Version 1.1.463) using the R-package mdatool.

3. Results and Discussion

3.1. ATR-FTR Analysis of Saliva/Salivette Dried Spots: Effect of Deproteinization Method

ATR-FTIR spectra were recorded on microspots "printed" from the dried spots on the ATR diamond window. The flat sample press tip (2 mm diameter) was employed to "stamp" the sample from the dried spots. After this, the PP sheet was removed. This procedure, not previously reported, allows in principle to prepare samples quickly onto a low-cost support and to obtain reliable and reproducible spectra using a microamount of sample. Using this method, at least three spectra can be recorded from three different areas of one single dried spot obtained from 50 µL.

Figure 2 shows a representative ATR-FTIR spectrum of a saliva dried spot. Figure 3 shows the spectra of all the analyzed samples before and after SNV normalization. The absorption bands of lipids, proteins, carbohydrates, and nucleic acids are evidenced. The IR spectrum of saliva is in fact a superposition of the absorption spectra of all these components in proportion to their concentration, following the Lambert–Beer law.

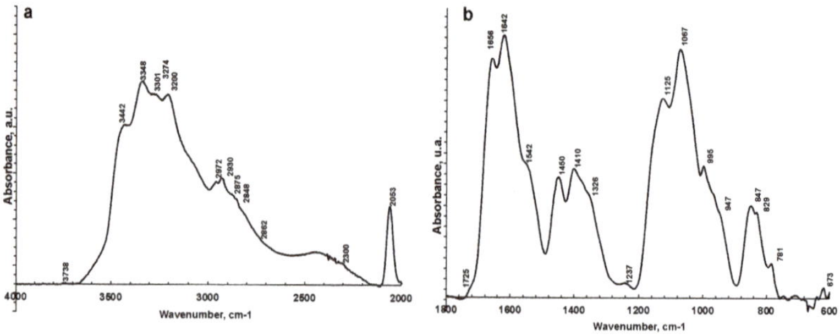

Figure 2. Representative ATR-FTIR spectra of saliva analyzed as is in 4000–2000 cm^{-1} (**a**) and 1800–600 cm^{-1} regions (**b**).

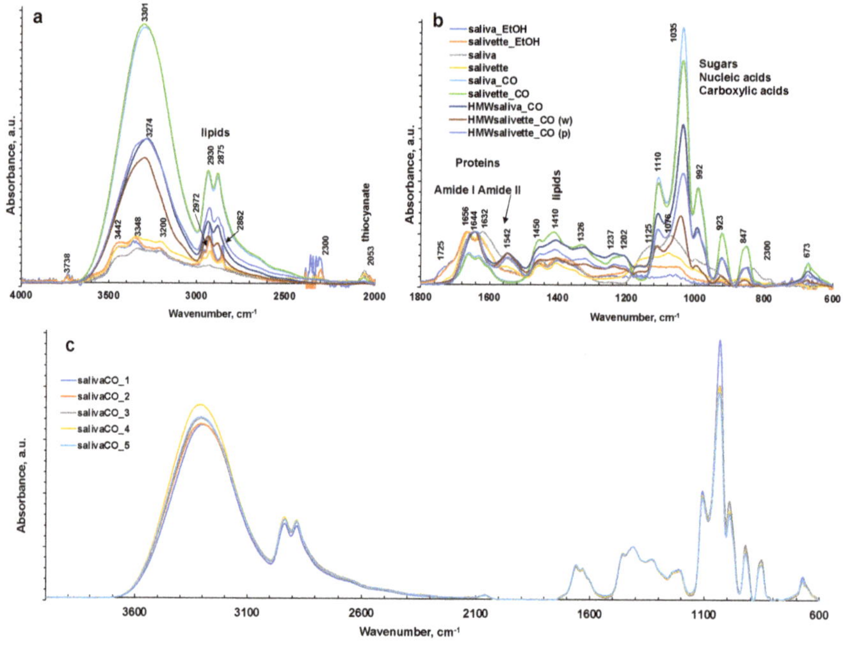

Figure 3. ATR-FTIR spectra of all representative saliva samples analyzed before and after deproteinization with ethanol (EtOH) and ultrafiltration with 3000 Da cut-off (CO) in 4000–2000 cm^{-1} (**a**) and 1800–600 cm^{-1} regions (**b**). (**c**) ATR-FTIR of N = 5 replicates of saliva sample after ultrafiltration with 3000 Da cut-off as example of reproducibility of the spectra. HMWsaliva_CO and HMWsalivette_CO refer to high-molecular-weight (HMW) compounds remaining in the upper part of 3 kDa cut-off filtering units (w = wiping, p = printing) as explained in the experimental part.

The sampling and the deproteinization method employed evidenced major changes in the FTIR spectra of dried spots in the 1750–600 cm^{-1} fingerprint region and in the N–H and OH stretching regions (3800–1600 cm^{-1}) and overlaid the latter in the region of C–H stretching in CH_2 and CH_3 (3000–2850 cm^{-1}).

The FTIR spectrum of almost all samples examined showed the characteristic FTIR features of biological samples: the peaks of proteins at 1656–1642 cm^{-1} (Amide I, C=O stretching), 1542 cm^{-1} (Amide II, N–H bending), and 1237 cm^{-1} (Amide III); nucleic acids (1100–850 cm^{-1}); P=O asymmetrical and symmetrical stretching vibrations of PO_2 phosphodiester groups from phosphorylated molecules (1125 cm^{-1} and 1076 cm^{-1}); and C–O stretching vibration coupled with C–O bending of the C–OH groups of carbohydrates (including glucose, fructose, and glycogen) at 1035 cm^{-1}. The absorptions typical of proteins (Amide I, II, and III) were not observed in the saliva_CO and salivette_CO samples, i.e., after deproteinization by 3 kDa cut-off filtering. The spectral region 1080–950 cm^{-1} also includes the sugar moieties of glycosylated proteins, (e.g., salivary amylase and mucins). Several authors report the assignment of specific bands in the fingerprint region to immunoglobins (1560–1464 cm^{-1} associated to IgG, 1420–1289 cm^{-1} and 1160–1028 cm^{-1} related to IgM, and 1285–1237 cm^{-1} designed to IgA) [28]. However, the salivary proteome is a complex protein mixture resulting from the activity of salivary glands and serum, from mucosal and/or immune cells, or from micro-organisms containing amylase (representing about 20% of total proteins), mucins (about 20%), 6% human serum albumin, 10% lysozyme, 10% IgA and IgG, lactoferrin, proline-rich proteins, histatins, cathelicidins, defensins, glycoproteins, lipoproteins, statherin, and matrix metalloproteases [2,52,53]. Human saliva contains also inorganic compounds (sodium, potassium, calcium, magnesium, chloride, and phosphate) and organic nonprotein components, such as bilirubin, creatinine, glucose, lactic and uric acids [2], and references therein.

The differences among the various sample groups, corresponding to different saliva preparation modes, were better evidenced, and the information from the spectra were extracted using principal component analysis (PCA). The results derived from the PCA on the FTIR spectra are shown in the PC1–PC2 score plots (Figure 4a), explaining 87.8% of the total variance. PC1 is responsible for the separation of samples deproteinized using 3 kDa cut-off, which show positive values of PC1 (Figure 4b, blue line) with respect to the other samples on the left side of the plot. Interestingly, the HMWsaliva_CO and HMWsalivette_CO samples (MW > 3 kDa) cluster between unprocessed samples and saliva_CO/salivette_CO samples, without significant differences if analyzed by wiping the tip onto ATR crystal (w) or by "printing" from dried spots (p). PC2 (Figure 4b, red line) separates all samples treated with EtOH that show positive values of PC2 with respect to all the others. Figures S1 and S2 show the PC1–PC3 and PC2–PC3 scores (a) and loading plots (b), explaining 67.2% and 30.9% of the total variance, respectively.

The PC1 loading plot (Figure 4b, blue line) clearly shows positive values of 4000–3100 cm^{-1} absorptions related to OH and NH stretching vibrations, negative values of Amide I and Amide II bands typical of proteins due to C=O and C–N stretching vibrations, respectively, of the bands assigned to unsaturated C=CH stretching of lipids (at 3000 cm^{-1}), symmetric -CH_3 stretching at 2922 cm^{-1} due primarily to proteins, and symmetric -CH_2 stretching at 2854 cm^{-1} due to lipids and proteins, and bending (at 1450 and 1378 cm^{-1}) of the CH_2 and CH_3 groups. In the region of 3600–2900 cm^{-1}, the absorption bands of the primary and secondary amines (-NH_2 and -NHR) are observed; the peaks at 3300–3200 cm^{-1} are assigned to O–H vibrations; N–H stretching is typically around 3364–3517 cm^{-1} and usually show a medium, somewhat broad signal (usually considerably less broad than a typical OH stretching). The positive values of PC1 at 3200–3300 cm^{-1} reflect the higher contents of water in all saliva and salivette samples after deproteinization with 3 kDa units. Another important region of the FTIR spectrum is the spectral range 1180–800 cm^{-1} that originates from various C–C/C–O stretching vibrations in sugar moieties, P–O stretching of phosphate groups in phosphorylated proteins, and nucleic acids and low-MW compounds. The 1032 cm^{-1} band is usually attributed to the C–O stretching vibration in glycogen, while lactic acid has peaks at 1032 and 916 cm^{-1}. Thus, the absorptions of low-MW metabolites in saliva/salivette spectra after 3 kDa cut-off ultrafiltration characterize PC1 components. The negative value in PC1, for these samples, of Amide I (1666–1622 cm^{-1}) and Amide II bands (1556 cm^{-1}), typical of proteins, also indicates that ultracentrifugation using 3 kDa cut-off is the only effective method for saliva deproteinization. The negative bands at 1137, 1078, 950, and 830 cm^{-1} of PC1 could be due to the removal of high-MW carbohydrates and nucleic acids from the saliva and salivette samples after cut off or the removal of phosphorylated molecules. The typical absorptions of high-MW compounds that characterize saliva and salivette samples are better evidenced in the negative components of PC3 (Figures S1 and S2, green line).

The PC2 loading plot shows remarkable positive values peaking at 3736, 3461, 3397 cm^{-1}, 3022sh, 2962, 2926, 2878sh, and 2857 cm^{-1}, characteristic of lipids. Positive values are also observed at 1750, 1719, and 1687 cm^{-1} and assigned to the C=O ester groups of lipids and cortisols and C=C stretching of cholesterol. These components are responsible for the clustering of the saliva_EtOH and salivette_EtOH samples. Among low-MW saliva components detected by FTIR, cortisol, phosphates, lactic acid, and urea are of interest from a medical point of view because their concentrations vary during physiological stress [44]. Our results suggest that the deproteinization in ethanol is not effective, in agreement with Araki, who reported that ethanol mostly precipitates non-protein nitrogen [54]. Table 2 shows with more detail the principal assignment of saliva MIR absorptions [7,10].

Negative values of the PC2 loading plot are observed at 1553, 1450, 1403, and 1321 cm^{-1}. The differences between the saliva and salivette samples mainly rely on marked negative peaks of PC2 (Figure 4b), i.e., the absorptions at 1553 cm^{-1} (amide II), 1042 with shoulders at 1137 and 1018 cm^{-1}, and 849 cm^{-1}. These absorptions, typical of C–O–C symmetric and asymmetric vibrations of sugar moieties of heavily glycosylated proteins (e.g., mucins [31]) (Table 2), let us hypothesize that the polymeric swab (Salivette®) may adsorb proteins characterized by HMW and/or high degrees of glycosylation.

Table 2. Principal Mid-Infrared (MIR) Bands of the Dataset and Chemical Assignments [7,10].

MIR Frequency	Band Tentative Assignment
~3736 cm^{-1}	stretching O–H
~3346 cm^{-1}	stretching N–H
~3275 cm^{-1}	stretching O–H symmetric
~3200–3550 cm^{-1}	symmetric and asymmetric vibrations attributed to water
~2968 cm^{-1}	CH3 stretching
~2930 cm^{-1}	stretching C–H
~2800–3000 cm^{-1}	C–H lipid region
~2100 cm^{-1}	combination of hindered rotation and O–H bending (water)
~1750 cm^{-1}	lipids: ν(C=C)
~1650 cm^{-1}	amide I: ν(C=O)
~1637 cm^{-1}	H–O–H scissoring
~1550 cm^{-1}	amide II: δ(N–H) coupled to ν(C–N)
~1450 cm^{-1}	methyl groups of proteins: δ CH2 and CH3 asymmetric
~1400 cm^{-1}	methyl groups of proteins: δ CH2 and CH3 symmetric
~1392–1396 cm^{-1}	fatty acids and polysaccharides
~1250–1260 cm^{-1}	amide III: ν(C–N)
~1155 cm^{-1}	carbohydrates: ν(C–O)
~1225 cm^{-1}	DNA and RNA: νas(PO2−)

Table 2. *Cont.*

MIR Frequency	Band Tentative Assignment
∼1080 cm^{-1}	DNA and RNA: νs(PO2−)
∼1030 cm^{-1}	glycogen vibration: νs(C−O)
∼992–986 cm^{-1}	ribose phosphate main chain and stretching vibration C–C of DNA backbone
∼971 cm^{-1}	nucleic acids and proteins: n(PO4)
∼960−966 cm^{-1}	C–O, C–C, deoxyribose νs = symmetric stretching; νas = asymmetric stretching; and δ = bending.

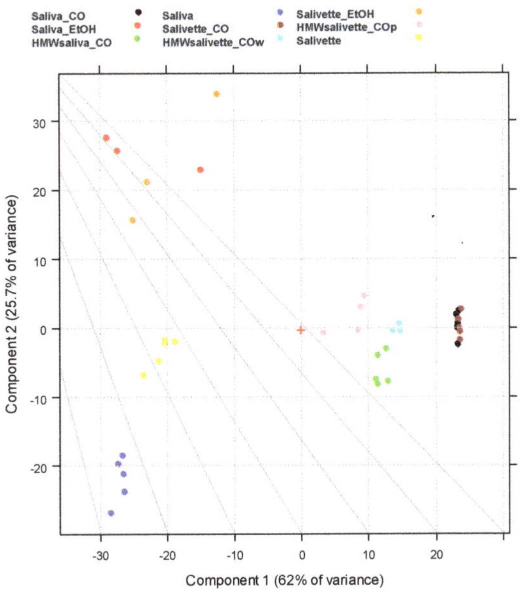

(**a**)

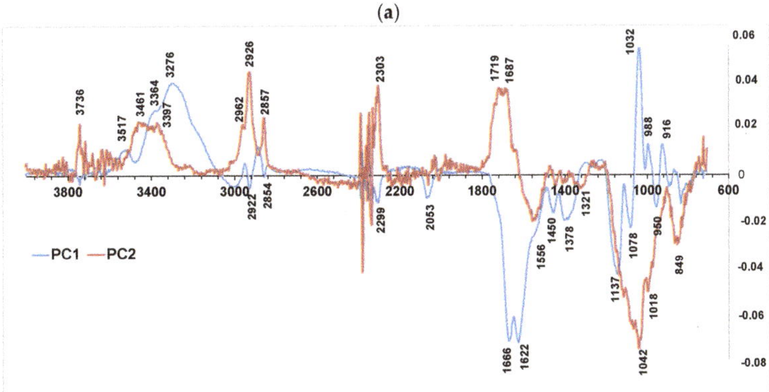

(**b**)

Figure 4. PCA results of SNV-normalized and centered ATR-FTIR spectra of saliva samples. (**a**) Score plot (87.7% of total variance); (**b**) loading plot of PC1 (blue line) and PC2 (red line). HMWsaliva_CO and HMWsalivette_CO refer to high-molecular-weight compounds (HMWCs) remaining in the upper part of 3 kDa cut-off filtering units (w = wiping, p = printing) as explained in the experimental part.

3.2. Choice of PP Support and Effect of Dried Spot Volume

Fifty μL was the optimized volume for the analysis of dried spots by FTIR that allowed to obtain "printed mini-spots" of suitable thickness to record high-quality FTIR spectra. If a smaller amount of sample is available for the analysis, e.g., 10 μL, the sample can be dried on PP and eventually gently scratched and microamounts analyzed by ATR-FTIR without significant changes in the spectra. The same experimental design performed on dried spots drop-casted onto aluminum foil did not gave satisfying, reproducible results likely because of the irregular thickness of the saliva dried spots or the rigidity of the aluminum foil. The good reproducibility of the saliva dried spots obtained on PP support may be also due to the hydrophobicity of the PP sheet itself. The ATR-FTIR measurements directly performed on the dried spots onto PP or aluminum foil have interference bands (data not shown for brevity) of the support employed unless higher volumes (≥50 μL) were used to obtain films of suitable thickness.

3.3. HPLC Analysis of Main Metabolites in Saliva/Salivette Samples

The concentrations of the main metabolites in saliva after the various sample handling procedures were determined by RP-HPLC-DAD [49]. Figure 5 shows the comparison of the concentration (mean and SD) of seven main metabolites determined in the saliva/salivette samples before and after deproteinization with 3 kDa cut-off filtration. The injection of the saliva_EtOH and salivette_EtOH samples did not give meaningful results likely because the precipitation in ethanol favors reaction/degradation of LMW metabolites (e.g., the decrease in the peak of uric acid and the increase in an unassigned peak at t_R = 4.348 min) and the disappearance of the peaks of pyruvic acid, valine (VAL), lactic acid, and propionic acid (Figure S3a,b). Figure S3c shows, as an example, UV/visible spectra of the peak at t_R = 5269 min (orange line) of the saliva_CO sample, which is due to uric acid, and UV/visible spectra of the peaks at t_R = 5.2599 (purple line) and 4.35 min (blue line) of the saliva_EtOH sample. Both these peaks have the absorption characteristics of uric acid, but only the peak at 5.2599 has the same retention time of uric acid standard solution.

The results show that for most of the metabolites the sampling by spitting or by swab does not affect their quantitation (lactic, propionic, uric acids, and valine). For other metabolites (creatinine and pyruvic acid), the salivette swab seems to partially adsorb the analyte. The filtering with cut-off filtration units instead does not affect their quantitation.

3.4. Raman Analysis on Saliva Dried Spots

Raman spectra were acquired from saliva dried spots on PP, glass, and aluminum foil-covered glass. The signals of PP strongly interfere with the analysis, while the spectra collected from samples onto glass were characterized by a poor S/N ratio. The deposition onto aluminum, as verified also by Bedoni and coworkers [55], is rather correlated with well-defined Raman bands, which are easily associable to the vibrational signatures of several biomolecules. Figure 6 shows the comparison of Raman spectra acquired at 785 nm of saliva before (Figure 6a) and after (Figure 6b) filtering with 3 kDa filters.

The characteristic features of proteins are clearly recognizable in the spectra of both saliva and salivette, dominating the investigated spectral region. In the spectra obtained after the cut-off at 3 kDa, the only signals related to proteins are the out-of-ring breathing of tyrosine (824 cm^{-1}), the C–C stretching of the proline ring (926 cm^{-1}), the C–C stretching of the protein β-sheet (978 cm^{-1}), and the band of Amide III (centered at 1255 cm^{-1}). Saliva treatment with filters to remove large biomolecules is thus necessary in Raman spectroscopy to obtain information from smaller metabolites. Protein precipitation with EtOH, instead, gives Raman spectra with high noise and low-intensity signals, and no reliable information could be deduced from them.

The PCA was applied to the preprocessed dataset acquired at 785 nm, obtaining a 95.6% of variance explained by the first two PCs (Figures S4 and S5). Saliva and salivette spectra cluster together and are clearly separated from the other samples along PC2. It appears, thus, that the Salivette® swab does not retain/release any compound at a significant concentration for Raman. The spectra of saliva_CO and salivette_CO are separated along PC1, while they appear indistinguishable along PC2, and a detailed analysis of the spectra revealed that salivette_CO samples show Raman signals at a lower intensity with respect to those of saliva_CO. As would be expected, the samples treated with EtOH form a close-packed cluster separated from the other groups.

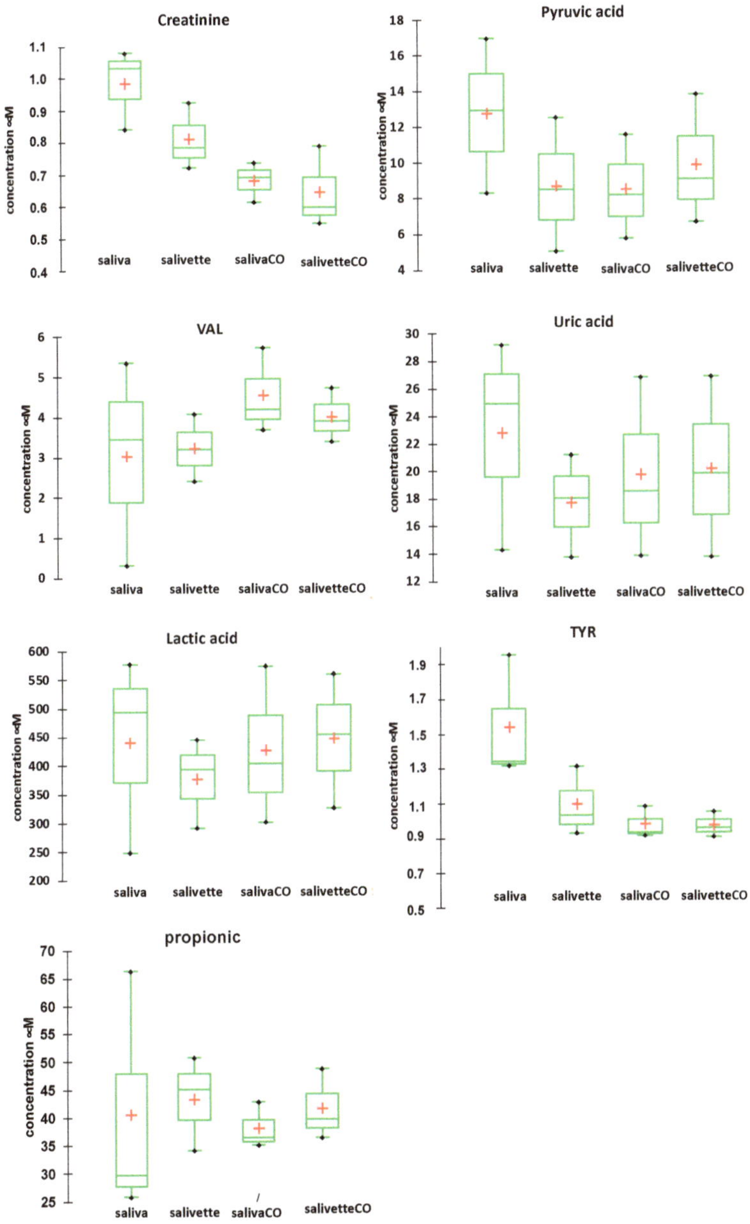

Figure 5. Box plots of the main metabolites determined in saliva/salivette samples before and after deproteinization with 3 kDa cut-off filtration. Red cross = mean value; black dots = minimum/maximum value; box = 1st quartile–3rd quartile range; bar = standard deviation.

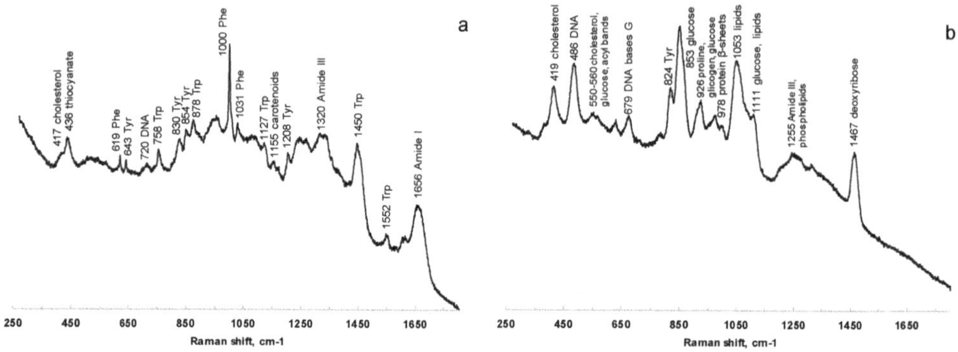

Figure 6. Comparison of Raman spectra at 785 nm of saliva before (**a**) and after (**b**) filtering with 3 kDa filters.

Spectra acquisition with a laser in the visible range is further complicated by molecular fluorescence. Specifically, we could not register any Raman working at 532 nm regardless of the processing protocol, while at 633 nm, protein removal with 3 kDa filters was necessary. In this case, the spectra of saliva_CO and salivette_CO mostly resemble those acquired at 785 nm, though the spectral bands are broader and less defined.

4. Conclusions

Vibrational spectroscopy (ATR-FTIR and Raman) of saliva in tandem with chemometrics is potentially a straightforward technique for pathology biomarker research and for personalized medicine screening to facilitate the diagnosis and follow up of patients during pharmacological therapies once biomarkers have been identified.

Multivariate analysis suggests that both Raman and FTIR spectral patterns are not affected by the saliva collection method (spitting or swab). The deproteinization method, instead, may affect the results of saliva-based vibrational spectroscopy, most of all because saliva contains nonprotein nitrogen that precipitates in ethanol [54]. Thus, the collection–processing protocol should be based on the biochemical component suitable to obtain differential diagnoses or to extract information on specific biomarkers [4]. As for the other spectrochemical approaches, FTIR is in fact advantageous for providing holistic information, but the extraction of information from the spectra is a key point to make this information useful for clinical purposes.

Although saliva collection by cotton swabs is not invasive, the spitting/drooling method is even easier and minimizes patient hassle, and it is cost-effective in repeated "personal monitoring" when the dynamics of salivary metabolites would be required. Raman analysis before and after protein removal with cut-off filters allows to obtain complementary information. It is not trivial or negligible to highlight that the development of methods based on vibrational spectroscopies, coupled with easy preanalytical steps (sampling/processing) and portable infrared and Raman spectrophotometers would in principle favor bedside applications. Lastly, the saliva deposition of multiple spots onto low-cost PP sheets and the acquisition of spectra on "printed" microamounts of SDSs transferred onto ATR diamond window is fast and novel, and the samples dry simultaneously, and it allows to obtain reproducible conditions and spectra, even when small amounts of sample are available.

Supplementary Materials: The following supporting information can be downloaded at: https://www.mdpi.com/article/10.3390/metabo13030393/s1, Figure S1. PCA results of SNV-normalized and centered (no scaling) ATR-FTIR spectra of saliva samples. (a) Score plot (69.6% of total variance); (b) loading plot PC1 (black line) and PC3 (green line). HMWsaliva_CO and HMWsalivette_CO refer to high-molecular-weight compounds (HMWCs) remaining in the upper part of 3 kDa cut-off filtering units (w = wiping, p = printing) as explained in the experimental part. Figure S2. PCA results of SNV-normalized and centered (no scaling) ATR-FTIR spectra of saliva samples. (a) Score plot (28.0% of total variance); (b) loading plot PC2 (red line) and PC3 (green line). HMWsaliva_CO and HMWsalivette_CO refer to high-molecular-weight compounds (HMWCs) remaining in the upper part of 3 kDa cut-off filtering units (w = wiping, p = printing) as explained in the experimental part. Figure S3. Absorbance HPLC chromatograms at 220 nm of saliva (a panel) and salivette (b panel) samples before and after deproteinization with 3 kDa cut-off filtration and precipitation in ethanol. (c) UV/visible spectra of the peak at tR = 5.2599 (purple line) and 4.35 min (blue line) of saliva_EtOH sample and tR = 5269 min (orange line) of saliva_CO sample. Figure S4. PC1 vs. PC2 score plots of Raman spectra acquired at 785 nm and preprocessed as described in Section 3.4. Legend: 1 (blue)—saliva; 2 (light blue)—salivette; 3 (green)—saliva_CO; 4 (yellow)—salivette_CO; 5 (orange)—saliva_EtOH; and 6 (red)—salivette_EtOH. Figure S5. PC1 (blue line) vs. PC2 (red line) loading plot of Raman spectra acquired at 785 nm and preprocessed as described in Section 3.4.

Author Contributions: Conceptualization, E.B. (Emilia Bramanti), B.C. and E.B. (Edoardo Benedetti); methodology, E.B. (Emilia Bramanti), B.C., M.O. and S.L.; validation, E.B. (Emilia Bramanti), B.C. and S.L.; resources, E.B. (Emilia Bramanti), B.C. and S.L.; writing—original draft preparation, E.B. (Emilia Bramanti); writing—review and editing, E.B. (Emilia Bramanti), E.B. (Edoardo Benedetti), B.C. and S.L. All authors have read and agreed to the published version of the manuscript.

Funding: This research received no external funding.

Institutional Review Board Statement: The study was conducted in accordance with the Declaration of Helsinki. Ethical review and approval are not applicable because all subjects were volunteers.

Informed Consent Statement: Informed consent was obtained from all subjects involved in the study. Written informed consent was obtained from volunteers to publish this paper.

Data Availability Statement: Data are available on request. Data is not publicly available due to privacy or ethical restrictions.

Conflicts of Interest: The authors declare no conflict of interest.

References

1. Sreebny, L.M. Saliva in health and disease: An appraisal and update. *Int. Dent. J.* **2000**, *50*, 140–161. [CrossRef] [PubMed]
2. Khurshid, Z.; Zohaib, S.; Najeeb, S.; Zafar, M.S.; Slowey, P.D.; Almas, K. Human Saliva Collection Devices for Proteomics: An Update. *Int. J. Mol. Sci.* **2016**, *17*, 846. [CrossRef] [PubMed]
3. Slowey, P.D. *Advances in Salivary Diagnostics*, 1st ed.; Streckfus, C.F., Ed.; Springer Berlin Heidelberg: Berlin/Heidelberg, Germany, 2015; ISBN 978-3-662-45398-8.
4. Paschotto, D.R.; PUPIN, B.; Bhattacharjee, T.T.; Soares, L.E.S. Saliva Preparation Method Exploration for ATR-FTIR Spectroscopy: Towards Bio-fluid Based Disease Diagnosis. *Anal. Sci.* **2020**, *36*, 1059–1064. [CrossRef]
5. Nonaka, T.; Wong, D.T.W. Saliva Diagnostics. *Annu. Rev. Anal. Chem.* **2022**, *15*, 107–121. [CrossRef]
6. Huang, Z.; Yang, X.; Huang, Y.; Tang, Z.; Chen, Y.; Liu, H.; Huang, M.; Qing, L.; Li, L.; Wang, Q.; et al. Saliva—A new opportunity for fluid biopsy. *Clin. Chem. Lab. Med.* **2023**, *61*, 4–32. [CrossRef] [PubMed]
7. Santos, M.C.D.; Morais, C.L.M.; Nascimento, Y.M.; Araujo, J.M.G.; Lima, K.M.G. Spectroscopy with computational analysis in virological studies: A decade (2006–2016). *Trends Anal. Chem.* **2017**, *97*, 244–256. [CrossRef]
8. Mitchell, A.L.; Gajjar, K.B.; Theophilou, G.; Martin, F.L.; Martin-Hirsch, P.L. Vibrational spectroscopy of biofluids for disease screening or diagnosis: Translation from the laboratory to a clinical setting. *J. Biophotonics* **2014**, *7*, 153–165. [CrossRef]
9. De Bruyne, S.; Speeckaert, M.M.; Delanghe, J.R. Applications of mid-infrared spectroscopy in the clinical laboratory setting. *Crit. Rev. Clin. Lab. Sci.* **2018**, *55*, 1–20. [CrossRef]
10. Movasaghi, Z.; Rehman, S.; Rehman, I.U. Fourier Transform Infrared (FTIR) Spectroscopy of Biological Tissues. *Appl. Spectrosc. Rev.* **2008**, *43*, 134–179. [CrossRef]
11. Naseer, K.; Ali, S.; Qazi, J. ATR-FTIR spectroscopy as the future of diagnostics: A systematic review of the approach using bio-fluids. *Appl. Spectrosc. Rev.* **2021**, *56*, 85–97. [CrossRef]

12. Anuradha, G.; Muthukrishnan, A.; Veeraraghavan, V.P. Saliva as a diagnostic fluid in diabetes mellitus—A review. *Drug Invent. Today* **2020**, *13*, 1100–1104.
13. Cameron, J.M.; Butler, H.J.; Palmer, D.S.; Baker, M.J. Biofluid spectroscopic disease diagnostics: A review on the processes and spectral impact of drying. *J. Biophotonics* **2018**, *11*, e201700299. [CrossRef] [PubMed]
14. Schultz, C.P.; Ahmed, M.K.; Dawes, C.; Mantsch, H.H. Thiocyanate Levels in Human Saliva: Quantitation by Fourier Transform Infrared Spectroscopy. *Anal. Biochem.* **1996**, *240*, 7–12. [CrossRef] [PubMed]
15. Khaustova, S.A.; Shkurnikov, M.U.; Grebenyuk, E.S.; Artyushenko, V.G.; Tonevitsky, A.G. Assessment of Biochemical Characteristics of the Saliva Using Fourier Transform Mid-Infrared Spectroscopy. *Bull. Exp. Biol. Med.* **2009**, *148*, 841–844. [CrossRef] [PubMed]
16. Sánchez-Brito, M.; Luna-Rosas, F.J.; Mendoza-González, R.; Mata-Miranda, M.M.; Martínez-Romo, J.C.; Vázquez-Zapién, G.J. A machine-learning strategy to evaluate the use of FTIR spectra of saliva for a good control of type 2 diabetes. *Talanta* **2021**, *221*, 121650. [CrossRef] [PubMed]
17. Sanchez-Brito, M.; Luna-Rosas, F.J.; Mendoza-Gonzalez, R.; Vazquez-Zapien, G.J.; Martinez-Romo, J.C.; Mata-Miranda, M.M. Type 2 diabetes diagnosis assisted by machine learning techniques through the analysis of FTIR spectra of saliva. *Biomed. Signal Process. Control* **2021**, *69*, 102855. [CrossRef]
18. Caixeta, D.C.; Aguiar, E.M.G.; Cardoso-Sousa, L.; Coelho, L.M.D.; Oliveira, S.W.; Espindola, F.S.; Raniero, L.; Crosara, K.T.B.; Baker, M.J.; Siqueira, W.L.; et al. Salivary molecular spectroscopy: A sustainable, rapid and non-invasive monitoring tool for diabetes mellitus during insulin treatment. *PLoS ONE* **2020**, *15*, e0223461. [CrossRef]
19. Lin, H.; Wang, Z.; Luo, Y.; Lin, Z.; Hong, G.; Deng, K.; Huang, P.; Shen, Y. Non/mini-invasive monitoring of diabetes-induced myocardial damage by Fourier transform infrared spectroscopy: Evidence from biofluids. *Biochim. Biophys. Acta Mol. Basis Dis.* **2022**, *1868*, 166445. [CrossRef]
20. Nogueira, M.S.; Barreto, A.L.; Furukawa, M.; Rovai, E.S.; Bastos, A.; Bertoncello, G.; e Silva, L.F.D.C. FTIR spectroscopy as a point of care diagnostic tool for diabetes and periodontitis: A saliva analysis approach. *Photodiagnosis Photodyn. Ther.* **2022**, *40*, 103036. [CrossRef]
21. Sanchez-Brito, M.; Vazquez-Zapien, G.J.; Luna-Rosas, F.J.; Mendoza-Gonzalez, R.; Martinez-Romo, J.C.; Mata-Miranda, M.M. Attenuated total reflection FTIR dataset for identification of type 2 diabetes using saliva. *Comput. Struct. Biotechnol. J.* **2022**, *20*, 4542–4548. [CrossRef]
22. Shaikh, S.; Yadav, D.K.; Rawal, R. Saliva based non invasive screening of Oral Submucous Fibrosis using ATR-FTIR spectroscopy. *J. Pharm. Biomed. Anal.* **2021**, *203*, 114202. [CrossRef]
23. Rodrigues, L.M.; Magrini Alva, T.D.; da Silva Martinho, H.; Almeida, J.D. Analysis of saliva composition in patients with burning mouth syndrome (BMS) by FTIR spectroscopy. *Vib. Spectrosc.* **2019**, *100*, 195–201. [CrossRef]
24. Ferreira, I.C.C.; Aguiar, E.M.G.; Silva, A.T.F.; Santos, L.L.D.; Cardoso-Sousa, L.; Araujo, T.G.; Santos, D.W.; Goulart, L.R.; Sabino-Silva, R.; Maia, Y.C.P. Attenuated Total Reflection-Fourier Transform Infrared (ATR-FTIR) Spectroscopy Analysis of Saliva for Breast Cancer Diagnosis. *J. Oncol.* **2020**, *2020*, 4343590. [CrossRef] [PubMed]
25. Paluszkiewicz, C.; Pięta, E.; Woźniak, M.; Piergies, N.; Koniewska, A.; Ścierski, W.; Misiołek, M.; Kwiatek, W.M. Saliva as a first-line diagnostic tool: A spectral challenge for identification of cancer biomarkers. *J. Mol. Liq.* **2020**, *307*, 112961. [CrossRef]
26. Barauna, V.G.; Singh, M.N.; Barbosa, L.L.; Marcarini, W.D.; Vassallo, P.F.; Mill, J.G.; Ribeiro-Rodrigues, R.; Campos, L.C.G.; Warnke, P.H.; Martin, F.L. Ultrarapid On-Site Detection of SARS-CoV-2 Infection Using Simple ATR-FTIR Spectroscopy and an Analysis Algorithm: High Sensitivity and Specificity. *Anal. Chem.* **2021**, *93*, 2950–2958. [CrossRef]
27. Nascimento, M.H.C.; Marcarini, W.D.; Folli, G.S.; da Silva Filho, W.G.; Barbosa, L.L.; de Paulo, E.H.; Vassallo, P.F.; Mill, J.G.; Barauna, V.G.; Martin, F.L.; et al. Noninvasive Diagnostic for COVID-19 from Saliva Biofluid via FTIR Spectroscopy and Multivariate Analysis. *Anal. Chem.* **2022**, *94*, 2425–2433. [CrossRef]
28. Martinez-Cuazitl, A.; Vazquez-Zapien, G.J.; Sanchez-Brito, M.; Limon-Pacheco, J.H.; Guerrero-Ruiz, M.; Garibay-Gonzalez, F.; Delgado-Macuil, R.J.; de Jesus, M.G.G.; Corona-Perezgrovas, M.A.; Pereyra-Talamantes, A.; et al. ATR-FTIR spectrum analysis of saliva samples from COVID-19 positive patients. *Sci. Rep.* **2021**, *11*, 19980. [CrossRef]
29. Kazmer, S.T.; Hartel, G.; Robinson, H.; Richards, R.S.; Yan, K.; van Hal, S.J.; Chan, R.; Hind, A.; Bradley, D.; Zieschang, F.; et al. Pathophysiological Response to SARS-CoV-2 Infection Detected by Infrared Spectroscopy Enables Rapid and Robust Saliva Screening for COVID-19. *Biomedicines* **2022**, *10*, 351. [CrossRef]
30. Das, T.; Harshey, A.; Srivastava, A.; Nigam, K.; Yadav, V.K.; Sharma, K.; Sharma, A. Analysis of the ex-vivo transformation of semen, saliva and urine as they dry out using ATR-FTIR spectroscopy and chemometric approach. *Sci. Rep.* **2021**, *11*, 11855. [CrossRef]
31. Hardy, M.; Kelleher, L.; de Carvalho Gomes, P.; Buchan, E.; Chu, H.O.M.; Goldberg Oppenheimer, P. Methods in Raman spectroscopy for saliva studies—A review. *Appl. Spectrosc. Rev.* **2022**, *57*, 177–233. [CrossRef]
32. Morais, C.L.M.; Paraskevaidi, M.; Cui, L.; Fullwood, N.J.; Isabelle, M.; Lima, K.M.G.; Martin-Hirsch, P.L.; Sreedhar, H.; Trevisan, J.; Walsh, M.J.; et al. Standardization of complex biologically derived spectrochemical datasets. *Nat. Protoc.* **2019**, *14*, 1546–1577. [CrossRef] [PubMed]
33. Bellagambi, F.G.; Lomonaco, T.; Salvo, P.; Vivaldi, F.; Hangouët, M.; Ghimenti, S.; Biagini, D.; Di Francesco, F.; Fuoco, R.; Errachid, A. Saliva sampling: Methods and devices. An overview. *Trends Anal. Chem.* **2020**, *124*, 115781. [CrossRef]

34. Wood, B.R.; Kochan, K.; Bedolla, D.E.; Salazar-Quiroz, N.; Grimley, S.L.; Perez-Guaita, D.; Baker, M.J.; Vongsvivut, J.; Tobin, M.J.; Bambery, K.R.; et al. Infrared Based Saliva Screening Test for COVID-19. *Angew. Chem. Int. Ed.* **2021**, *60*, 17102–17107. [CrossRef] [PubMed]
35. Scott, D.A.; Renaud, D.E.; Krishnasamy, S.; Meric, P.; Buduneli, N.; Cetinkalp, S.; Liu, K.Z. Diabetes-related molecular signatures in infrared spectra of human saliva. *Diabetol. Metab. Syndr.* **2010**, *2*, 48. [CrossRef]
36. Bel'skaya, L.V.; Solomatin, D.V. Influence of surface tension on the characteristics of FTIR spectra on the example of saliva. *J. Mol. Liq.* **2021**, *335*, 116173. [CrossRef]
37. Bel'skaya, L.V.; Sarf, E.A.; Solomatin, D.V. Age and Gender Characteristics of the Infrared Spectra of Normal Human Saliva. *Appl. Spectrosc.* **2020**, *74*, 536–543. [CrossRef]
38. Bel'skaya, L.V.; Sarf, E.A. Biochemical composition and characteristics of salivary FTIR spectra: Correlation analysis. *J. Mol. Liq.* **2021**, *341*, 117380. [CrossRef]
39. Saranya, K.K.N.; Maradi, A.; Chithresan, K.; Janakiram, S.H.; Krishna, P. Molecular signatures in infrared spectra of saliva in healthy, chronic and aggressive periodontitis. *Vib. Spectrosc.* **2020**, *111*, 103179. [CrossRef]
40. Chrimatopoulos, C.; Pavlou, E.; Kourkoumelis, N.; Sakkas, V. Discriminating the salivary profile of athletes using ATR-FTIR spectroscopy and chemometrics. *Chemom. Intell. Lab. Syst.* **2022**, *230*, 104660. [CrossRef]
41. da Silva, J.B.; de Carvalho, A.E.V.; Schneider, C.; Corbellini, V.A. Saliva may predict quality of life in psoriasis as measured by Fourier transform infrared spectroscopy (FTIR) and chemometrics. *Photodiagnosis Photodyn. Ther.* **2022**, *39*, 103017. [CrossRef]
42. Vieira, C.; Pupin, B.; Bhattacharjee, T.T.; Sakane, K.K. Infrared Spectroscopy Based Study of Biochemical Changes in Saliva during Maximal Progressive Test in Athletes. *Anal. Sci.* **2021**, *37*, 1157–1163. [CrossRef]
43. Sultana, R.R.; Zafarullah, S.N.; Kirubamani, N.H. Analysis of saliva and serum of normal and anomalies pregnant women—Folic acid deficiency using FTIR spectroscopy. *Indian J. Sci. Technol.* **2014**, *7*, 367–373. [CrossRef]
44. Khaustova, S.; Shkurnikov, M.; Tonevitsky, E.; Artyushenko, V.; Tonevitsky, A. Noninvasive biochemical monitoring of physiological stress by Fourier transform infrared saliva spectroscopy. *Analyst* **2010**, *135*, 3183–3192. [CrossRef]
45. Pellegrini, D.; Onor, M.; Degano, I.; Bramanti, E. Development and validation of a novel derivatization method for the determination of lactate in urine and saliva by liquid chromatography with UV and fluorescence detection. *Talanta* **2014**, *130*, 280–287. [CrossRef]
46. Biagini, D.; Lomonaco, T.; Ghimenti, S.; Fusi, J.; Cerri, E.; De Angelis, F.; Bellagambi, F.G.; Oger, C.; Galano, J.M.; Bramanti, E.; et al. Saliva as a non-invasive tool for monitoring oxidative stress in swimmers athletes performing a VO2max cycle ergometer test. *Talanta* **2020**, *216*, 120979. [CrossRef]
47. Ammazzini, S.; Onor, M.; Pagliano, E.; Mester, Z.; Campanella, B.; Pitzalis, E.; Bramanti, E.; D'Ulivo, A. Determination of thiocyanate in saliva by headspace gas chromatography-mass spectrometry, following a single-step aqueous derivatization with triethyloxonium tetrafluoroborate. *J. Chromatogr. A* **2015**, *1400*, 124–130. [CrossRef]
48. Campanella, B.; Lomonaco, T.; Benedetti, E.; Onor, M.; Nieri, R.; Marmorino, F.; Cremolini, C.; Bramanti, E. Fast, Direct Dihydrouracil Quantitation in Human Saliva: Method Development, Validation, and Application. *Int. J. Environ. Res. Public Heal.* **2022**, *19*, 6033. [CrossRef]
49. Campanella, B.; Lomonaco, T.; Benedetti, E.; Onor, M.; Nieri, R.; Bramanti, E. Validation and Application of a Derivatization-Free RP-HPLC-DAD Method for the Determination of Low Molecular Weight Salivary Metabolites. *Int. J. Environ. Res. Public Health* **2020**, *17*, 6158. [CrossRef]
50. Campanella, B.; Onor, M.; Lomonaco, T.; Benedetti, E.; Bramanti, E. HS-SPME-GC-MS approach for the analysis of volatile salivary metabolites and application in a case study for the indirect assessment of gut microbiota. *Anal. Bioanal. Chem.* **2019**, *411*, 7551–7562. [CrossRef]
51. Goodson, J.M.; Kantarci, A.; Hartman, M.-L.; Denis, G.V.; Stephens, D.; Hasturk, H.; Yaskell, T.; Vargas, J.; Wang, X.; Cugini, M.; et al. Metabolic Disease Risk in Children by Salivary Biomarker Analysis. *PLoS ONE* **2014**, *9*, e98799. [CrossRef]
52. Scarano, E.; Fiorita, A.; Picciotti, P.M.; Passali, G.C.; Calò, L.; Cabras, T.; Inzitari, R.; Fanali, C.; Messana, I.; Castagnola, M.; et al. Proteomics of saliva: Personal experience. *Acta Otorhinolaryngol. Ital.* **2010**, *30*, 125–130. [PubMed]
53. Fábián, T.K.; Fejérdy, P.; Csermely, P. Saliva in Health and Disease, Chemical Biology of. In *Wiley Encyclopedia of Chemical Biology*; John Wiley & Sons, Inc: Hoboken, NJ, USA, 2008; pp. 1–9. ISBN 9780470048672.
54. ARAKI, Y. Nitrogenous substances in saliva. I. Protein and non-protein nitrogens. *Jpn. J. Physiol.* **1951**, *2*, 69–78. [CrossRef] [PubMed]
55. Carlomagno, C.; Banfi, P.I.; Gualerzi, A.; Picciolini, S.; Volpato, E.; Meloni, M.; Lax, A.; Colombo, E.; Ticozzi, N.; Verde, F.; et al. Human salivary Raman fingerprint as biomarker for the diagnosis of Amyotrophic Lateral Sclerosis. *Sci. Rep.* **2020**, *10*, 10175. [CrossRef] [PubMed]

metabolites

Article

Simultaneous Quantification of Steroid Hormones Using hrLC-MS in Endocrine Tissues of Male Rats and Human Samples

Guillermo Bordanaba-Florit [1,*], Sebastiaan van Liempd [2], Diana Cabrera [2], Félix Royo [1,3] and Juan Manuel Falcón-Pérez [1,2,3,4,*]

[1] Exosomes Laboratory, Center for Cooperative Research in Biosciences (CIC bioGUNE), Basque Research and Technology Alliance (BRTA), 48160 Derio, Spain; froyo@cicbiogune.es
[2] Metabolomics Platform, Center for Cooperative Research in Biosciences (CIC bioGUNE), Basque Research and Technology Alliance (BRTA), 48160 Derio, Spain; smvanliempd@cicbiogune.es (S.v.L.); dcabrera@cicbiogune.es (D.C.)
[3] Centro de Investigación Biomédica en Red de Enfermedades Hepáticas y Digestivas (Ciberehd), 28029 Madrid, Spain
[4] Ikerbasque, Basque Foundation for Science, 48011 Bilbao, Spain
* Correspondence: gbordanaba@cicbiogune.es (G.B.-F.); jfalcon@cicbiogune.es (J.M.F.-P.)

Abstract: Steroid hormones play a vital role in the regulation of cellular processes, and dysregulation of these metabolites can provoke or aggravate pathological issues, such as autoimmune diseases and cancer. Regulation of steroid hormones involves different organs and biological compartments. Therefore, it is important to accurately determine their levels in tissues and biofluids to monitor changes after challenge or during disease. In this work, we have developed and optimized the extraction and quantification of 11 key members of the different steroid classes, including androgens, estrogens, progestogens and corticoids. The assay consists of a liquid/liquid extraction step and subsequent quantification by high-resolution liquid chromatography coupled time-of-flight mass spectrometry. The recoveries range between 74.2 to 126.9% and 54.9 to 110.7%, using a cell culture or urine as matrix, respectively. In general, the signal intensity loss due to matrix effect is no more than 30%. The method has been tested in relevant steroidogenic tissues in rat models and it has also been tested in human urine samples. Overall, this assay measures 11 analytes simultaneously in 6 min runtime and it has been applied in adrenal gland, testis, prostate, brain and serum from rats, and urine and extracellular vesicles from humans.

Keywords: liquid chromatography–mass spectrometry; time-of-flight; steroid hormones; androgens; urinary extracellular vesicles; hormone-dependent disease; metabolomics

Citation: Bordanaba-Florit, G.; Liempd, S.v.; Cabrera, D.; Royo, F.; Falcón-Pérez, J.M. Simultaneous Quantification of Steroid Hormones Using hrLC-MS in Endocrine Tissues of Male Rats and Human Samples. *Metabolites* **2022**, *12*, 714. https://doi.org/10.3390/metabo12080714

Academic Editor: Joana Pinto

Received: 28 June 2022
Accepted: 28 July 2022
Published: 30 July 2022

Publisher's Note: MDPI stays neutral with regard to jurisdictional claims in published maps and institutional affiliations.

1. Introduction

Steroid hormones are involved in a wide range of physiological processes and their production and delivery is regulated via the hypothalamus–pituitary–adrenal gland and –gonadal axes (Figure 1) [1]. Regulation is, amongst other things, subject to circadian rhythm, stress and sex. There are five classes of steroid hormones, namely glucocorticoids, mineralocorticoids, progestogens, androgens and estrogens. These different classes have distinct biological functions. The glucocorticoids are involved in the stress and immune response, while the mineralocorticoids are more related the maintenance of cell homeostasis [1,2]. In addition, the androgens and estrogens highly regulate cellular proliferation, development and differentiation. Hence, dysregulation of the steroid signal cascades often results in hormone-dependent pathologies. For instance, the carcinogenesis of breast and prostate cancer (PCa) are strongly influenced by the systemic presence of active estrogens [3] and androgens [4,5], respectively. Specifically, in PCa, the androgen receptor triggers the tumorigenic growth at a molecular level. The active steroid hormones, such as

5-α dihydrotestosterone (DHT), are the major ligands in this molecular pathway and cause the progression of PCa at early stages [4,6].

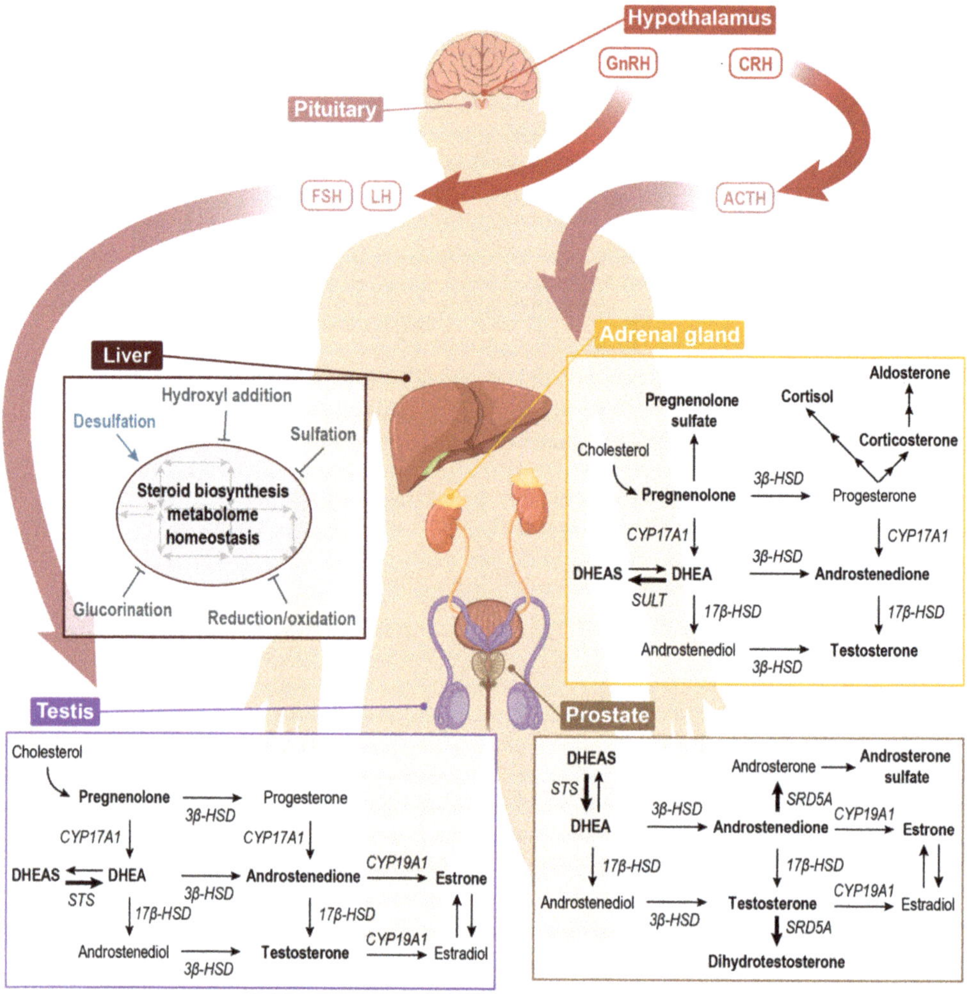

Figure 1. Schematic representation of the steroid hormone biosynthesis pathway in relevant organs and its regulation. CRH stimulates the release of ACTH from the pituitary gland. ACTH stimulates the production of cortisol (exerts negative feedback on CRH and ACTH) and DHEAS in adrenal glands. Pulses of GnRH from hypothalamic neurons stimulate pulses of LH as well as FSH. LH stimulates testosterone production in testis. Liver maintains pathway's homeostasis and several processes may happen: sulf desulfation makes metabolites available to feed the pathway while processes indicated with a flat end arrow inactivate metabolites that are in circulation. Bold arrows indicate a higher activity of the specific reaction. In bold, the metabolites that are majorly produced in each specific organ are represented. ACTH: adrenocorticotropin; CRH: corticotropin-releasing hormone; FSH: follicle stimulating hormone; GnRH: gonadotropin-releasing hormone; LH: luteinizing hormone; CYP17A1: Steroid 17-alpha-monooxygenase; CYP19A1: aromatase; SULT: hydroxysteroid sulfotransferase; STS: steroid sulfatase; 3β-HSD: 3β-Hydroxysteroid dehydrogenase; 17β-HSD: 17β-Hydroxysteroid dehydrogenase; DHEA: dehydroepiandrosterone; DHEAS: DHEA sulfate.

In mammals, the precursor of sterol biosynthesis is cholesterol, which is further utilized in the adrenal glands, gonads and sexual-derived tissues to produce steroid hormones. There are 99 metabolites involved in the steroid hormone biosynthesis pathway and over 100 reactions are catalyzed by 61 different enzymes [7,8]. All of the steroid compounds share a sterane backbone structure. The physiological role of each individual steroid hormone is primarily defined by the layout of double bonds, hydroxyl and keto groups around this basic sterane backbone structure [1]. The main structural difference between the classes is the carbon atom arrangement i.e., the androgens are C-19, the estrogens are C-18, the progestogens are C-20 and the corticoids are C-21.

In the first step of the steroid hormone biosynthesis, cholesterol is internalized into the mitochondria where it is fed as a substrate to produce pregnenolone (Figure S1, Supplementary Materials). This is the main precursor for steroid hormones produced de novo [4] inside the mitochondria. Pregnenolone can be converted to progesterone or dehydroepiandrosterone (DHEA), which can be further metabolized to glucocorticoids and mineralocorticoids (C-21) or to androgens (C-19), such as testosterone, DHT or androsterone and estrogens (C-18), respectively (Figure S1, Supplementary Materials). Interestingly, this metabolic network is tissue-dependent. Different organs are specialized on particular modules of the pathway that are physiologically relevant to perform their function. For instance, the adrenal glands are the producers of C-21 hormones, while prostate shows a high SRD5A activity, which catalyzes the conversion of testosterone to DHT (Figure 1).

Indeed, this is an intricate network of metabolites. Many of these metabolites participate as ligands in a wide span of signaling cascades and biological processes, and their levels vary strongly between different biological compartments. While cholesterol is the unique de novo precursor in steroid hormone biosynthesis, there exists an interchange between cells and tissues that anaplerotically feeds the pathway at the intermediate steps [9]. This means that the compounds upstream of the pathway can be provided by the cell environment. In this line, sulfated steroids are of interest since they are, unlike their unsulfated counterparts, readily soluble in the cytoplasm and in biofluids, such as blood or urine. Notably, the sulfates of steroids are considered endogenous and active neurosteroids [9,10]. Over the past few decades, it has been established that sulfonation is not only a process to inactivate and excrete steroid hormones; it also acts as a systemic reservoir for peripheral or local steroidogenesis in non-steroidogenic tissues, i.e., the brain or prostate [9,11]. In addition, it has been reported that the secreted vesicles, also known as extracellular vesicles (EVs), participate in many of the physiological processes [12,13] and they can contain a wide variety of cargos, such as lipids, proteins, metabolites, sugars and even DNA [12–15]. The hormone steroids and related cargos are transported by the blood and other body fluids as sulfated species, but they could also be transported by EVs to reach the target tissues.

The steroid hormone metabolism and the consequences of dysregulation have gained interest within the biomedical community to understand and diagnose hormone-dependent diseases, rather than the historic usage of steroid hormones in therapeutics. Indeed, a number of methods to detect and quantify steroid hormones have been reported during the last two decades. Many of the studies describe methodologies to detect steroids from several biological sources: cell cultures [3,16,17]; urine samples [18–20]; animal tissues [21–23]; human serum [24–26]; human hair [27] and waste water [28,29]. In general, steroid metabolomics methodologies focus on profiling a specific set of metabolites of interest in targeted tissues (or in circulation) rather than analyzing steroidogenesis status in a system of organs and related fluids. The methods are usually developed for similar non-sulfated steroids that efficiently ionize in the same mode, avoiding the exploration of the detection and quantification of many different steroids simultaneously [16,23,25,26]. Methodologically, these studies describe a variety of extraction, separation and detection methods. In particular, the solid phase extraction (SPE) and reversed phase liquid chromatographic-based methods are deployed in the isolation and separation of these compounds. The detection is mostly performed with triple quadrupole instruments. In addition, gas chromatography-coupled

MS methods was also utilized in a few of the studies. All of these methods have their advantages and disadvantages.

We describe a method for the detection of endogenous steroid hormones and their intermediates, using liquid/liquid extraction and ultra-performance liquid chromatography (UPLC), coupled with high resolution time-of-flight mass spectrometry (hrLCMS). UPLC provides fast cycling times and a high chromatographic resolution. The high mass resolution obtained with time-of-flight mass spectrometry results in high specificity, while the sensitivities are on par with triple quadrupole methods. This method was applied to metabolically profile several animal tissues and urinary EVs (uEVs). Different biological matrices, including prostate, adrenal gland, testicles, brain and liver of Wistar male rats but also human urinary samples, were tested in this assay. To our knowledge, the present work presents for the first time a reliable and optimized hrLCMS assay to analyze the key endogenous steroid hormones in endocrine tissue, bioliquids and EVs.

2. Materials and Methods

2.1. Tissue and Biofluid Samples

The tissues and serum were obtained from three wild-type (Wistar, RjHan:WI) rats obtained from Janvier Labs, Le Genest-Saint-Isle, France. All of the urine samples were obtained from a healthy male on either the morning or the afternoon. uEVs were obtained by ultracentrifuging urine samples as described elsewhere [5]. Urine samples and uEVs were characterized in several physicochemical parameters and protein markers, respectively. For a more detailed information on sample collection, preparation and characterization refer to Figure S1 (Supplementary Materials).

2.2. Chemicals and Standards

The DHEA, DHT, cortisol (in methanol solution) and the sodium salt of androsterone sulfate were obtained from Cerilliant Corporation (Round Rock, TX, USA). Supelco (Bellefonte, PA, USA) procured androstenedione. The sodium salts of DHEAS and pregnenolone were obtained from Avanti Polar Lipids, Inc. (Alabaster, AL, USA). The testosterone, aldosterone, corticosterone, estrone, pregnenolone 3-sulfate (sodium salt form), leucine-enkephalin (Leu-Enk), chloroform (>99.8% pure; of chromatography grade) and ammonia solution were purchased from Sigma-Aldrich (St. Louis, MO, USA). The LC-MS grade water, acetonitrile, formic acid and methanol were purchased from Fisher Chemical (Fair Lawn, NJ, USA).

2.3. LCMS Sample Preparation

The steroid metabolites were extracted by liquid–liquid extraction using a methanol/water mixture and chloroform as extraction liquids. The EV fractions were sonicated for 15 min in a total volume of 400 μL 50% *v*/*v* methanol/water mixture containing 1 mM ammonia to lysate EVs. The cell culture (DU145 cell line), fixed on culture well plates, was scrapped after 5 min incubation with 500 μL 50% *v*/*v* methanol/water mixture containing 1 mM ammonia. Tissue aliquots—approximately 50 mg—were lysed, using 1.4 mm zirconium oxide beads into standard 2 mL homogenizer tubes (Precellys, Montigny, France). Each sample was homogenized in 500 μL 50% *v*/*v* methanol/water mixture containing 1 mM ammonia by performing two cycles of 40 s at 6000 rpm in a FastPrep-24TM 5G bead beating grinder (MP Biomedicals, Solon, OH, USA). After lysis, 400 μL of the homogenate—either tissue, EV fraction or DU145 cell culture—was transferred to a clean Eppendorf® tube. Subsequently, 400 μL of LCMS grade chloroform was added on top of the 400 μL of any lysated sample and shaken for 60 min at 1400 rpm at 4 °C. Then, the samples were centrifuged for 30 min at 14,000 rpm at 4 °C in order to precipitate the proteins and to separate the organic from the aqueous phases.

The aqueous (top) and organic (bottom) phases were separated. The protein fraction was precipitated on the meniscus between these two immiscible phases. Then, 250 μL of each fraction was transferred to the clean Eppendorf® tubes and evaporated using a centrifugal vacuum concentrator. The pellets from the organic fraction were dissolved in 100 µL pure methanol and the pellets from the aqueous fractions were dissolved in 50% *v*/*v* methanol/water. All of the resuspended pellets were centrifuged for 30 min at 13,000 rpm and 4 °C. Finally, 80 µL of the resuspended pellets were transferred to deactivated glass vials or 96-well plates for injection into the hrLCMS system.

2.4. Ultra-High Performance Liquid Chromatography (UPLC)

The chromatographic separation of the analytes was performed with an ACQUITY UPLC I-Class PLUS System (Waters Inc., Milford, MA, USA). This system was equipped with a cooled (10 °C) Process Sample Manager with a sample loop of 10 µL and a Sample Organizer, a Binary Solvent Manager and a High Temperature Column Heater. A reversed-phased 1.0 mm × 100 mm BEH C18 column (Waters Inc., Milford, MA, USA), thermostated at 40 °C, was used for separating the analytes. The samples were injected from either 2 mL deactivated glass vials or 700 µL round 96-well polypropylene plates.

The chromatographic behavior was optimized with respect to the peak intensity and an adequate separation of the 11 analytes along the run. The gradient elution was accomplished with an aqueous mobile phase (eluent A) consisting of 99.9% water with 0.1% formic acid and an organic mobile phase (eluent B) consisting of 99.9% acetonitrile with 0.1% formic acid. The flow rate was 140 µL per min. Several gradients were tested during the optimization process (Table S1, Supplementary Materials) in order to avoid break-through (elution of analyte in the injection peak) and to obtain a good peak separation. The optimal gradient was as follows: start at 30% B; a linear increase to 80% B in 3.8 min.; a step increase from 80% to 99%; constant at 99% for 1.0 min and back to 30% B in 0.2 min. The total cycle time from injection to injection was 6 min. The injection volume for all of the samples was 2 µL.

2.5. Mass Spectrometry

A time-of-flight mass spectrometer SYNAPT G2-S (Waters Inc.) was utilized for the detection of the analytes. The instrument was operated in either positive (ESI+) or negative (ESI-) electrospray ionization mode and in full-scan mode with a scan range between 50 Da and 1200 Da and scan time of 0.2 s.

The z-spray source parameters: temperatures; gas flows; capillary position and voltages were tuned, as detailed elsewhere [30]. The optimal source parameters for this assay in either ESI+ or ESI− are summarized in Table S2 (Supplementary Materials). The ion optics were fine-tuned by spraying Leu-Enk (100 ppb), at a rate of 10 µL per min, to a resolution over 20,000 (FWHM) for m/z 556.2771. The same Leu-Enk solution was sprayed as a lock mass to correct for m/z fluctuations along the assay. The lock mass solution was introduced into the source every 90 s using a second ESI probe and it was recorded for 0.5 s. Mass spectrometer spectra was corrected according to fluctuations detected in the lock mass.

2.6. Statistical Analysis

2.6.1. Analyte Recovery Study

The extraction step efficiency was assessed by performing a recovery assay with various mixtures of organic solvents and water. Five different extraction buffers were tested in this assay: 25/75% *v*/*v* and 50/50% *v/v* of methanol/water mixture; 25/74.9/0.1% *v/v/v* and 50/49.9/0.1% *v/v*/*v* of methanol/water/formic acid mixture and 50/50% *v/v* of methanol/water mixture with 1mM ammonia. To compare and calculate the recoveries of 10 different analytes, a culture of a prostate cancer cell line-DU145-was spiked with the analyte standards. Each well containing $5 \cdot \times 10^5$ cells was spiked with a mix of standards at 2 µM before lysis (pre-spiked) and at the resuspension stage (post-spiked) with a standard mix at 10 µM. Thus, the pre-spikes contained 1 nmol in 500 µL and post-spikes (aqueous and organic fractions) contained the same total amount in 100 µL, which would be the theoretical maximum absolute if there was no loss during the extraction. In addition, for each extraction solution, the non-spiked samples were prepared in order to correct for endogenous metabolites in the matrix. The samples for the pre-spiked, post-spiked and non-spiked conditions and the five different extraction buffers were prepared in biological triplicates.

Only the absolute peak areas were taken into consideration to establish the recovery efficiency in the extraction step. The average peak areas were obtained by mean smoothing the raw signals of triplicates. The recovery (*R*) was determined by dividing the corrected pre-spike average by the corrected post-spike average and represented as a percentage (Equation (1)). Both the pre-spiked and post-spiked raw signals ought to be corrected by subtracting the endogenous analytes signal in the DU145 culture matrix ($S_{non\text{-}spike}$). However, as the $S_{non\text{-}spike}$ of DU145 culture matrix was less than 0.05% of the signal, endogenous correction was neglected during the calculation. Importantly, the pre-spikes were corrected with respect to analyte loss (α) during the extraction procedure. Moreover, the raw signals of each sample did not have to be corrected by the amount of initial samples, because every well contained the same amount of cells.

$$R\ (\%) = \frac{\alpha\left(S_{pre-spike} - S_{non-spike}\right)}{S_{post-spike} - S_{non-spike}} \times 100 \quad (1)$$

2.6.2. Study of Matrix Effect in Analyte Quantification

In order to assess the matrix effect (*ME*) in the quantification of the analytes, the post-spiked raw signal was compared to an equivalent raw signal of a mixture of analytes (10 μM) in solution. The post-spiked raw signals were corrected by subtracting the endogenous analytes detected in the non-spiked DU145 culture samples. Then, the numerator was divided by the average peak areas of the standards and expressed as a percentage (Equation (2)):

$$ME\ (\%) = \frac{S_{post-spike} - S_{non-spike}}{S_{standards}} \times 100 \quad (2)$$

2.6.3. Analyte Semi-Quantification

In this work, a calibration curve was prepared in solution with 50% *v/v* methanol/water for the semi-quantification of the analytes. This calibration curve consisted of a serially diluted mixture containing all of the analytes, starting at a concentration of 10 μM. The initial concentration was diluted to half concentration twice, resulting in 5 μM and 2.5 μM concentration in the curve. Then, this set of triplets was diluted in five decades; it resulted in the following 15 different concentrations per analyte: 10; 5; 2.5; 1; 0.5; 0.25; 0.1; 0.05; 0.025; 0.01; 0.005; 0.0025; 0.001; 0.0005 and 0.00025 μM. The calibration samples were injected at the beginning and at the end of each experiment; the average of these two points was used to semi-quantify the metabolites in the tissues.

The limit of detection (LOD) for each analyte was set to be the lowest concentration at which the signal-to-noise (S/N) ratio was above three. The LOQ was defined as the lowest concentration at which the S/N ratio was above 10. The highest quantifiable concentration was the highest concentration per analyte that fits the calibration curve with an acceptable accuracy and precision (CV $\leq$ 15%) [16].

In general, the data of a calibration curve range over several orders of magnitude, the data are not linear and tend to be heteroscedastic [31]. For this reason, the relation between the peak area and the sample concentration was determined by power-fitting [30]. The power fitting resulted in a calibration curve (Equation (3)) with α and b as the fitted parameters. Once the sample concentrations were calculated using a calibration method in solution, the amount (in nanomole) per gram of tissue weight was estimated:

$$Peak\ area = \alpha[concentration]^b \quad (3)$$

3. Results

3.1. Liquid Chromatography and Mass Spectrometry Method

We compared six different chromatographic methodologies (Table S1, Supplementary Materials) to satisfactorily separate the analytes. The gradient 6 (30% B to 80% B in 3.8 min; detailed steps in Table S2, Supplementary Materials) showed the best peak separation along this run time compared to other tested gradients (data available in [32]). Due to the nature of the stationary phase, analytes elute in order of increasing hydrophobicity. The resulting extracted ion current (XIC) chromatograms of a standard mixture at 10 µM are depicted in Figure S2 (Supplementary Materials). In brief, aldosterone (m/z 361.2015; ESI+) elutes at 0.99 min, cortisol (m/z 363.2171; ESI+) at 1.20 min, DHEAS (m/z 367.1579; ESI−) at 1.60 min, corticosterone (m/z 347.2222; ESI+) at 1.68 min, androsterone sulfate (m/z 369.1736; ESI−) at 1.85 min, pregnenolone sulfate (m/z 395.1892; ESI−) at 2.23 min, estrone (m/z 271.1698; ESI+) at 2.39 min, androstenedione (m/z 287.2011; ESI+) and DHEA (m/z 289.2168; ESI+) co-elute at 2.40 min, DHT (m/z 291.2324; ESI+) at 2.65 min, pregnenolone (m/z 317.2481; ESI+) at 3.25 min.

Regarding the mass spectrometry method, the Leu-Enk signal (m/z 556.2771) was aimed at a resolution of over 20,000 (FWHM) and provided the necessary mass accuracy to evaluate assay analytes. Isotope pattern matching and the use of chemical standards confirming elution times further ensured the specificity. In general, the mass accuracies for the analytes in solution were between −1 to 1 mDa. It is noteworthy that several analytes were not adequately separated during the chromatographic elution. The corticosterone and DHEAS elute at similar retention times—1.60 min and 1.68 min-, however, the MS could properly distinguish them by their m/z difference and their fragmentation pattern. Moreover, the DHEAS was not detected with a high intensity signal in ESI+ mode. For this reason, the corticosterone was measured in ESI+ and the DHEAS in ESI− mode. Likewise, estrone, DHEA and androstenedione eluted in approximately 2.40 min. In this case, one could only rely on the MS sensitivity (estrone m/z 271.1698, DHEA m/z 289.2168, androstenedione m/z 287.2011) and on a fragmentation pattern that was sensitive enough to distinguish and quantify them separately.

3.2. Analyte Recovery Optimization

Afterwards, we evaluated the recovery of 11 analytes using a biphasic liquid–liquid method and analyzed them with the optimized hrLCMS method. The extraction was performed, using the DU145 cell line as a matrix. Five different mixtures of organic solvents and water, containing either formic acid or ammonia to modify the pH of the extraction buffer or no pH modifier, were assessed (Table S3, Supplementary Materials). The addition of formic acid strived for lowering the pH approximately to three, while 1mM ammonia modified the extraction buffer to pH 8–9 in order to chemically neutralize the functional groups of the steroid compounds. From the previous experiments in our metabolomics platform, we observed that in liquid–liquid extraction requires at least 25% organic solvent during the extraction step to precipitate the proteins. This is important to avoid clogging the chromatographic system [30]. Moreover, the effectivity of tissue homogenization using beads has been reported as high and does not differ much from the homogenization of other matrices, such as urine or cell cultures [30,33]. Therefore, the calculated recoveries are ultimately dependent on the extraction buffer utilized, regardless of the homogenization methodology.

During the optimization process, it was determined that the steroid sulfate compounds were recovered completely in the aqueous fraction, whilst steroids without sulfate group were found in the organic fraction. Notably, only cortisol was detected systematically in both of the fractions (Figure S3, Supplementary Materials); however, it was majorly recovered in the organic (80% or higher) rather than in the aqueous (approximately 20%) fraction. Moreover, the addition of formic acid to the extraction buffer led to a dramatic decrease in the recoveries of the sulfate compounds and a slight decrease in the rest of the steroid analytes (Figure S3, Supplementary Materials). One can infer that the presence of protons in the buffer do not stabilize steroid charges and severely hampers the extraction of sulfate steroids in a polar environment. The supplementation of 1mM ammonia outperformed the extraction in terms of recovery and robustness, compared to the other extraction liquids. Notably, the recovery values using different percentages of methanol in the extraction buffer do not differ much. However, the extraction efficiency of the sulfate compounds using 25% *v/v* methanol underperforms 50% *v/v* methanol, with a recovery loss of 40 to 50%.

In Table 1, the recoveries of the 11 selected analytes, using a mixture of 50/50% *v/v* methanol/water with 1mM ammonia as the extraction buffer, are reported. In general, the present methodology is able to recover and detect over 90% of the initially spiked analyte. Only DHT was detected in a lower percentage; approximately 80% of the initially spiked DHT was recovered. As expected in a biphasic extraction, the hormone steroids were retrieved in an apolar environment and the sulfated steroids in a polar solvent. Besides cortisol, pregnenolone sulfate was also reported in both of the fractions; it was mainly recovered in the more polar solvent and a derisory amount in the organic fraction. Using this methodology, the recoveries for 10 µM of analyte ranged from 74.2% to 126.9%. These values are acceptable for routine muti-analyte hrLCMS analysis since all of the results are reproducible [34]. Thus, extraction using 50/50% *v/v* of methanol/water mixture with 1 mM ammonia was selected for further experiments in different biological matrices.

Table 1. Summary of the optimized method characteristics. The recoveries (±standard deviation) and matrix effect as signal loss (±standard deviation) of the extraction procedure in two different biological matrices ($n = 6$; biological matrix: DU145 cell) are reported. In addition, LOD and LOQ values of the analytes in the adequate fraction are compiled. LOD: Limit of detection; LOQ: Limit of quantification.

Analyte	Fraction	Recovery (%)	Matrix Effect (%)	LOD (nM)	LOQ (nM)
Pregnenolone	Organic	97.2 (±1.9)	25.2 (±3.1)	2.5 nM	10 nM
	Aqueous	-	24.0 (±2.8)		
DHEA	Organic	122.7 (±2.9)	37.7 (±5.7)	5.0 nM	50 nM
	Aqueous	-	28.0 (±6.2)	-	-
Androstenedione	Organic	102.2 (±3.2)	30.8 (±4.6)	0.25 nM	0.5 nM
	Aqueous	-	23.2 (±4.5)		
Estrone	Organic	103.7 (±3.8)	25.5 (±4.8)	5.0 nM	10 nM
	Aqueous	-	25.7 (±4.0)		
DHT	Organic	74.2 (±3.4)	23.1 (±3.9)	0.25 nM	1.0 nM
	Aqueous	-	23.4 (±2.9)		
Cortisol	Organic	114.3 (±3.8)	25.9 (±4.2)	0.5 nM	1.0 nM
	Aqueous	22.28 (±4.5)	17.6 (±4.7)		
Aldosterone	Organic	99.8 (±1.77)	18.7 (±4.3)	0.5 nM	2.5 nM
	Aqueous	-	17.7 (±5.1)		
Corticosterone	Organic	109.4 (±3.1)	25.1 (±3.6)	0.25 nM	1.0 nM
	Aqueous	-	20.2 (±3.2)		
Testosterone	Organic	126.9 (±1.7)	14.3 (±1.9)	0.25 nM	0.25 nM
	Aqueous	-	8.0 (±2.1)		
Pregnenolone sulfate	Organic	6.9 (±2.7)	25.2 (±3.1)	0.25 nM	1.0 nM
	Aqueous	94.8 (±1.9)	24.0 (±2.8)		
DHEAS	Organic	-	42.6 (±1.1)	0.25 nM	0.5 nM
	Aqueous	108.0 (±1.4)	42.5 (±0.1)		

Furthermore, the performance of the optimized methodology was tested, using urine as the matrix since it has a high interest for clinical applications. Six samples of urine from a male individual were pooled and aliquoted in different two volumes to assess the matrix effect on the recovery efficiency. In Table 2, the recoveries of the 10 analytes are reported; DHEA recovery has not been retrieved, because its peak was masked by testosterone's signal. In general, over 85% of the initially spiked analyte is recovered and detected in 50 µL urine matrix. Importantly, the sulfated steroids are not recovered with the same efficiency; DHEAS and pregnenolone sulfate report a recovery efficiency of 75.7% and 54.9%, respectively. The recoveries of the analytes using 250 µL urine as matrix describes a slight decrease in the non-sulfated steroids while the efficiency decay is dramatic in the sulfated species.

Table 2. Summary of the recoveries using the optimized methodology in urine matrix. The recoveries (±standard deviation) of two different volumes (50 µL and 250 µL) of pre-pooled urine are reported ($n = 3$).

Analyte	Urine Volume	Recovery (%)
Pregnenolone	50 µL 250 µL	92.4 (±3.6) 99.3 (±4.8)
Androstenedione	50 µL 250 µL	93.0 (±3.9) 79.3 (±3.8)
Estrone	50 µL 250 µL	94.2 (±3.3) 84.8 (±4.8)
DHT	50 µL 250 µL	76.3 (±4.1) 71.2 (±3.76)
Cortisol	50 µL 250 µL	87.0 (±3.0) 72.4 (±3.6)
Aldosterone	50 µL 250 µL	110.7 (±2.9) 103.1 (±3.2)
Corticosterone	50 µL 250 µL	96.2 (±2.8) 84.3 (±3.6)
Testosterone	50 µL 250 µL	104.1 (±2.1) 96.3 (±5.1)
Pregnenolone sulfate	50 µL 250 µL	54.9 (±1.5) 25.5 (±1.2)
DHEAS	50 µL 250 µL	75.7 (±2.5) 44.0 (±4.2)

3.3. Matrix Effect

It is well known that the phospholipids and other lipids, typically enriched in biological matrices, such as tissues, body fluids or cell cultures, can cause ion suppression in mass spectrometry, thereby hampering the analyte signal [35,36]. This phenomenon negatively influences the detection of the analytes and may underestimate their quantification. For a specific matrix, the higher the ion suppression effect is, the higher the signal loss. Therefore, the conclusions drawn by detecting and quantifying the analytes under these conditions could be misleading.

The matrix effect of each analyte was defined as the signal loss measured at the resuspension step (sample spiked with 10 µM analyte mix) compared to 10 µM of each analyte in solution. The signal loss was calculated in five different extraction procedures, because they can influence ion suppression. The matrix effect reported in this work was estimated for a prostate cancer cell line (DU145) culture and urine samples. To note, signal loss is specific for each matrix and each independent experiment. In further experiments, in which quantification is required, the matrix effect should be calculated in every particular assay. From our optimization experiments, one can infer that the matrix effect is fraction-dependent, because there is a significant difference between signal loss comparing organic and aqueous fractions (Figure S4, Supplementary Materials). This phenomenon is likely observed due to a differential extraction of the phosphatidylcholine (or other lipid) compounds [30,35]. Strikingly, this fraction dependency was not observed upon the addition of ammonia to the extraction liquid. Moreover, the presence of ammonia resulted in a signal loss of up to half-fold compared to extraction liquids with acidic modifier or no pH modifier addition. This suggests that the ammonia impairs the extraction of the lipidic compounds from the biological matrix, hence, decreasing the ion suppression phenomenon in mass spectrometry.

In Table 1, the matrix effect (expressed as signal loss (%)) of a DU145 culture of 11 selected analytes, using a 50/50% *v*/*v* of methanol/water mixture with 1mM ammonia for extraction, is reported. In general, the present methodology loses approximately 15 to 40% of the signal of non-sulfated analytes but it mainly lays between 20 to 30% loss. On the other hand, the sulfated steroids display a 40 to 50% loss of signal, regardless of the extraction fraction. The signal loss of the 10 µM analytes spiked in DU145 cell line were: 25.2% for pregnenolone, 37.7% for DHEA, 30.8% for androstenedione, 25.5% for estrone, 23.1% for DHT, 25.9% and 20.2 % for cortisol in the organic and aqueous fraction, respectively, 18.6% for aldosterone, 25.0% for corticosterone, 46.1% for pregnenolone sulfate and 42.5% for DHEAS. All of the analytes are majorly recovered back in a particular fraction of the extraction procedure, which is the one selected to report the matrix effect. Signal loss of sulfate compounds refer to aqueous fraction measurement and the other steroids refer to signal loss in organic fraction.

3.4. *Semi-Quantitation of Steroids in Animal Tissues*

The hrLCMS method was most sensitive in detecting androstenedione, DHT, corticosterone, pregnenolone sulfate and DHEAS with a LOD (S/N > 3) of 250 pM in a 50/50% *v*/*v* methanol/water solution. The detection limit for cortisol and aldosterone was 0.5 nM, and a LOD of 2.5 nM was determined for pregnenolone. The least responsive ions were those for DHEA and estrone with a LOD of 5.0 nM. With regards to the quantification limits, androstenedione and DHEAS were the most sensitive compounds, with a LOQ (S/N > 10) of 0.5 nM in solution. The cortisol, corticosterone, pregnenolone sulfate and DHT were in the second group of the most quantifiable ions showing a LOQ of 1.0 nM. The quantitation limit for aldosterone was 2.5 nM, while a LOQ of 0.01 µM was estimated for pregnenolone and estrone. The DHEA was the compound with the highest quantitation threshold (0.05 µM).

We found that the concentration range of the steroid hormones is typically low in tissues, ranging from pico- to nanomole per gram of tissue, and cannot be detected in some tissues (Table 3). Only pregnenolone, androstenedione, DHT, corticosterone, cortisol and testosterone were detected in the tissues or serum of Wistar rats. Pregnenolone and cortisol are only quantified in the adrenal gland tissue, however, pregnenolone is also detected in the brain and testicles. Adrenal gland and testicles reported picomole amounts of androstenedione per gram of tissue. Moreover, DHT was quantified in the prostate, adrenal gland and testicles. In prostate, the amount of DHT was two-fold the quantitation in the other tissues. The testosterone and corticosterone were quantified in all of the measured rat samples. In general, they were reported in the picomole per gram range in tissues. In serum, they were quantified in the nM range. Interestingly, the adrenal gland described nanomole per gram concentrations of corticosterone. Furthermore, testosterone was found in a one order of magnitude higher amount in the adrenal gland and testicles compared to prostate and brain.

Table 3. Quantitation of three independent Wistar rat tissues: adrenal gland, prostate and brain. Adrenal glands of the same animal were titered independently, also, the prostate lobes of each rat. The averages in nmol per gram of tissue, standard deviations and coefficients of variation (%) of the three groups of samples are reported.

Analyte	Quantification (nmol/g Tissue)	Adrenal Gland	Prostate	Brain	Testicle	Serum (nM)
Pregnenolone	Amount	7.04	-	Detected	Detected	-
	St. dev.	3.74				
	%cv	53				
Androstenedione	Amount	5.97×10^{-3}	-	Detected	1.45×10^{-3}	Detected
	St. dev.	3.35×10^{-3}			1.38×10^{-3}	
	%cv	56			95	
DHT	Amount	3.47×10^{-3}	7.57×10^{-3}	Detected	2.70×10^{-3}	Detected
	St. dev.	1.02×10^{-3}	2.40×10^{-3}		7.92×10^{-4}	
	%cv	29	31		29	
Corticosterone	Amount	18.89	4.01×10^{-3}	2.42×10^{-2}	1.25×10^{-3}	28.01
	St. dev.	10.05	5.15×10^{-3}	7.04×10^{-3}	7.98×10^{-4}	3.31
	%cv	53	128	29	63	12
Cortisol	Amount	0.45	-	-	-	-
	St. dev.	0.19				
	%cv	43				
Testosterone	Amount	4.53×10^{-3}	6.92×10^{-4}	7.02×10^{-4}	9.18×10^{-3}	0.20
	St. dev.	1.47×10^{-3}	2.36×10^{-4}	4.29×10^{-4}	4.53×10^{-3}	0.02
	%cv	32	34	60	49	

The standard deviations and coefficients of the variation are rather large, indicating an important variability among the samples obtained from the same strain but independent animals. One could expect this biological variation and it suggests that treatments, stress or any procedure applied to animals can potentially influence the outcome in further experiments.

3.5. *Quantitation of Steroid Hormones in Human Urinary Samples*

Six different urine samples were characterized in several physicochemical parameters (Table S4, Supplementary Materials) to examine whether the sample collection resulted in homogenous sample groups, regardless of the metabolomics' analysis. No blood, ketone bodies or glucose were detected in the urine sample, and the pH value and density of the urine were similar in all of the samples. The urine samples were centrifuged in two serial steps at 10,000× *g* for 30 min to isolate the so-called P10K fraction—typically containing vesicles of 150 to 200 nm diameter and above—followed by a 100,000× *g* centrifugation for 90 min to isolate the so-called P100K—typically containing vesicles of 100 to 150 nm diameter and below (up to 50 nm) [37]. The supernatant of the second centrifugation was also analyzed and referred to as SN100K.

In this set of urine samples, the current methodology is able to detect and quantify androstenedione, cortisol and DHEAS (Table 4). The other steroids of the panel were below the LOQ and, in general, also below the LOD. The androstenedione and cortisol were detected only in the urine and SN100K. It was not possible to detect them associated with the EVs, and they are majorly solubilized in the urine. The androstenedione was found in lower concentrations compared to cortisol and the variability between the collection days was high (40 to 60%) regardless of the collection time. Concerning cortisol, the variability was extremely high between the morning collection days (approximately 50 to 85%) whilst the concentration of the afternoon collected samples was stable (approximately 2% variation). DHEAS was the compound detected in the highest concentration (μM range) soluble in urine, compared to androstenedione and cortisol (nM range). Similar to androstenedione, the DHEAS showed a high variability over independent collection days at both the morning and afternoon collection times. To note, DHEAS was the only metabolite detectable in the EV fraction. In Table 4, the absolute amount (μmol) in 50 mL of urine is reported but also the relative amount (in ppm) of the total detected metabolite that is associated with the EVs. Importantly, DHEAS was not quantifiable (S/N < 10) in all of the samples collected at morning time, but it was detectable in all of the cases (S/N > 3). According to our analysis, a range of 0.5 to 3.0 ppm of DHEAS was associated with the EVs in the urine samples (Table 4; detailed calculations available in [32]).

Table 4. Quantitation of urine human samples (n = 6, U001 to U006, Table S4, Supplementary Materials). The isolated EV fraction are also included in the table. In the table, the three analytes detected in the urine-derived samples.

Sample	Collection Time	Androstenedione (nM)	Cortisol (nM)	DHEAS		
				Conc. (μM)	EV-Associated DHEAS (μM)	EV-Associated DHEAS in Urine (ppm)
Urine	Morning	2.25 (±0.92)	40.1 (±33.5)	0.36 (±0.16)	-	-
Urine	Afternoon	1.95 (±0.78)	35.7 (±0.7)	1.27 (±0.87)	-	-
SN100K	Morning	2.31 (±1.53)	29.9 (±14.7)	1.33 (±0.94)	-	-
SN100K	Afternoon	1.82 (±0.64)	31.4 (±0.7)	0.87 (±0.92)	-	-
P10K	Morning	-	-	-	1.75	0.90
P10K	Afternoon	-	-	-	0.76 (±0.08)	0.79 (±0.41)
P100K	Morning	-	-	-	6.17	3.19
P100K	Afternoon	-	-	-	0.74 (±0.01)	

Concentration (±standard deviation) of the analytes in urine and supernatant fraction of both morning and afternoon collected urine is shown. Absolute amount and relative amount (±standard deviation) of DHEAS is calculated in 50 mL of initial sample of both morning and afternoon collected urine.

The isolation of the EVs in the pellet fractions was confirmed with the presence of typical EV markers by Western blotting (Figure S5, Supplementary Materials). Typical urine exosome markers, such as CD9, CD63 and AQP2, were intensified in P100K fractions, confirming that this fraction is enriched in EVs. However, they are sample-dependent and were detected in various amounts. In addition, LAMP2A and CD10 were detected only in the P100K fraction of U003-derived EVs preparation. Annexin V and AQP2 were found in both P100K and P10K, but also in different amounts among urine samples.

4. Discussion

This work describes a fast and simple hrLCMS methodology, able to detect and quantify 11 key metabolites of the steroid hormones biosynthesis in several biological matrices. Their importance in diseases, such as PCa and other steroid-dependent diseases, spotlights this assay as a powerful tool to study the role of steroid hormones in the development and progression of hormone-dependent diseases and to assess the metabolic status of patients via liquid biopsy analysis. In brief, this method identifies and quantifies 11 steroids, including corticoids, androgens and metabolic intermediates, in a high-throughput method of 6 min. Although testosterone and androsterone sulfate were not included in the recovery experiment, the methodology is able to separate, identify and quantify them.

All of the steroid hormones are primarily derived from cholesterol, which provides the sterane ring structure shared by all of these compounds (Figure S1, Supplementary Materials). Subtle chemical differences, unique to each steroid hormone, significantly complicate the separation of such structurally similar molecules. Furthermore, the structure of the steroids and position of the functional groups determine their preferred ionization mode and efficiency [18,24]. For instance, testosterone and DHEA—with the same molecular formula—display different ionization efficiencies. DHT or androstenedione are readily ionized in positive mode, in contrast to DHEA or pregnenolone, which are not strongly ionized due to the presence of keto groups in the ionizable region (Figure S1, Supplementary Materials). In order to increase the signal intensity, the MS could be operated in enhanced duty-cycle (EDC) mode; this is a more appropriate approach in targeted analyses, where the analyte empirical formulas are known. In this strategy, the MS signals of a given retention time are measured in separate scan functions to enhance the *m/z* of the selected analyte. Measuring in EDC instead of full-scan mode may increase by several fold the S/N ratio of a given metabolite [30,38,39]. Therefore, the EDC mode is an option to consider for those samples in which the analytes S/N ratio falls above the LOD, but are not always quantifiable.

An LCMS method is usually evaluated in terms of efficiency, accuracy and sensitivity of the measurement. The process efficiency is a combination of recovery efficiency and matrix effect of each metabolite [40], and the sensitivity is evaluated with the LOD and LOQ of each metabolite. Different studies identifying and quantifying steroid compounds in biological matrices report a wide range of efficiency recoveries. For example, in PCa cell cultures, a recovery range of 54.7% to 78.1% was reported [16] while in breast cancer cell cultures, recoveries ranging 95.7 to 102.0% were reported [16]. Our data, with recoveries ranging from approximately 75% to 125%, suggest that a cell culture as matrix does not impair the extraction of the steroid metabolites. The urine matrix does not impair the extraction of the non-sulfated steroids but the sulfated species suffer a recovery efficiency decay. To note, the studies measuring steroids in urine and tissues, as biological matrices report recovery efficiencies of over 100% in some of the cases [17–19]. An explanation for this phenomenon might be that the metabolites can be either free in solution or tethered to other molecules, such as membranal lipids during the extraction process. For this reason, the organic and aqueous phase recoveries are not adding up to 100% in this assay. In case of detecting a metabolite in two fractions, the addition of both of the signals is perhaps a better approach to quantify that specific metabolite. However, our assay is very convenient, since all of the metabolites (except cortisol) are recovered in only one fraction. This permits a faster measurement of the steroid hormones in different biological matrices.

The existing quantitation methods for steroid hormone compounds have a wide span of LOQ, ranging from 0.002 to 10 ng per mL. However, it is highly dependent on the analyzed matrix, i.e., a urine matrix shows a range from 0.002 to 0.2 ng per mL [18,19], whilst the cell matrices display a higher LOQ up to 10 ng per mL [16]. This suggests that the matrix effect also depends on the specific matrix where the metabolites are contained. Comparing these studies, the cell matrices report a lower sensitivity compared to urine; this is important when applying this method in future experiments or assays. In fact, this observation spotlights the major limitation of this study: the quantitation has been performed semi-quantitively. Ion suppression in mass spectrometry negatively affects the analyte signal, and subsequently underestimates its quantitation, or it simply hampers its detection. Moreover, ion suppression may be limiting the detection of certain steroid compounds in several matrices, i.e., EV preparations. In consequence, this method should be utilized in matrices that facilitate the detection of the steroids. A matrix-spiked calibration is usually the appropriate method to quantify the absolute amounts of analytes in samples [30]. In this work, a calibration curve of the analyte standards was prepared in solution with 50% *v/v* methanol/water as a solvent. Such an approach cannot compute the absolute amounts of the analytes in tissue, since the matrix effect is not considered, however, a semi-quantitative approximation of the metabolites in tissues can be calculated. In this assay, the reported LOQ range lies between 0.50 and 50 nM (equivalent to 0.14 and 14.42 ng per mL) in solution, similar to previous studies. However, it is advised to use matrix-spiked curves in further experiments using this assay.

The time required to perform the chromatographic separation is typically long in the literature; they report runtimes from over 10 min up to 45 min [3,5,18–22,27]. Only the work of Quanson et al. [16] and Indapurkar et al. [17] described a methodology with a short runtime (4 to 5 min); however, they tested and applied the method solely in cell matrices: PCa and induced pluripotent stem cell lines, respectively. Indapurkar et al. [17] developed a methodology specific for estradiol-related metabolites and Quanson et al. [16] measured androgenic steroids using an ultra-performance convergence chromatography.

In 2012, Maeda et al. accomplished the separation, detection and quantification of a panel of steroids in rat organs except in the liver, but using an HPLC system. For this reason, their sample preparation strategy demanded high volumes of extraction buffer—15 mL of acetonitrile per sample—and required a total run time of 11 min. In this work, the volumes are lower than 1 mL and the run time for different types of samples is lower than 10 min.

In order to test the performance of our methodology, we have measured steroid hormone analytes from several rat tissues: adrenal glands; testis; prostate; liver and brain. The data shown in Table 3 are in accordance with the fact that the pathway is tissue-dependent in regular physiological conditions. Two metabolites upstream of the pathway, pregnenolone and androstenedione, were quantified in the adrenal glands, but could not be quantified in prostate or brain. This hints that the adrenal glands are in charge of the conversion of cholesterol into the steroid compounds in complex organisms, such as rats; this is in line with previous findings in the literature [41–43]. Likewise, the adrenal glands are known to produce corticoid hormones. Our data confirms this, since corticosterone is quantified in a higher amount—three to four orders of magnitude—when compared to the prostate, brain and testicles. The adrenal glands also seem to accumulate androgens (Table 3); however, the presence of active androgens (DHT) is two-fold higher in the prostate compared to other tissues. Importantly, the ratio DHT/testosterone, which are the active and non-active paired androgens, was approximately 11 in prostate, while the adrenal gland and testis were below 1. Because the presence of the active androgen plays a physiological role in prostate, the ratio of DHT/testosterone was also higher in this tissue.

Since the first urinary metabolomics attempts to analyze urinary samples and other biofluids, several methodologies have been developed during the last few years [18–20]. Nevertheless, none of the reported methodologies was optimal to assess the steroids in the EV sample preparations, tissues or body fluids in a fast and simple manner. Up to date, many of the studies have shown metabolomics in EVs [5,20,44], but none of them has reported the detection of steroid hormones in a targeted approach. A plausible explanation is that the identification and detection of compounds similar in molecular mass—even the same one in some cases—hampers the allocation of mass signals with the corresponding chromatographic peak. For those steroids, i.e., DHEA and testosterone, which share an empirical formula, the identification of each specific compound remains challenging using MS and the identification relies on chromatographic separation.

Importantly, we have been able to quantify the steroid hormones in urine samples and derived uEV in a fast and simple manner. However, only one DHEAS was detected in the uEVS and cortisol, androstenedione and DHEAS were detected in the urine samples. These EVs were isolated by ultracentrifugation, including a washing step to avoid any contamination from the soluble fraction. The urine samples from a healthy man were collected on different days and different time of collection (morning and afternoon). The time collection was a parameter to be assessed from a metabolomics perspective, but we found out that inter-day variability also had a high impact on the analysis. Morning samples are considered to contain a higher concentration of steroid analytes coming from the prostate, possibly due to accumulation and leakage towards the urinary tract during the night. However, this trend was not described in our morning samples. The reason may be that urine sample U003 (Table S4, Supplementary Materials) was not available for metabolomics analysis; the analysis of the soluble fractions of urine (after uEV isolation), which includes U003, in the morning samples had a higher concentration of DHEAS. This highlights the importance of analyzing a larger cohort to obtain significant results non-dependent on a unique highly concentrated sample.

In the end, this is a fast and sensitive method that was successfully applied for the detection and quantification of a panel of steroid hormone compounds in biological samples in 6 min runtime per sample. The sensitivity of this method makes it ideally suited for multiple in vivo applications. In this manuscript, we explored the analysis of steroids in several rat tissues and also in human urine and uEV samples. This has evident applications in profiling the metabolic status of patients suffering any hormone-dependent disease. It should be noted that the assay requires a longer cleanse step to wash the column out of the lipids and peptides when running a long experiment with many tissue samples. To our knowledge, this is the first hrLCMS-based method able to detect and quantify steroid hormones associated with EVs isolated from body fluids in a targeted approach.

Supplementary Materials: The following supporting information can be downloaded at: https://www.mdpi.com/article/10.3390/metabo12080714/s1, Supporting Information S1: Additional experimental details related to sample collection and characterization; Supporting Information S2: Table S1–S4 Supplementary tables with method optimization data and urine characterization; Supporting Information S3: Figure S1–S5 Supplementary figures including metabolomics network, method optimization results and urine characterization.

Author Contributions: Conceptualization and methodological design, G.B.-F., S.v.L. and D.C.; experiments, data analysis and original draft writing, G.B.-F.; supervision, S.v.L., D.C., F.R. and J.M.F.-P.; writing—review and editing, G.B.-F., S.v.L., D.C., F.R. and J.M.F.-P. All authors have read and agreed to the published version of the manuscript.

Funding: The authors of this study were supported by funds from the European Union's Horizon 2020 research and innovation programme under grant agreement no. 860303. This research was funded by the Spanish Ministry of Economy and Competitiveness MINECO, grant number RTI2018-094969-B-I00.

Institutional Review Board Statement: All animal experimentation was conducted in accordance with Spanish guidelines for the care and use of laboratory animals, and protocols were approved by the CIC bioGUNE Institute and the regional Basque Country ethical committee (ref. P-CBG-CBBA-0219). All efforts were made to minimize the suffering of the animals. The animal study protocol was approved by the Institutional Review Board (or Ethics Committee) of Diputación Foral de Bizkaia (protocol code P-CBG-CBBA-0219 and date of approval).

Informed Consent Statement: Informed consent was obtained from the subject involved in the study.

Data Availability Statement: All data which support the reported results have been uploaded to figshare (https://figshare.com/articles/dataset/_/20231493) (accessed on 27 June 2022). The reference to this data was added as [32] in the text.

Acknowledgments: We thank Arkaitz Carracedo's lab for kindly donating DU145 cell line used for the experiments. We also thank Exosomes lab staff at CIC bioGUNE for experiment assistance and guidance.

Conflicts of Interest: The authors declare no conflict of interest.

References

1. Miller, W.L.; Auchus, R.J. The Molecular Biology, Biochemistry, and Physiology of Human Steroidogenesis and Its Disorders. *Endocr. Rev.* **2011**, *32*, 81–151. [CrossRef] [PubMed]
2. Bereshchenko, O.; Bruscoli, S.; Riccardi, C. Glucocorticoids, Sex Hormones, and Immunity. *Front. Immunol.* **2018**, *9*, 1332. [CrossRef] [PubMed]
3. Poschner, S.; Zehl, M.; Maier-Salamon, A.; Jäger, W. Simultaneous Quantification of Estrogens, Their Precursors and Conjugated Metabolites in Human Breast Cancer Cells by LC–HRMS without Derivatization. *J. Pharm. Biomed. Anal.* **2017**, *138*, 344–350. [CrossRef]
4. Cardoso, H.J.; Carvalho, T.M.A.; Fonseca, L.R.S.; Figueira, M.I.; Vaz, C.V.; Socorro, S. Revisiting Prostate Cancer Metabolism: From Metabolites to Disease and Therapy. *Med. Res. Rev.* **2021**, *41*, 1499–1538. [CrossRef] [PubMed]
5. Clos-Garcia, M.; Loizaga-Iriarte, A.; Zuñiga-Garcia, P.; Sánchez-Mosquera, P.; Rosa Cortazar, A.; González, E.; Torrano, V.; Alonso, C.; Pérez-Cormenzana, M.; Ugalde-Olano, A.; et al. Metabolic Alterations in Urine Extracellular Vesicles Are Associated to Prostate Cancer Pathogenesis and Progression. *J. Extracell. Vesicles* **2018**, *7*, 1470442. [CrossRef] [PubMed]
6. Lonergan, P.; Tindall, D. Androgen Receptor Signaling in Prostate Cancer Development and Progression. *J. Carcinog.* **2011**, *10*, 20. [CrossRef] [PubMed]
7. Kanehisa, M.; Goto, S. KEGG: Kyoto Encyclopedia of Genes and Genomes. *Nucleic Acids Res.* **2000**, *28*, 27–30. [CrossRef]
8. Kanehisa, M.; Furumichi, M.; Sato, Y.; Ishiguro-Watanabe, M.; Tanabe, M. KEGG: Integrating Viruses and Cellular Organisms. *Nucleic Acids Res.* **2021**, *49*, D545–D551. [CrossRef]
9. Mueller, J.W.; Gilligan, L.C.; Idkowiak, J.; Arlt, W.; Foster, P.A. The Regulation of Steroid Action by Sulfation and Desulfation. *Endocr. Rev.* **2015**, *36*, 526–563. [CrossRef]

10. Foster, P.A.; Mueller, J.W. Insights into Steroid Sulfation and Desulfation Pathways. *J. Mol. Endocrinol.* **2018**, *61*, T271–T283. [CrossRef]
11. Langford, R.; Hurrion, E.; Dawson, P.A. Genetics and Pathophysiology of Mammalian Sulfate Biology. *J. Genet. Genom.* **2017**, *44*, 7–20. [CrossRef] [PubMed]
12. Bordanaba-Florit, G.; Royo, F.; Kruglik, S.G.; Falcón-Pérez, J.M. Using Single-Vesicle Technologies to Unravel the Heterogeneity of Extracellular Vesicles. *Nat. Protoc.* **2021**, *16*, 3163–3185. [CrossRef] [PubMed]
13. Yáñez-Mó, M.; Siljander, P.R.-M.; Andreu, Z.; Zavec, A.B.; Borràs, F.E.; Buzas, E.I.; Buzas, K.; Casal, E.; Cappello, F.; Carvalho, J.; et al. Biological Properties of Extracellular Vesicles and Their Physiological Functions. *J. Extracell. Vesicles* **2015**, *4*, 27066. [CrossRef] [PubMed]
14. Kalra, H.; Simpson, R.J.; Ji, H.; Aikawa, E.; Altevogt, P.; Askenase, P.; Bond, V.C.; Borràs, F.E.; Breakefield, X.; Budnik, V.; et al. Vesiclepedia: A Compendium for Extracellular Vesicles with Continuous Community Annotation. *PLoS Biol.* **2012**, *10*, 8–12. [CrossRef]
15. Pathan, M.; Fonseka, P.; Chitti, S.V.; Kang, T.; Sanwlani, R.; Van Deun, J.; Hendrix, A.; Mathivanan, S. Vesiclepedia 2019: A Compendium of RNA, Proteins, Lipids and Metabolites in Extracellular Vesicles. *Nucleic Acids Res.* **2019**, *47*, D516–D519. [CrossRef]
16. Quanson, J.L.; Stander, M.A.; Pretorius, E.; Jenkinson, C.; Taylor, A.E.; Storbeck, K.H. High-Throughput Analysis of 19 Endogenous Androgenic Steroids by Ultra-Performance Convergence Chromatography Tandem Mass Spectrometry. *J. Chromatogr. B Anal. Technol. Biomed. Life Sci.* **2016**, *1031*, 131–138. [CrossRef]
17. Indapurkar, A.; Hartman, N.; Patel, V.; Matta, M.K. Simultaneous UHPLC-MS/MS Method of Estradiol Metabolites to Support the Evaluation of Phase-2 Metabolic Activity of Induced Pluripotent Stem Cell Derived Hepatocytes. *J. Chromatogr. B Anal. Technol. Biomed. Life Sci.* **2019**, *1126–1127*, 121765. [CrossRef] [PubMed]
18. Im, E.; Lew, B.L.; Lee, M.Y.; Lee, J.; Paeng, K.J.; Chung, B.C. Simultaneous Determination of Androgens and Prostaglandins in Human Urine Using Ultra-High-Performance Liquid Chromatography–Tandem Mass Spectrometry. *J. Chromatogr. B Anal. Technol. Biomed. Life Sci.* **2019**, *1109*, 45–53. [CrossRef]
19. Qin, F.; Zhao, Y.Y.; Sawyer, M.B.; Li, X.F. Hydrophilic Interaction Liquid Chromatography-Tandem Mass Spectrometry Determination of Estrogen Conjugates in Human Urine. *Anal. Chem.* **2008**, *80*, 3404–3411. [CrossRef]
20. Lee, Y.R.; Im, E.; Kim, H.; Lew, B.L.; Sim, W.Y.; Lee, J.; Oh, H.B.; Paeng, K.J.; Hong, J.; Chung, B.C. Untargeted Metabolomics and Steroid Signatures in Urine of Male Pattern Baldness Patients after Finasteride Treatment for a Year. *Metabolites* **2020**, *10*, 131. [CrossRef]
21. Maeda, N.; Tanaka, E.; Suzuki, T.; Okumura, K.; Nomura, S.; Miyasho, T.; Haeno, S.; Yokota, H. Accurate Determination of Tissue Steroid Hormones, Precursors and Conjugates in Adult Male Rat. *J. Biochem.* **2013**, *153*, 63–71. [CrossRef]
22. Fernandes, D.; Navarro, J.C.; Riva, C.; Bordonali, S.; Porte, C. Does Exposure to Testosterone Significantly Alter Endogenous Metabolism in the Marine Mussel Mytilus Galloprovincialis? *Aquat. Toxicol.* **2010**, *100*, 313–320. [CrossRef] [PubMed]
23. Li, Y.; Yan, Z.; Li, X.; Yin, X.; Li, K. Uplc–Tof–Ms Method for Simultaneous Quantification of Steroid Hormones in Tissue Homogenates of Zebrafish with Solid-Phase Extraction. *Molecules* **2021**, *26*, 6213. [CrossRef]
24. Ceglarek, U.; Kortz, L.; Leichtle, A.; Fiedler, G.M.; Kratzsch, J.; Thiery, J. Rapid Quantification of Steroid Patterns in Human Serum by On-Line Solid Phase Extraction Combined with Liquid Chromatography-Triple Quadrupole Linear Ion Trap Mass Spectrometry. *Clin. Chim. Acta* **2009**, *401*, 114–118. [CrossRef]
25. Koal, T.; Schmiederer, D.; Pham-Tuan, H.; Röhring, C.; Rauh, M. Standardized LC-MS/MS Based Steroid Hormone Profile-Analysis. *J. Steroid Biochem. Mol. Biol.* **2012**, *129*, 129–138. [CrossRef] [PubMed]
26. Yuan, T.F.; Le, J.; Wang, S.T.; Li, Y. An LC/MS/MS Method for Analyzing the Steroid Metabolome with High Accuracy and from Small Serum Samples. *J. Lipid Res.* **2020**, *61*, 580–586. [CrossRef]
27. Dong, Z.; Wang, C.; Zhang, J.; Wang, Z. A UHPLC-MS/MS Method for Profiling Multifunctional Steroids in Human Hair. *Anal. Bioanal. Chem.* **2017**, *409*, 4751–4769. [CrossRef] [PubMed]
28. Isobe, T.; Shiraishi, H.; Yasuda, M.; Shinoda, A.; Suzuki, H.; Morita, M. Determination of Estrogens and Their Conjugates in Water Using Solid-Phase Extraction Followed by Liquid Chromatography-Tandem Mass Spectrometry. *J. Chromatogr. A* **2003**, *984*, 195–202. [CrossRef]
29. Reddy, S.; Iden, C.R.; Brownawell, B.J. Analysis of Steroid Conjugates in Sewage Influent and Effluent by Liquid Chromatography-Tandem Mass Spectrometry. *Anal. Chem.* **2005**, *77*, 7032–7038. [CrossRef]
30. Van Liempd, S.; Cabrera, D.; Mato, J.M.; Falcon-Perez, J.M. A Fast Method for the Quantitation of Key Metabolites of the Methionine Pathway in Liver Tissue by High-Resolution Mass Spectrometry and Hydrophilic Interaction Ultra-Performance Liquid Chromatography. *Anal. Bioanal. Chem.* **2013**, *405*, 5301–5310. [CrossRef]
31. Schepers, U.; Ermer, J.; Preu, L.; Wätzig, H. Wide Concentration Range Investigation of Recovery, Precision and Error Structure in Liquid Chromatography. *J. Chromatogr. B* **2004**, *810*, 111–118. [CrossRef]
32. Bordanaba-Florit, G. Simultaneous Quantification of Steroid Hormones Using hrLC-MS in Endocrine Tissues of Male Rats and Human Samples (Dataset). 2022. Available online: https://figshare.com/articles/dataset/_/20231493 (accessed on 27 June 2022).
33. Liang, X.; Ubhayakar, S.; Liederer, B.M.; Dean, B.; Ran-Ran Qin, A.; Shahidi-Latham, S.; Deng, Y. Evaluation of Homogenization Techniques for the Preparation of Mouse Tissue Samples to Support Drug Discovery. *Bioanalysis* **2011**, *3*, 1923–1933. [CrossRef]

34. Method Validation and Quality Control Procedures for Pesticide Residues Analysis in Food and Feed. 2012. Available online: https://www.eurl-pesticides.eu/library/docs/allcrl/AqcGuidance_Sanco_2011_12495.pdf (accessed on 27 June 2022).
35. Ghosh, C.; Shinde, C.P.; Chakraborty, B.S. Influence of Ionization Source Design on Matrix Effects during LC-ESI-MS/MS Analysis. *J. Chromatogr. B Anal. Technol. Biomed. Life Sci.* **2012**, *893–894*, 193–200. [CrossRef] [PubMed]
36. Zhou, W.; Yang, S.; Wang, P.G. Matrix Effects and Application of Matrix Effect Factor. *Bioanalysis* **2017**, *9*, 1839–1844. [CrossRef] [PubMed]
37. Bobrie, A.; Colombo, M.; Krumeich, S.; Raposo, G.; Thery, C. Diverse Subpopulations of Vesicles Secreted by Different Intracellular Mechanisms Are Present in Exosome Preparations Obtained by Differential Ultracentrifugation. *J. Ayurveda Integr. Med.* **2012**, *1*, 18397. [CrossRef] [PubMed]
38. Giles, K.; Pringle, S.D.; Worthington, K.R.; Little, D.; Wildgoose, J.L.; Bateman, R.H. Applications of a Travelling Wave-Based Radio-Frequency-Only Stacked Ring Ion Guide. *Rapid Commun. Mass Spectrom.* **2004**, *18*, 2401–2414. [CrossRef]
39. Kenny, D.J.; Worthington, K.R.; Hoyes, J.B. Scanwave: A New Approach to Enhancing Spectral Data on a Tandem Quadrupole Mass Spectrometer. *J. Am. Soc. Mass Spectrom.* **2010**, *21*, 1061–1069. [CrossRef]
40. Taylor, P.J. Matrix Effects: The Achilles Heel of Quantitative High-Performance Liquid Chromatography-Electrospray-Tandem Mass Spectrometry. *Clin. Biochem.* **2005**, *38*, 328–334. [CrossRef]
41. Campbell, I. Adrenocortical Hormones. *Anaesth. Intensive Care Med.* **2008**, *9*, 420–423. [CrossRef]
42. Rosol, T.J.; Yarrington, J.T.; Latendresse, J.; Capen, C.C. Adrenal Gland: Structure, Function, and Mechanisms of Toxicity. *Toxicol. Pathol.* **2016**, *29*, 41–48. [CrossRef]
43. Rainey, W.E.; Nakamura, Y. Regulation of the Adrenal Androgen Biosynthesis. *J. Steroid Biochem. Mol. Biol.* **2008**, *108*, 281–286. [CrossRef] [PubMed]
44. Puhka, M.; Takatalo, M.; Nordberg, M.E.; Valkonen, S.; Nandania, J.; Aatonen, M.; Yliperttula, M.; Laitinen, S.; Velagapudi, V.; Mirtti, T.; et al. Metabolomic Profiling of Extracellular Vesicles and Alternative Normalization Methods Reveal Enriched Metabolites and Strategies to Study Prostate Cancer-Related Changes. *Theranostics* **2017**, *7*, 3824–3841. [CrossRef] [PubMed]

metabolites

Article

Mapping of Urinary Volatile Organic Compounds by a Rapid Analytical Method Using Gas Chromatography Coupled to Ion Mobility Spectrometry (GC–IMS)

Giulia Riccio [1,2], Silvia Baroni [1,2], Andrea Urbani [1,2] and Viviana Greco [1,2,*]

[1] Department of Basic Biotechnological Sciences, Intensivological and Perioperative Clinics, Università Cattolica del Sacro Cuore, 00168 Rome, Italy
[2] Department of Diagnostic and Laboratory Medicine, Unity of Chemistry, Biochemistry and Clinical Molecular Biology, Fondazione Policlinico Universitario A. Gemelli IRCCS, 00168 Rome, Italy
* Correspondence: viviana.greco@unicatt.it

Citation: Riccio, G.; Baroni, S.; Urbani, A.; Greco, V. Mapping of Urinary Volatile Organic Compounds by a Rapid Analytical Method Using Gas Chromatography Coupled to Ion Mobility Spectrometry (GC–IMS). *Metabolites* **2022**, *12*, 1072. https://doi.org/10.3390/metabo12111072

Academic Editor: Joana Pinto

Received: 21 September 2022
Accepted: 2 November 2022
Published: 5 November 2022

Abstract: Volatile organic compounds (VOCs) are a differentiated class of molecules, continuously generated in the human body and released as products of metabolic pathways. Their concentrations vary depending on pathophysiological conditions. They are detectable in a wide variety of biological samples, such as exhaled breath, faeces, and urine. In particular, urine represents an easily accessible specimen widely used in clinics. The most used techniques for VOCs detections are expensive and time-consuming, thus not allowing for rapid clinical analysis. In this perspective, the aim of this study is a comprehensive characterisation of the urine volatilome by the development of an alternative rapid analytical method. Briefly, 115 urine samples are collected; sample treatment is not needed. VOCs are detected in the urine headspace using gas chromatography coupled to ion mobility spectrometry (GC–IMS) by an extremely fast analysis (10 min). The method is analytically validated; the analysis is sensitive and robust with results comparable to those reported with other techniques. Twenty-three molecules are identified, including ketones, aldehydes, alcohols, and sulphur compounds, whose concentration is altered in several pathological states such as cancer and metabolic disorders. Therefore, it opens new perspectives for fast diagnosis and screening, showing great potential for clinical applications.

Keywords: GC–IMS; metabolomics; urine; volatile organic compounds; volatilomics

1. Introduction

Volatilomics is a recent and promising branch of metabolomics that focuses on the study of small molecules and volatile organic compounds (VOCs) with significant potential for biomarker discovery and screening [1].

Specifically, VOCs are a large and highly differentiated class of molecules, continuously produced in the human body and released as intermediates or products of cellular metabolic pathways. They include ketones, aldehydes, alcohols, sulphur compounds, esters, aromatic hydrocarbons, and terpenes, whose concentrations vary depending on pathophysiological conditions, and are detectable in a wide variety of biological samples (exhaled breath, urine, blood, faeces, and skin).

In recent years, the diagnostic potential of VOCs has been strongly recognised. There is an increasingly evident correlation between the profile of VOCs and various diseases, including diabetes [2], irritable bowel syndrome, asthma [3], and, above all, cancer [4].

Compared to other types of metabolites, which have to be extracted from tissues or body fluids prior to analysis, VOCs are directly accessible in the gas phase (headspace), thus requiring minimal sample preparation and enabling noninvasive, real-time monitoring.

Consequently, headspace analyses may find easy applicability in the clinical setting.

As for the biological matrix, in addition to breath, urine is the most used fluid for the detection of VOCs. It is a biological fluid easy to collect with a noninvasive sampling, less complex than other fluids [5], and available in large volumes so VOCs can be detected even in high concentrations.

Therefore, it represents a well-suited source for VOCs metabolomics investigation.

Moreover, urinary VOCs can vary both in concentration and in the types of molecules depending on several variables such as diet, therapies, genetic factors, and smoking habits, which must be taken into account during analysis [6].

Gas chromatography coupled to mass spectrometry (GC–MS) is the gold standard technique used to detect urinary VOCs. GC–MS is an extremely useful tool; however, it is also extremely expensive and time-consuming, and it requires highly skilled personnel and is not portable. Therefore, it is not a suitable technique to be implemented in the clinical setting [1,6,7].

As a result, there is an urgent need for fast and non-invasive innovative methodologies for VOCs analysis that can be implemented in clinical early diagnosis applications.

In this context, the aim of this study is to develop an alternative analytical method using a high-sensitivity gas chromatographic system coupled to an ion mobility spectrometer (GC–IMS) for the rapid detection of urinary VOCs.

To the best of our knowledge, GC–IMS has already been applied to detect different VOCs profiles in breath samples and to distinguish between diagnostic groups related to inflammatory bowel disease (IBD) [8]. Furthermore, IMS is finding great application in the analysis of exhaled breath samples of lung cancer patients [9]. Recently, the potential of VOCs profiling in the urine of lung cancer patients to differentiate them from healthy subjects is also being evaluated with GC–IMS and an electronic nose (e-nose) [10]. The main advantages of this technology were highlighted, including non-invasiveness, portability, ease of use, and cost-effectiveness.

The implementation of this method could open up new perspectives for extremely rapid diagnosis and screening, showing great potential for clinical applications.

2. Materials and Methods

2.1. Chemicals and Materials

The ketone mix was composed of six ketones (2-butanone, 2-pentanone, 2-hexanone, 2-heptanone, 2-octanone, and 2-nonanone) (S.C.A.T. Europe GmbH, Walldorf, Germany). Chemical standards, such as 4-heptanone, were of analytical grade (Thermo Fisher Scientific, Waltham, MA, USA). In addition, 20 mL headspace vials (screw top, rounded bottom, clear glass vial (vial size: 22.5 × 75.5 mm)) and caps (screw cap 18 mm, argent magnetic, PTFE/silicone septum, septum thickness 1.5 mm) were sterile (Thermo Fisher Scientific, Waltham, MA, USA). Needles (calibre 21 G, colour green, size: 0.8 × 50 mm) were purchased from Agani Needle (Terumo Europe N. V., Leuven, Belgium) and a 5 mL Luer Lock Solo syringe was purchased from Injekt B. Braun (B. Braun, Melsungen, Germany). MilliQ water was prepared using the Elix® 70 water purification system (Merk, Dramstadt, Germany).

2.2. Analytical Method Validation

For column normalisation and internal calibration, a standard mixture of six ketones (S_0 as defined in Table 1) was analysed. It included 2-butanone, 2-pentanone, 2-hexanone, 2-heptanone, 2-octanone, and 2-nonanone (mixed volume ratio 1:1:1:1:1:1). Seven different solutions (M_1, M_2, M_3, M_4, M_5, M_6, and M_7) were prepared at the concentrations outlined in Table 1. An amount of 2 mL of each solution was put in a screw vial and left to settle for 10 min to allow the transition of VOCs to the gas phase in the headspace. Then, 3 mL of vial headspace was withdrawn and injected in the instrument. Each measurement was performed in triplicate after the blank in the experimental condition.

Table 1. Concentration values of ketone mixture standard solutions used for column normalisation and for calibration.

Compound	S_0 ppm (g/mL)	M_1 ppb (µg/L)	M_2 ppb (µg/L)	M_3 ppb (µg/L)	M_4 ppb (µg/L)	M_5 ppb (µg/L)	M_6 ppb (µg/L)	M_7 ppb (µg/L)
2-Butanone	0.135	216	162	108	70.2	54	27	10.8
2-Pentanone	0.135	216	162	108	70.2	54	27	10.8
2-Hexanone	0.135	216	162	108	70.2	54	27	10.8
2-Heptanone	0.137	218.4	163.8	109.2	71	54	27.3	10.9
2-Octanone	0.137	218.4	163.8	109.2	71	54	27.3	10.9
2-Nonanone	0.137	218.4	163.8	109.2	71	54	27.3	10.9

Linearity was calculated with standard solutions of 4-heptanone in the range of concentrations of 0–160 ppb, plotting IMS peaks intensity (y-axis) against the 4-heptanone concentration (x-axis). Slope regression was calculated with a linear regression analysis. The minimum concentration value for which an IMS signal is measured, corresponding to the detection limit (limit of detection, LOD), was calculated from the slope regression in order to evaluate the sensitivity of the method.

2.3. Sample Collection

Urine samples were collected at Clinical Chemistry, Biochemistry, and Molecular Biology Operations Unit (UOC), Fondazione Policlinico Universitario A. Gemelli IRCCS (Rome, Italy). All the investigations were performed on the residual sample aliquots after the conclusions of all clinical procedures. Samples were stored at room temperature for no more than six hours in order to avoid the degradation. The pH of urine samples was in the range of 5.0–7.5.

2.4. Sample Preparation

An amount of 2 mL of urine sample was withdrawn from the residual urine and immediately put in 20 mL glass screw vials. Vials were closed with the appropriate screw cap equipped by a Silicon/PTFE septum to allow for picking the gas phase from the headspace. Samples were incubated at 37 °C for 10 min before the analysis, facilitating the transition and the stabilisation of VOCs between the liquid phase and vial headspace. An amount of 3 mL of headspace air was withdrawn with a sterile syringe from the vial and injected through a Luer adapter into the system. Samples were directly injected without any pre-concentration or extraction.

2.5. GC–IMS Analysis

Samples were analysed by a GC–IMS system (G.A.S., Dortmund, Germany), a combination of a gas chromatograph and an ion mobility mass spectrometer. Volatile chemical compounds, which are contained in the vial headspace, are physically pre-separated by GC and detected by IMS after a second separation in a drift tube, allowing for analysis of complex mixtures with the concentration at the parts per billion level (ppb/µg/L). Technical features are shown in Table 2. Briefly, GC–IMS is equipped with a gas recycling flow unit (CGFU) to purify ambient air, used as a carrier gas at 40 °C in GC and as a drift gas at 45 °C in IMS. The flow rate of carrier gas is set at 5 mL for the first 30 s and increased to 30 mL/min within 10 min, while the drift gas flow rate is set at 150 mL/min. A capillary DB wax column, thermostated at 40 °C, is used. VOCs ionise through a β-radiation tritium (^3H) source with 300 MBq of activity in positive ion mode. After a soft chemical-ionisation, ions move to a 10 cm drift tube driven by a ±5000 V electric field. Drift gas molecules enter in the drift tube and collide with analytes accelerated by the electric field, whose separation depends on the molecular weight, charge, and spatial structure. They reach a Faraday plate where the ion current is measured as a function of time. The overall time of analysis is 10 min.

Table 2. Experimental conditions of GC–IMS device. Technical parameters have been schematised both for chromatographic elution (column, carrier gas, flow control, injection volume, and sampling) and ion mobility mass spectrometry (ionisation, model, drift gas, and detector).

GC-IMS Technical Parameters	
Gas Chromatograph	
Column	Capillary, DB wax
Carrier Gas	Air, CGFU Circular Gas Flow Unit
Flow Control	Electronic pressure controller
Injection Volume	3 mL
Sampling	Heated 6-port-valve incl. sample pump
Ion Mobility Spectrometer	
Ionisation	API, ^{3}H-Tritium Source (<380 MBq)
Model	Time-of-flight/10 cm tube, ±5000 V
Drift Gas	Air, CGFU Circular Gas Flow Unit
Detection	Faraday Plate

2.6. Data Analysis

Spectrum visualisation, organisation of data measurement, and setting of experimental conditions were enabled by VOCal software (v0.1.3, G.A.S., Dortmund, Germany). Column normalisation was carried out by analysing the standard mixture of six ketones with increasing molecular weight and retention indexes (Ri), or Kovats indexes were calculated by an algorithm of libraries of the software VOCal based on the formula:

$$I = 100 \times \left[n + (N - n) \frac{\log(Rt_{unknown}) - \log(Rt_n)}{\log(Rt_N) - \log(Rt_n)} \right]$$

where the variables are as follows:

I, the Kovats retention index of the peak;
n, the carbon number of the shorter alkane;
N, the carbon number of the longer alkane;
Rt, the retention time registered.

The retention time, Rt, and the drift time, Dt, are the two main values recognised by the device. In particular, Rt is defined as the time in seconds that a compound spends in the column after being injected. Dt is the time an ionised compound takes to reach the detector during an acceleration due to an electric field in a drift tube. The spectra obtained are a three-dimensional pseudo-colour representation reporting the Rt on the y-axis and Dt on the x-axis.

After all acquisitions, the areas of the most relevant peaks are highlighted and selected using the VOCal software. The identification of VOC species is based on the Ri and Dt of each peak calculated from those of standard ketones using the IMS database of GC/IMS Library Search tool software (NIST2014 db wax).

Calibration was performed by analysing the ketone mix at seven different concentrations. Afterwards, the quantification was carried out for the ketone mix compounds as well as for urine samples.

3. Results

3.1. Analytical Method Validation

Before the analysis of biological samples, an analytical validation of instrumental parameters is carried out. First, in order to identify VOCs, column normalisation is carried out by analysing a mixture of ketones including compounds with different molecular weights (2-butanone, 2-pentanone, 2-hexanone, 2-heptanone, 2-octanone, and 2-nonanone). Their Rt and Dt cover a range of our interest, in which most of the common volatile compounds contained in the human biological samples are included and detectable with this device. A typical spectrum of the ketone mix at the concentration of 108 ppb is shown in Figure 1 (Figure 1a).

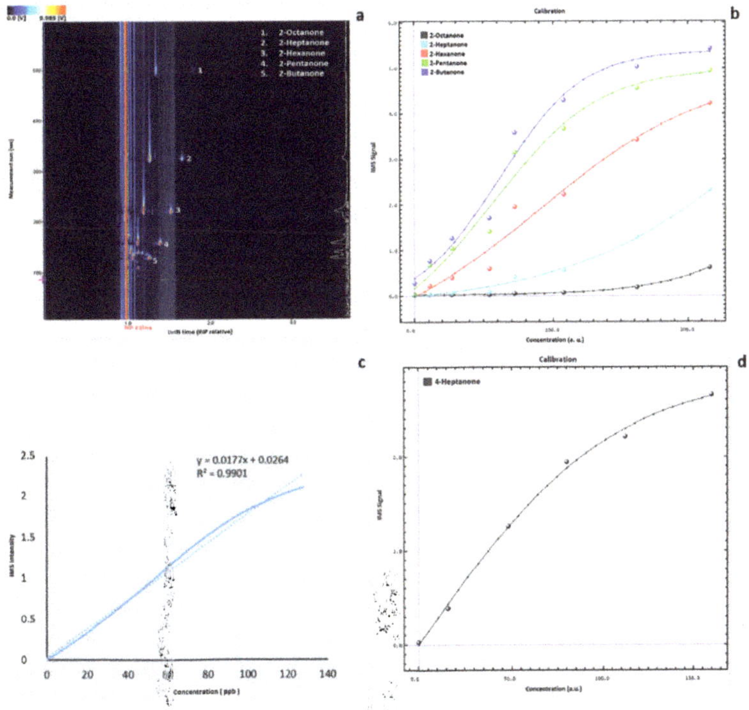

Figure 1. (**a**) Example of GC–IMS output of the ketone mix profile at the concentration of 108 ppb. The detected compounds have been highlighted. Each compound's Dt has been normalised by means of the software application to the signal of the reaction ion peak (RIP). It represents the total number of ions available for ionisation, and therefore, it is used as the reference signal. The colour representation corresponds to a three-dimensional spectrum. An increasing concentration of VOCs is outlined by the colour change from blue to red. (**b**) Calibration curve obtained with the VOCal software by measuring the ketones mixture at seven different concentrations in the range of 218.4–10.8 ppb. Each colour corresponds to a detected compound in the ketone mix (blue = 2-butanone; green = 2-pentanone, red = 2-hexanone; light blue = 2-heptanone; black = 2-octanone). Dots represent the signal intensity for the concentrations analysed (expressed as arbitrary unit, a.u.); lines show the fit of the calibration curve. (**c**) Linearity range for 4-heptanone analysed by means of GC–IMS. The linearity curve and the regression line are reported for the concentration range of 0–128 ppb. (**d**) Calibration curve obtained with the VOCal software by measuring 4-heptanone at five different concentrations in the range of 0–160 ppb.

As reported in the method section (Table 1), seven solutions of the ketones mixture at different concentrations are analysed in triplicate to obtain a calibration curve. The 2-nonanone signal is extremely low; thus, it is not shown (Figure 1b).

In particular, 4-heptanone is selected to assess the linearity range. The standard solutions of this VOC at different concentrations (8, 16, 48, 80, 112, and 160 ppb) are analysed. Specifically, the curve for 4-heptanone is linear and statistically acceptable

($R^2 \geq 0.9901$) in the concentration range of 0–128 ppb, while for higher concentrations (>128 ppb), the linearity is slightly lower ($R^2 \geq 0.9802$) (Figure 1c).

To assess the sensitivity of the method, LOD is calculated from the regression slope. The value found from the regression slope is 4.66 ppb, which is close to the experimentally detectable LOD value by analysing 4-heptanone solutions at low concentration in the range of 1–5.5 ppb as reported by the instrumental features (Figure 1d).

3.2. VOCs Analysis in Urine Samples

In order to obtain a comprehensive urinary VOCs profiling, 115 urine samples are analysed by the GC–IMS device as described in the methods section. Our test does not require any sample treatment, thus greatly reducing the analysis time. Samples are directly injected and analysed by GC–IMS.

For each sample, a three-dimensional GC–IMS spectrum is obtained (Figure 2). Seven main classes of volatile compounds are identified as reported in Table 3. These include ketones, sulphur compounds, esters, aldehydes, alcohols, and aromatic hydrocarbons, terpenes.

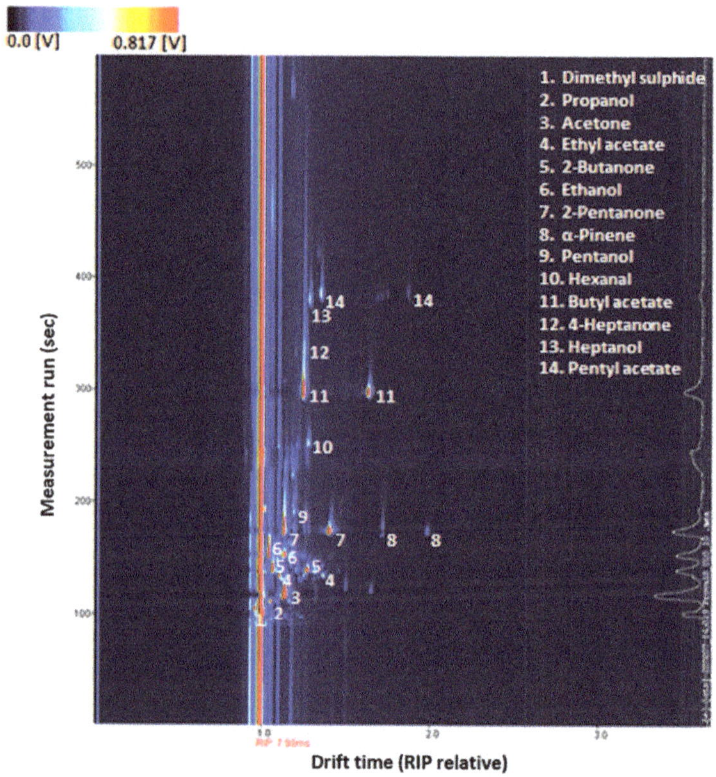

Figure 2. Example of GC–IMS spectrum of a urine sample. Detected VOCs have been highlighted. Increasing concentrations of VOCs are outlined by the colour change from blue to red.

Table 3. Summary of VOCs detected across the population in urine samples. (a) class of molecule to which the VOC belongs; (b) list of detected VOCs; (c) retention time (Rt) to which the VOC was eluted; (d) retention index (Ri) of VOC calculated by the VOCal software; (e) percentage of the population in which VOCs were detected.

Class (a)	VOCs (b)	Rt [s] (c)	Ri (d)	% (e)
Ketones	Acetone	119	812	100
	2-butanone	141	897	100
	2-pentanone	177	979	97
	4-heptanone	334	1125	16
	2-hexanone	256	1070	0.87
Aldehydes	Propanal	112	763	87
	Pentanal	176	977	44
	Hexanal	255	1070	28
	3-methylbutanal	159	945	13
	Heptanal	385	1152	11
Sulphur compounds	Dimethyl sulphide	107	718	21
	Diallyl sulphide	405	1161	0.87
Alcohols	Ethanol	154	934	100
	Propanol	217	1033	16
	Pentanol	574	1226	13
	2-methyl-1-propanol	271	1083	13
	2-methyl-1-butanol	527	1209	1.74
	2-hexanol	528	1210	0.87
Esters	Butyl acetate	281	1091	72
	Pentyl acetate	338	1153	51
	Ethyl acetate	141	901	16
Aromatic Hydrocarbons	Toluene	228	1045	13
Terpen	α-pinene	175	974	16

The molecules identified occur heterogeneously within the population (Figure 3).

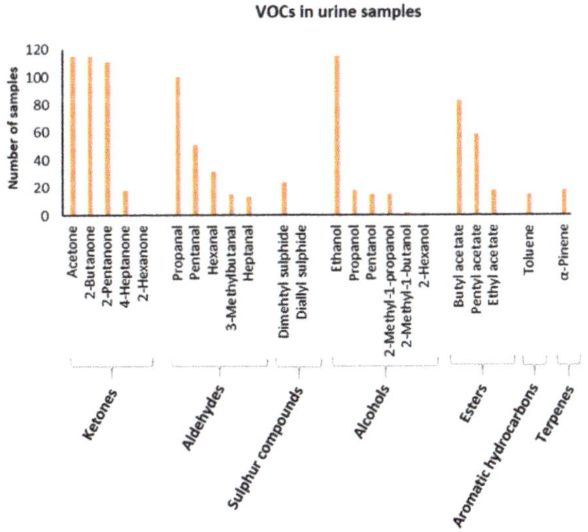

Figure 3. Graphical representation of VOCs profile in urine samples. The x-axis shows detected VOCs grouped in classes of molecules, and the y-axis shows the number of samples showing that VOC.

In particular, ketones represent the main compounds. Among these, acetone and 2-butanone are detected in the entire sample population. Then, 2-pentanone is found in 97% of the population, 4-heptanone is detected in 16% of the population, and finally 2-hexanone is found in only one sample (0.87%).

Among the aldehydes class, propanal is found in 87% of the population, pentanal is found in 44%, hexanal in 28%, 3-methyl butanal in 13%, and heptanal in 11%.

Sulphur compounds, such as dimethyl sulphide and diallyl sulphide, are found in 21 and 0.87%, respectively.

The class of alcohols is the most abundant in number of detected compounds: ethanol is present in the entire population, propanol in 16% of the population, pentanol in 13%, 2-methyl-1-propanol in 13%, 2-methyl-1-butanol in 1.74%, and 2-hexanol in 0.87%.

Regarding the class of esters, butyl acetate is the most abundant and is found in 72% of the population, pentyl acetate is found in 51%, and ethyl acetate in 16%.

Among the aromatic hydrocarbons, toluene is found in 13% of the population and, among the terpenes, α-pinene is found in 16%.

3.3. VOCs Identification in a Sub-Population of Urine Samples

The presence of some exclusive VOCs is related to a specific subpopulation of urinary samples. This group includes 15 samples characterised by a value of ketone bodies higher than 60 mg/dL. We dwell on their analysis.

Specifically, six classes of VOCs are identified. Among these, most overlap those identified in all other samples. However, some specifically distinguish these samples, including 2-hexanone, 3-methylbutanal, pentanol, 2-methyl-1-propanol, and 2-hexanol. In particular, 3-methylbutanal (aldehydes class), pentanol, and 2-methyl-1-propanol (alcohol group) are detected in all the subpopulation. Some details on the possible origin of the detected VOCs are reported in Table 4 [11].

Table 4. Summary of VOCs detected in urine with excess of ketone bodies. [(a)] class of molecule to which the VOC belongs; [(b)] list of detected VOCs; [(c)] number of samples that contain the VOC; [(d)] putative origin of the detected VOCs (Endo = VOC endogenously produced; Exo = VOC resulting from exogenous sources (food, environment, an medication); M = VOC from microbial metabolism; D = VOC from drug metabolism, as reported by Porto-Figueira et al. [11]).

Class [(a)]	VOCs [(b)]	Number of Samples [(c)]	Origin [(d)] [11]
Ketones	Acetone	15	Endo, M
	2-pentanone	14	Exo (Food)
	2-butanone	12	Endo
	4-Heptanone	7	Endo
	2-hexanone	1	Endo
Aldehydes	3-methylbutanal	15	Unknown
Sulphur compounds	Dimethyl sulphide	15	Endo/Exo (D, M)
Alcohols	Ethanol	15	Unknown
	Propanol	15	Unknown
	Pentanol	15	Unknown
	2-methyl-1-propanol	15	Unknown
	2-hexanol	1	Unknown
Esters	Butyl acetate	2	Unknown
Terpenes	α-pinene	3	Endo/Exo (Food)

4. Discussion

To the best of our knowledge, to date, the most commonly used sampling procedures for VOCs analysis are Solid-Phase Micro Extraction (SPME) for the headspace and Stir Bar Sorptive Extraction (SBSE), N,O-Bis (trimethylsilyl)trifluoroacetamide (BSTFA) derivatisation, or centrifugation for the liquid phase [12,13]. These are followed by metabolomics analysis based on GC–MS, High-Performance Liquid Chromatography with an Electrospray Ionisation source and a Time-of-Flight Mass Spectrometry detector (HPLC–ESI–TOF), Selected-Ion Flow-Tube Mass Spectrometry (SIFT-MS), and sensors (e.g., Electronic Nose, e-Nose) [14,15]. Although these are considered the gold-standard techniques for urinary VOCs detection, they are extremely expensive and time-consuming and are, thus, not suitable for fast clinical applications. In this perspective, we develop and validate an innovative analytical method to overcome some of the limits reported so far. The main strengths of our method are its ease of use and rapid results. In particular, our analysis is performed on a GC–IMS. Both the dual-physical separation of VOCs and the high sensitivity of the IMS allow identification of compounds at the ppb level. In parallel, we use a simple device, which allows for the direct introduction of the sample in the equipment, avoiding the alteration of the analytes concentration due to extraction or pre-concentration methods. This method provides results in 10 min. The extremely low time and cost of analysis make it a particularly useful technique for fast initial screening.

Based on our results, a good level of sensitivity is achieved and a linearity range is supplied at the concentration of interest (from 5 to 130 ppb).

In order to obtain a comprehensive and fast mapping of urine volatilome, this method is applied to a first cohort of 115 urine samples from a heterogeneous population of patients without a specific preselection. Twenty-three VOCs related to seven different classes of molecules are detected. As shown in Table 3, their origin can be diverse, including endogenous synthesis and/or production resulting from microbial metabolism and external sources [11]. Ketones are one of the major classes of molecules detected in urine samples. As reported [16], they are common in urine of both healthy and ill subjects. In addition, acetone, 2-butanone, 2-pentanone, and 4-heptanone are the major ketones detected in our samples. Acetone is present in all samples, and it is the most abundant VOCs. This endogenous compound can derive from two different metabolic pathways: from the glucose metabolism through the β-oxidation of acetoacetic acid or from the hydrogenation of isopropanol [17]. At physiological concentrations (133 ppb–6 ppm) [18], acetone is related to the energy metabolism. Conversely, at higher concentration, acetone is considered as a biomarker for diabetes mellitus and type I diabetes [19]. 2-butanone and 2-pentanone are possible biomarkers for lung [11,20] and bladder [21] cancer. In these above-mentioned studies, VOCs (acetone, 2-butanone, and 2-pentanone) are detected using GC–MS analysis after a solid-phase micro-extraction (SPME) [21,22]. With our method, we are able to identify these molecules by reducing the analysis time, which emphasises its potential for clinical studies.

4-heptanone is a common volatile constituent of human urine; it is of unknown origin and it may arise from in vivo decarboxylation of an oxoacid (3-oxo-2-ethylhexanoic acid) from plasticisers with a similar process to acetone from acetoacetic acid [23]. Different research studies, based on headspace solid-phase micro-extraction (HS-SPME) coupled with the GC–MS technique, report 4-heptanone as a possible biomarker for bladder [21], breast [24], lung [11], and renal cell [22] carcinoma.

Among the volatile sulphur compounds, dimethyl sulphide is highly present in urine and is a major contributor to their odour [1]. This VOC is considered as a biomarker for the lung and colorectal cancer [11,25]. To the best of our knowledge, no data have been collected on diallyl sulphide.

Esters are not common urinary VOCs. Among them, only ethyl acetate is shown as a putative biomarker for lung cancer. It has been detected in urine by a headspace GC equipped with a programmed temperature vaporiser and mass spectrometry detector (HS–PTV–GC–MS) [26]. Aldehydes can be produced from the oxygen free-radical-mediated lipid peroxidation of fatty acids. Hexanal is one of the most common aldehydes found in urine [27]. It has been detected with SPME–GC–MS [20,21], Needle Trap Micro-Extraction (NTME) GC–MS [11], and HS–GC–MS [26] and is considered a potential biomarker for many types of cancer such as bladder [21], colorectal [25], leukaemia [16], prostate [28], and especially for lung cancer [29]. Heptanal is the second most found aldehyde in urine samples. In particular, a decrease in its concentration is related to lung [29], colorectal, leukaemia, and lymphoma cancer [16], while an increase in its levels is related to head and neck cancer [30].

The most widely used technique for detecting aldehydes is SPME–GC–MS. A study performed by Khalid et al. identified pentanal as a biomarker for prostate cancer [31]. Propanal is also detected in all our samples, but no other evidence has been collected so far.

Alcohols can have different origins such as the reduction of fatty acids in the gastrointestinal tract [32]. To the best of our knowledge, ethanol, n-propanol, and n-butanol are the most common alcohols in urine and their concentration increases for diabetic patients [33]. Many of the other compounds could be produced by exogenous sources such as food.

Taking into account all the results, although the number of VOCs detectable by other techniques are higher than ours, our method is able to overlap the detection of many compounds. As an example, in the recent study of Taunk et al. [34], the authors showed a volatilomic urinary profile for patients with lung cancer compared to healthy controls using the headspace solid-phase microextraction technique combined with the GC–MS methodology. Interestingly, many VOCs related to clinical differences, such as acetone, 2-butanone, 2-pentanone, 4-heptanone, and toluene, are also detected by our approach.

In parallel, propanal, hexanal, 3-methylbutanal, 2-butanone, and 4-heptanone are widely related to different types of cancer, as reported by Pinto et al.'s study [35]. In addition, Silva et al. [24] described the urinary volatilomic composition of patients with breast cancer and healthy individuals to detect possible VOCs biomarkers. These include some VOCs detectable by our approach, including acetone, 2-butanone, 2-pentanone, hexanal, ethyl acetate, and toluene.

Finally, we focus our attention on a specific class of urine samples characterised by an excess of ketone bodies (>60 mg/dL). Compared to the larger population, more alcohols are found in the 15 samples, many of which are present in all of them (Table 4). Among the detected aldehydes, the compounds differ from the rest of the population. With regard to ketones, 2-hexanone is found in addition to the others previously detected and mentioned. Volatile compounds such as acetone, dimethyl sulphide, 3-methylbutanal, propanol, pentanol, and ethanol are found in all our samples as shown in the gallery plot of the main peak areas (Figure 4).

In conclusion, this study aimed to comprehensively profile urinary VOCs by rapid GC–IMS analysis. Based on our results, this methodological approach promises to discriminate VOCs in clinically well-classified patient groups.

We are aware that our study shows some limitations. First, this approach does not allow the quantification of identified VOCs. This would require the development of a more accurate analytical protocol, with the use of specific internal standards, and further investigation of analytical parameters such as precision (repeatability and intermediate precision), limit of quantitation (LOQ), robustness, and recovery [36].

Furthermore, this preliminary study does not take into account contributory factors that may influence both the synthesis and the concentration of VOCs themselves. The latter factors include the clinical features of the population analysed, such as demographic characteristics, diet, alcohol consumption, smoking, and various environmental factors [7,37].

As is well known, the assessment of both pre-analytical and analytical factors is a critical point for the research of biomarkers in biological fluids [38–40].

All these important issues, which will be explored in subsequent studies, are beyond the objective of the present manuscript, which, as mentioned, is to obtain a qualitative mapping of the urinary VOCs profile with a rapid screening method.

The overlap of our results with those of other studies mentioned above strengthens the reliability of our proposed method. In this context, GC–IMS stands as a powerful, robust, and easy-to-use technique for separating and detecting VOCs for a rapid, nontargeted screening approach.

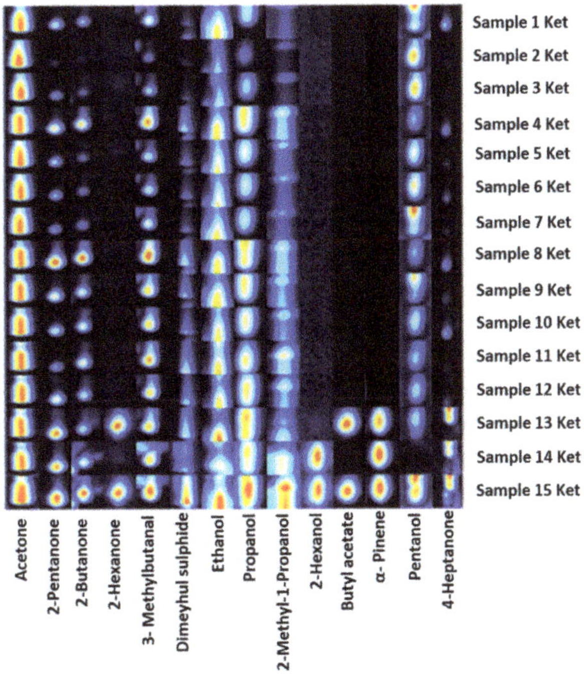

Figure 4. Gallery plot of GC–IMS signals of 14 VOCs species detected in 15 urine samples with ketone bodies value over 60 mg/dL.

5. Conclusions

Although GC–MS remains the gold-standard technique for detecting urinary VOCs, it is also extremely time-consuming and expensive. Therefore, it is not a suitable technique to be implemented in the context of fast clinical screening.

With this in mind, we propose an analytically validated alternative method based on the use of GC–IMS for the rapid detection of VOCs in urine, biological fluid widely used in the clinic. This method is not intended to replace more sensitive techniques and must be coupled to analysis for VOCs quantification. However, based on our results, it can represent a first step for rapidly obtaining a profile of urinary VOCs useful for clinical applications.

Author Contributions: Conceptualization, V.G. and A.U.; methodology, G.R., S.B. and V.G.; software, G.R.; validation, G.R. and V.G.; formal analysis, G.R. and V.G.; investigation, G.R. and V.G.; resources, G.R., S.B. and V.G.; data curation, G.R. and V.G.; writing—original draft preparation, G.R. and V.G.; writing—review and editing, V.G., A.U. and S.B.; visualization, G.R. and V.G.; supervision, V.G. and A.U.; project administration, V.G. and A.U.; funding acquisition, A.U. All authors have read and agreed to the published version of the manuscript.

Funding: This research received no external funding.

Institutional Review Board Statement: This study was approved by the Ethics Committee at Fondazione Policlinico A. Gemelli IRCCS- Rome, Italy on 23 September 2022 (MeSolVOCs, protocol number ID5271). All investigations were carried out on the residual aliquots of the samples following the conclusions of all clinical procedures. No clinical indication has been provided as a result of these molecular investigations.

Informed Consent Statement: The patient's consent was not required for this study, as all investigations were carried out on the residual aliquots of samples from anonymous subjects following the conclusions of all clinical procedures.

Data Availability Statement: The data presented in this study are available in article.

Conflicts of Interest: The authors declare no conflict of interest.

References

1. da Costa, B.R.B.; De Martinis, B.S. Analysis of Urinary VOCs Using Mass Spectrometric Methods to Diagnose Cancer: A Review. *Clin. Mass Spectrom.* **2020**, *18*, 27–37. [CrossRef] [PubMed]
2. Behera, B.; Joshi, R.; Vishnu, G.A.; Bhalerao, S.; Pandya, H.J. Electronic Nose: A Non-Invasive Technology for Breath Analysis of Diabetes and Lung Cancer Patients. *J. Breath Res.* **2019**, *13*, 024001. [CrossRef] [PubMed]
3. Van Vliet, D.; Smolinska, A.; Jöbsis, Q.; Rosias, P.; Muris, J.; Dallinga, J.; Dompeling, E.; Van Schooten, F.J. Can Exhaled Volatile Organic Compounds Predict Asthma Exacerbations in Children? *J. Breath Res.* **2017**, *11*, 016016. [CrossRef] [PubMed]
4. Shigeyama, H.; Wang, T.; Ichinose, M.; Ansai, T.; Lee, S.W. Identification of Volatile Metabolites in Human Saliva from Patients with Oral Squamous Cell Carcinoma via Zeolite-Based Thin-Film Microextraction Coupled with GC-MS. *J. Chromatogr. B. Analyt. Technol. Biomed. Life Sci.* **2019**, *1104*, 49–58. [CrossRef] [PubMed]
5. Greco, V.; Piras, C.; Pieroni, L.; Ronci, M.; Putignani, L.; Roncada, P.; Urbani, A. Applications of MALDI-TOF Mass Spectrometry in Clinical Proteomics. *Expert Rev. Proteom.* **2018**, *15*, 683–696. [CrossRef]
6. Schmidt, K.; Podmore, I. Current Challenges in Volatile Organic Compounds Analysis as Potential Biomarkers of Cancer. *J. Biomark.* **2015**, *2015*, 981458. [CrossRef]
7. Tyagi, H.; Daulton, E.; Bannaga, A.S.; Arasaradnam, R.P.; Covington, J.A. Urinary Volatiles and Chemical Characterisation for the Non-Invasive Detection of Prostate and Bladder Cancers. *Biosensors* **2021**, *11*, 437. [CrossRef]
8. Tiele, A.; Wicaksono, A.; Kansara, J.; Arasaradnam, R.P.; Covington, J.A. Breath Analysis Using Enose and Ion Mobility Technology to Diagnose Inflammatory Bowel Disease—A Pilot Study. *Biosensors* **2019**, *9*, 55. [CrossRef]
9. Westhoff, M.; Litterst, P.; Freitag, L.; Urfer, W.; Bader, S.; Baumbach, J.I. Ion Mobility Spectrometry for the Detection of Volatile Organic Compounds in Exhaled Breath of Patients with Lung Cancer: Results of a Pilot Study. *Thorax* **2009**, *64*, 744–748. [CrossRef]
10. Gasparri, R.; Capuano, R.; Guaglio, A.; Caminiti, V.; Canini, F.; Catini, A.; Sedda, G.; Paolesse, R.; Di Natale, C.; Spaggiari, L. Volatolomic Urinary Profile Analysis for Diagnosis of the Early Stage of Lung Cancer. *J. Breath Res.* **2022**, *16*, 046008. [CrossRef]
11. Porto-Figueira, P.; Pereira, J.A.M.; Câmara, J.S. Exploring the Potential of Needle Trap Microextraction Combined with Chromatographic and Statistical Data to Discriminate Different Types of Cancer Based on Urinary Volatomic Biosignature. *Anal. Chim. Acta* **2018**, *1023*, 53–63. [CrossRef]
12. Aggarwal, P.; Baker, J.; Boyd, M.T.; Coyle, S.; Probert, C.; Chapman, E.A. Optimisation of Urine Sample Preparation for Headspace-Solid Phase Microextraction Gas Chromatography-Mass Spectrometry: Altering Sample PH, Sulphuric Acid Concentration and Phase Ratio. *Metabolites* **2020**, *10*, 482. [CrossRef] [PubMed]
13. Aggio, R.B.M.; Mayor, A.; Coyle, S.; Reade, S.; Khalid, T.; Ratcliffe, N.M.; Probert, C.S.J. Freeze-Drying: An Alternative Method for the Analysis of Volatile Organic Compounds in the Headspace of Urine Samples Using Solid Phase Micro-Extraction Coupled to Gas Chromatography-Mass Spectrometry. *Chem. Cent. J.* **2016**, *10*, 9. [CrossRef] [PubMed]
14. Shen, C.; Sun, Z.; Chen, D.; Su, X.; Jiang, J.; Li, G.; Lin, B.; Yan, J. Developing Urinary Metabolomic Signatures as Early Bladder Cancer Diagnostic Markers. *Omics* **2015**, *19*, 1–11. [CrossRef] [PubMed]
15. Brooks, S.W.; Moore, D.R.; Marzouk, E.B.; Glenn, F.R.; Hallock, R.M. Canine Olfaction and Electronic Nose Detection of Volatile Organic Compounds in the Detection of Cancer: A Review. *Cancer Invest.* **2015**, *33*, 411–419. [CrossRef]
16. Silva, C.L.; Passos, M.; Cmara, J.S. Investigation of Urinary Volatile Organic Metabolites as Potential Cancer Biomarkers by Solid-Phase Microextraction in Combination with Gas Chromatography-Mass Spectrometry. *Br. J. Cancer* **2011**, *105*, 1894–1904. [CrossRef]
17. Matin, A.A.; Maleki, R.; Farajzadeh, M.A.; Farhadi, K.; Hosseinzadeh, R.; Jouyban, A. Headspace SPME-GC Method for Acetone Analysis and its Biomedical Application. *Chromatographia* **2007**, *66*, 383–387. [CrossRef]
18. Mochalski, P.; Unterkofler, K. Quantification of Selected Volatile Organic Compounds in Human Urine by Gas Chromatography Selective Reagent Ionization Time of Flight Mass Spectrometry (GC-SRI-TOF-MS) Coupled with Head-Space Solid-Phase Microextraction (HS-SPME). *Analyst* **2016**, *141*, 4796–4803. [CrossRef]
19. Yamada, K.; Ohishi, K.; Gilbert, A.; Akasaka, M.; Yoshida, N.; Yoshimura, R. Measurement of Natural Carbon Isotopic Composition of Acetone in Human Urine. *Anal. Bioanal. Chem.* **2015**, *408*, 1597–1607. [CrossRef]
20. Hanai, Y.; Shimono, K.; Matsumura, K.; Vachani, A.; Albelda, S.; Yamazaki, K.; Beauchamp, G.K.; Oka, H. Urinary Volatile Compounds as Biomarkers for Lung Cancer. *Biosci. Biotechnol. Biochem.* **2012**, *76*, 679–684. [CrossRef]
21. Cauchi, M.; Weber, C.M.; Bolt, B.J.; Spratt, P.B.; Bessant, C.; Turner, D.C.; Willis, C.M.; Britton, L.E.; Turner, C.; Morgan, G. Evaluation of Gas Chromatography Mass Spectrometry and Pattern Recognition for the Identification of Bladder Cancer from Urine Headspace. *Anal. Methods* **2016**, *8*, 4037–4046. [CrossRef]
22. Wang, D.; Wang, C.; Pi, X.; Guo, L.; Wang, Y.; Li, M.; Feng, Y.; Lin, Z.; Hou, W.; Li, E. Urinary Volatile Organic Compounds as Potential Biomarkers for Renal Cell Carcinoma. *Biomed. Rep.* **2016**, *5*, 68–72. [CrossRef] [PubMed]
23. Walker, V.; Mills, G.A. Urine 4-Heptanone: A β-Oxidation Product of 2-Ethylhexanoic Acid from Plasticisers. *Clin. Chim. Acta* **2001**, *306*, 51–61. [CrossRef]

24. Silva, C.L.; Perestrelo, R.; Silva, P.; Tomás, H.; Câmara, J.S. Implementing a Central Composite Design for the Optimization of Solid Phase Microextraction to Establish the Urinary Volatomic Expression: A First Approach for Breast Cancer. *Metabolomics* **2019**, *15*, 64. [CrossRef]
25. Porto-Figueira, P.; Pereira, J.; Miekisch, W.; Câmara, J.S. Exploring the Potential of NTME/GC-MS, in the Establishment of Urinary Volatomic Profiles. Lung Cancer Patients as Case Study. *Sci. Rep.* **2018**, *8*, 13113. [CrossRef]
26. Santos, P.M.; del Nogal Sánchez, M.; Pozas, Á.P.C.; Pavón, J.L.P.; Moreno-Cordero, B. Determination of Ketones and Ethyl Acetate—A Preliminary Study for the Discrimination of Patients with Lung Cancer. *Anal. Bioanal. Chem.* **2017**, *409*, 5689–5696. [CrossRef]
27. Broza, Y.Y.; Mochalski, P.; Ruzsanyi, V.; Amann, A.; Haick, H. Hybrid Volatolomics and Disease Detection. *Angew. Chem. Int. Ed.* **2015**, *54*, 11036–11048. [CrossRef]
28. Lima, A.R.; Pinto, J.; Azevedo, A.I.; Barros-Silva, D.; Jerónimo, C.; Henrique, R.; de Lourdes Bastos, M.; Guedes de Pinho, P.; Carvalho, M. Identification of a Biomarker Panel for Improvement of Prostate Cancer Diagnosis by Volatile Metabolic Profiling of Urine. *Br. J. Cancer* **2019**, *121*, 857–868. [CrossRef]
29. Guadagni, R.; Miraglia, N.; Simonelli, A.; Silvestre, A.; Lamberti, M.; Feola, D.; Acampora, A.; Sannolo, N. Solid-Phase Microextraction–Gas Chromatography–Mass Spectrometry Method Validation for the Determination of Endogenous Substances: Urinary Hexanal and Heptanal as Lung Tumor Biomarkers. *Anal. Chim. Acta* **2011**, *701*, 29–36. [CrossRef]
30. Opitz, P.; Herbarth, O. The Volatilome-Investigation of Volatile Organic Metabolites (VOM) as Potential Tumor Markers in Patients with Head and Neck Squamous Cell Carcinoma (HNSCC). *J. Otolaryngol. Head Neck Surg.* **2018**, *47*, 42. [CrossRef]
31. Khalid, T.; Aggio, R.; White, P.; De Lacy Costello, B.; Persad, R.; Alkateb, H.; Jones, P.; Probert, C.S.; Ratcliffe, N. Urinary Volatile Organic Compounds for the Detection of Prostate Cancer. *PLoS ONE* **2015**, *10*, e0143283. [CrossRef] [PubMed]
32. Garner, C.E.; Smith, S.; de Lacy Costello, B.; White, P.; Spencer, R.; Probert, C.S.J.; Ratcliffem, N.M. Volatile Organic Compounds from Feces and Their Potential for Diagnosis of Gastrointestinal Disease. *FASEB J.* **2007**, *21*, 1675–1688. [CrossRef] [PubMed]
33. Liebich, H.; Buelow, H.; Kallmayer, R. Quantification of Endogenous Aliphatic Alcohols in Serum and Urine. *J. Chromatogr. A* **1982**, *239*, 343–349. [CrossRef]
34. Taunk, K.; Porto-Figueira, P.; Pereira, J.A.M.; Taware, R.; da Costa, N.L.; Barbosa, R.; Rapole, S.; Câmara, J.S. Urinary Volatomic Expression Pattern: Paving the Way for Identification of Potential Candidate Biosignatures for Lung Cancer. *Metabolites* **2022**, *12*, 36. [CrossRef] [PubMed]
35. Pinto, J.; Amaro, F.; Lima, A.R.; Carvalho-Maia, C.; Jerónimo, C.; Henrique, R.; Bastos, M.D.L.; Carvalho, M.; de Pinho, P.G. Urinary Volatilomics Unveils a Candidate Biomarker Panel for Noninvasive Detection of Clear Cell Renal Cell Carcinoma. *J. Proteome Res.* **2021**, *20*, 3068–3077. [CrossRef] [PubMed]
36. Guy, R.I.H.T. *Guideline, Validation of Analytical Procedures: Text and Methodology Q2 (R1)*; International Conference on Harmonisation: Geneva, Switzerland, 2005; pp. 11–12.
37. Bannaga, A.S.; Tyagi, H.; Daulton, E.; Covington, J.A.; Arasaradnam, R.P. Exploratory Study Using Urinary Volatile Organic Compounds for the Detection of Hepatocellular Carcinoma. *Molecules* **2021**, *26*, 2447. [CrossRef] [PubMed]
38. Greco, V.; Piras, C.; Pieroni, L.; Urbani, A. Direct Assessment of Plasma/Serum Sample Quality for Proteomics Biomarker Investigation. *Serum/Plasma Proteom.* **2017**, *1619*, 3–21. [CrossRef]
39. Greco, V.; Pieragostino, D.; Piras, C.; Aebersold, R.; Wiltfang, J.; Caltagirone, C.; Bernardini, S.; Urbani, A. Direct Analytical Sample Quality Assessment for Biomarker Investigation: Qualifying Cerebrospinal Fluid Samples. *Proteomics* **2014**, *14*, 1954–1962. [CrossRef]
40. Stevens, V.L.; Hoover, E.; Wang, Y.; Zanetti, K.A. Pre-Analytical Factors that Affect Metabolite Stability in Human Urine, Plasma, and Serum: A Review. *Metabolites* **2019**, *9*, 156. [CrossRef]

metabolites

Article

A Sample Preparation Method for the Simultaneous Profiling of Signaling Lipids and Polar Metabolites in Small Quantities of Muscle Tissues from a Mouse Model for Sarcopenia

Yupeng He [1], Marlien van Mever [1], Wei Yang [1], Luojiao Huang [1], Rawi Ramautar [1], Yvonne Rijksen [2,3], Wilbert P. Vermeij [2,3], Jan H. J. Hoeijmakers [2,3,4,5], Amy C. Harms [1], Peter W. Lindenburg [1,6] and Thomas Hankemeier [1,*]

1 Metabolomics and Analytics Centre, Leiden Academic Centre for Drug Research, Faculty of Science, Leiden University, Einsteinweg 55, 2333 CC Leiden, The Netherlands
2 Princess Máxima Center for Pediatric Oncology, 3584 CS Utrecht, The Netherlands
3 Oncode Institute, 3521 AL Utrecht, The Netherlands
4 Department of Molecular Genetics, Erasmus MC Cancer Institute, Erasmus University Medical Center Rotterdam, 3015 GD Rotterdam, The Netherlands
5 Institute for Genome Stability in Aging and Disease, Cologne Excellence Cluster for Cellular Stress Responses in Aging-Associated Diseases (CECAD), University of Cologne, 50931 Cologne, Germany
6 Research Group Metabolomics, Leiden Center for Applied Bioscience, University of Applied Sciences Leiden, 2333 CK Leiden, The Netherlands
* Correspondence: hankemeier@lacdr.leidenuniv.nl; Tel.: +31-71-527-1340

Abstract: The metabolic profiling of a wide range of chemical classes relevant to understanding sarcopenia under conditions in which sample availability is limited, e.g., from mouse models, small muscles, or muscle biopsies, is desired. Several existing metabolomics platforms that include diverse classes of signaling lipids, energy metabolites, and amino acids and amines would be informative for suspected biochemical pathways involved in sarcopenia. The sample limitation requires an optimized sample preparation method with minimal losses during isolation and handling and maximal accuracy and reproducibility. Here, two developed sample preparation methods, BuOH-MTBE-Water (BMW) and BuOH-MTBE-More-Water (BMMW), were evaluated and compared with previously reported methods, Bligh-Dyer (BD) and BuOH-MTBE-Citrate (BMC), for their suitability for these classes. The most optimal extraction was found to be the BMMW method, with the highest extraction recovery of 63% for the signaling lipids and 81% for polar metabolites, and an acceptable matrix effect (close to 1.0) for all metabolites of interest. The BMMW method was applied on muscle tissues as small as 5 mg (dry weight) from the well-characterized, prematurely aging, DNA repair-deficient *Ercc1*$^{\Delta/-}$ mouse mutant exhibiting multiple–morbidities, including sarcopenia. We successfully detected 109 lipids and 62 polar targeted metabolites. We further investigated whether fast muscle tissue isolation is necessary for mouse sarcopenia studies. A muscle isolation procedure involving 15 min at room temperature revealed a subset of metabolites to be unstable; hence, fast sample isolation is critical, especially for more oxidative muscles. Therefore, BMMW and fast muscle tissue isolation are recommended for future sarcopenia studies. This research provides a sensitive sample preparation method for the simultaneous extraction of non-polar and polar metabolites from limited amounts of muscle tissue, supplies a stable mouse muscle tissue collection method, and methodologically supports future metabolomic mechanistic studies of sarcopenia.

Keywords: metabolomics extraction; signaling lipids; polar metabolites; muscle tissue; muscle ageing and sarcopenia

Citation: He, Y.; van Mever, M.; Yang, W.; Huang, L.; Ramautar, R.; Rijksen, Y.; Vermeij, W.P.; Hoeijmakers, J.H.J.; Harms, A.C.; Lindenburg, P.W.; et al. A Sample Preparation Method for the Simultaneous Profiling of Signaling Lipids and Polar Metabolites in Small Quantities of Muscle Tissues from a Mouse Model for Sarcopenia. *Metabolites* **2022**, *12*, 742. https://doi.org/10.3390/metabo12080742

Academic Editor: Joana Pinto

Received: 15 July 2022
Accepted: 10 August 2022
Published: 12 August 2022

1. Introduction

Sarcopenia is characterized by the age-related loss of muscle mass and function, constitutes a major health problem, and is associated with a high loss of quality of life [1,2].

Globally, 11–50% of those aged 80 or above suffer from sarcopenia [3], and this number is increasing with the rapid growth of the ageing population, thereby creating an enormous socioeconomic and health care burden. The molecular mechanisms underlying sarcopenia are still not well understood and effective medication is lacking [4]. Metabolomics is a powerful approach for obtaining molecular insight into complex diseases and for the discovery of disease biomarkers [5]. Previous muscle function metabolomics studies revealed that dysregulation of signaling lipids (i.e., oxylipins, free fatty acids, oxidative stress markers) [6–8], energy metabolites (i.e., ATP, citrate, pyruvate) [9–11], and amino acids and amines [10,12,13] were highly associated with weak muscle contractile function. Therefore, a systematic metabolomics mechanistic study of these non-polar (signaling lipids) and polar (energy metabolites, amino acids and amines) metabolites is needed for understanding the biochemistry behind sarcopenia and for the identification of biomarkers for the diagnosis, prevention, and treatment of sarcopenia. Mice deficient in the DNA excision-repair gene, *Ercc1* ($Ercc1^{\Delta/-}$), show numerous age-related pathologies and accelerated ageing features [14–16], and are widely used in the studies of ageing and age-related diseases, including muscle wasting and sarcopenia [17–20]. Moreover, this mouse mutant is an excellent model for several rare, but very severe progeroid human DNA repair syndromes, including Cockayne syndrome, xeroderma pigmentosum, Fanconi anemia, and XFE1 syndrome [21–23]. As the $Ercc1^{\Delta/-}$ mice exhibit early cessation of growth, only small amounts (i.e., 5–50 mg dry weight) of (skeletal) muscle can be collected, necessitating the development of a single sensitive, reproducible sample preparation method suitable for analysis by multiple metabolomics platforms, thereby allowing for the analysis of non-polar and polar metabolites.

The Bligh and Dyer (BD) method is a traditional sample preparation method for the extraction of non-polar and polar components, which is able to non-selectively extract a wide range of metabolites [24,25]. Medina et al. evaluated sample extraction methods with isopropanol and 1-butanol:methanol for simultaneous extraction of 584 non-polar and 116 polar metabolites; however, the method mainly focused on the metabolome analysis of human plasma samples and some of our targeted signaling lipids—oxylipins and bile acids were not covered [26]. Löfgren developed an automated butanol:methanol extraction method for lipids, however, the method mainly focused on the plasma lipid classes, i.e., cholesterol, triacylglycerol, phosphatidylcholine, sphingomyelin, and lysophospholipids [27]. BuOH-MTBE-Citrate (BMC) is a sensitive sample preparation method for sample limited applications; Di Zazzo et al. applied this method for the analysis of oxylipins, oxidative stress markers, endocannabinoids, and bile acids for ocular surface cicatrizing conjunctivitis, and identified 9S-hydroxy octadecatrienoic acid (9S-HOTrE) and 5-hydroxy eicosapentaenoic acid (5-HEPE) as potential diagnostic biomarker candidates. However, the performance of BMC on a small amount of muscle tissues still remains unknown, and because of the addition of a non-volatile (citric acid/phosphate) buffer, the extracted aqueous phase was not compatible with mass spectrometric detection [28].

In this work, we report the development of a sample preparation method that allows for the simultaneous extraction of targeted non-polar and polar metabolites from biomass-limited mouse muscle tissues (i.e., 5–50 mg dry weight). With this approach, we would like to obtain more insight into the etiology of sarcopenia using a metabolomics approach. For this purpose, two extraction methods based on BMC [28] were developed and compared with Di Zazzo et al.'s BMC [28] and BD methods [29]. The optimal method with the highest extraction recovery and acceptable matrix effect was applied to muscle tissues of $Ercc1^{\Delta/-}$ mice to study the effect of the muscle tissue isolation speed on metabolite stability. Overall, this work yielded a sensitive sample preparation method for the simultaneous extraction of non-polar and polar metabolites from limited amounts of muscle tissues, supplied a reference method for an existing sarcopenia samples collection, and methodologically supports the metabolomic analysis of sarcopenia.

2. Materials and Methods

2.1. Chemicals

Methanol and chloroform were purchased from Biosolve Chimime SARL (Dieuze, France). The 1-butanol was purchased from Acros Organics (Geel, Belgium). Butylated hydroxytoluene (BHT) and methyl tert-butyl ether (MTBE), and citric acid and sodium dihydrogen phosphate dehydrate were obtained from Sigma-Aldrich (Steinheim, Germany). MilliQ water was obtained from a Millipore high-purity water dispenser (Billerica, MA, USA). All solvents were HPLC grade or higher.

For internal standards (ISTDs), deuterium-, carbon-, and/or nitrogen-labelled metabolites were used. Labelled oxylipins, fatty acids, and endocannabinoids ISTDs were acquired from Cayman Chemicals (Ann Arbor, MI, USA). Labelled lysophospholipids, sphingolipids, and bile acid and steroid ISTDs were ordered from Avanti Polar Lipids (Alabaster, AL, USA). Labelled amino acids and amine ISTDs were ordered from Cambridge Isotope Laboratories (Andover, MA, USA), and labelled ATP, AMP, and UTP were purchased from Sigma-Aldrich (Steinheim, Germany).

2.2. ISTDs Preparation

For lipid ISTDs, the stock solution was prepared in MeOH in a stated concentration (Table S1) containing 0.4 mg/mL BHT. This includes the classes of oxylipins, fatty acids, endocannabinoids, bile acids and steroids, lysophospholipids, and sphingolipids. For the stock solution of amino acids and amine ISTDs, 9 kinds of ISTDs (Table S2) were prepared in MilliQ water with a concentration of 0.5 mg/mL. Stock solutions of ATP ($^{13}C_{10}$,$^{15}N_5$), AMP ($^{13}C_{10}$,$^{15}N_5$), and UTP ($^{13}C_9$,$^{15}N_2$) were prepared in MilliQ water at 10 mg/mL (Table S3).

2.3. Muscle Samples

The development and evaluation of extraction methods were performed on pig muscle tissues that serve as a uniform source for multiple experiments and as a surrogate for mouse tissue, which was only available in scarce quantities. The pig muscle tissue was stored at −80 °C before extraction. Muscle tissue from mice deficient in the DNA excision-repair gene *Ercc1* (*Ercc1*$^{\Delta/-}$) was utilized for the study of effect of sample isolation speed on metabolite stability for sarcopenia. The generation and characterization of *Ercc1*$^{\Delta/-}$ mice is described in [15,16,20]. Three kinds of muscle types, gastrocnemius + soleus (Gas + Sol), quadriceps (Quadr), and extensor digitorum longus + tibialis anterior (EDL + TA), were collected at the animal facility of the Erasmus Medical Center, Rotterdam, Netherlands. All above experiments were performed in accordance with the Principles of Laboratory Animal Care and with the guidelines approved by the Dutch Ethical Committee (permit Nos. 139-12-13 and 139-12-18) in full accordance with European legislation.

Fast and delayed (15-min delayed) muscle tissue collection procedures were applied to study the effects of sample isolation speed on metabolite stability. Briefly, mice were anaesthetized using CO_2. For fast sample isolation, a large piece of Quadr tissue was dissected immediately and rapidly frozen in liquid nitrogen, and EDL + TA and Gas + Sol tissue were carefully isolated as described in [30]. Following dissection, the muscles were immediately frozen in liquid-nitrogen-cooled isopentane and stored at −80 °C [19]. For delayed sample isolation, the Quadr, EDL + TA, and Gas + Sol tissues from the other hind leg of the same mouse were kept for 15 min at room temperature, then were isolated and frozen as described above for the fast isolation. All samples were stored at −80 °C until analysis.

2.4. Extraction Methods

For the development of an extraction method yielding high extraction efficiency for both polar metabolites and signaling lipids, four extraction methods were compared and evaluated using pig muscle tissues, i.e., the Bligh-Dyer (BD), BuOH-MTBE-Citrate (BMC), BuOH-MTBE-Water (BMW), and BuOH-MTBE-more-Water (BMMW) extraction methods. Thirty mg (±20%) of frozen wet pig muscle tissue was lyophilized in a VaCo I freeze-dryer (Zirbus, Bad Grund, Germany; connected to a E2M12 high vacuum pump, Edwards, Crawley, England) for 24 h and weighed. To homogenize muscle tissues thoroughly, a dry-homogenization method was used by adding 100 mg (±10%) of zirconium oxide beads (0.5 mm; Next Advance, Averill Park, NY, USA) to the freeze-dried tissue, and homogenized in a Bullet Blender (BBX24; Next Advance, Averill Park, NY, USA) for 15 min at speed 9 [29]. Labelled ISTDs (10 µL amino acids & amines, 10 µL ATP & AMP & UTP, 10 µL lipids stock solution) were spiked in the muscle samples before and after extraction for the evaluation of the four extraction methods.

2.4.1. Bligh-Dyer Extraction (BD)

A previously reported Bligh-Dyer extraction was utilized for the polar and non-polar analyte extraction [29]. Briefly, 400 µL of cold MeOH and 125 µL of cold MilliQ water were added to the muscle tissues and homogenized by using the Bullet Blender for 15 min at speed 9. Then, 450 µL of homogenate was transferred to a new tube after centrifugation (500× *g*, 5 min, 4 °C), and vortexed with cold chloroform (450 µL), water (250 µL), and MeOH (50 µL) for 2 min. The samples were next left on ice for 10 min to partition, and centrifuged (2000× *g*, 10 min, 4 °C) to obtain a clear biphasic mixture. The 500 µL of upper aqueous/polar phase and 400 µL of lower organic/non-polar phase were collected separately by using positive-displacement Microman pipettes (Gilson, Middleton, WI, USA) without disturbing the layer between both phases. These were then evaporated in a SpeedVac Vacuum concentrator (Thermo Savant SC210A, Waltham, MA, USA) and reconstituted in 50 µL of MeOH for the organic phase and 100 µL of 50%-MeOH–50%-MilliQ water for the aqueous phase.

2.4.2. BuOH-MTBE-Citrate Extraction (BMC)

A reported lipid extraction method [28], BuOH-MTBE-Citrate extraction (BMC), was tested for the muscle samples. In this method, 5 µL of antioxidant solution (0.4 mg/mL BHT:EDTA = 1:1), 150 µL of 0.2 M citric acid-0.4 M disodium hydrogen phosphate buffer at pH 4.5, and 1 mL of extraction solution (BuOH: MTBE = 1:1, *v*/*v*) were added to all samples and allowed to settle on ice for 20 min before homogenization in the Bullet Blender for 15 min at speed 9. Then, the homogenized samples were centrifuged (2000× *g*, 4 °C) for 10 min, and 900 µL of the upper organic phase was collected, evaporated, and reconstituted using the same method described in 3.4.1 BD method.

2.4.3. BuOH-MTBE-Water Extraction (BMW)

The extraction procedure for the BMW method is similar to the BMC method, but the 150 µL of citric acid/phosphate buffer was replaced with 150 µL of cold MilliQ water. After collection of the upper organic phase, 500 µL more of ice-cold MilliQ water was added to more easily collect the lower aqueous phase. After vortexing and centrifugation at 2000× *g* at 4 °C for 10 min, 350 µL of the lower aqueous phase was then collected.

2.4.4. BuOH-MTBE-More-Water Extraction (BMMW)

A larger aqueous phase volume (400 µL of cold MilliQ water) was utilized in the BMMW method instead of the 150 µL of cold MilliQ water used in the BMW method. Two-hundred µL of the lower aqueous phase was directly collected after collection of the upper organic phase.

2.5. LC/CE-MS Quality Control

Some extra extracted pig muscle tissues were pooled together as quality control (QC) samples. A QC sample was injected once each 6–8 samples to evaluate and correct for changes in the sensitivity of the instruments. The metabolites with a relative standard deviation (RSD) of quality control (QC) samples less than 30% were used for statistical analysis.

2.5.1. Lipid Metabolite Analysis

The signaling lipid metabolites were measured according to a validated ultra-performance liquid chromatography tandem mass spectrometry (UPLC-MS/MS) method in our lab [28]. Briefly, each sample was measured with two complementary reverse phase methods using mobile phases with a different pH.

The low-pH run utilized an Acquity BEH C18 column (50 × 2.1 mm, 1.7 µm; Waters, Milford, USA) on a Shimadzu LC-30AD (Kyoto, Japan) hyphenated to a SCIEX Q-Trap 6500+ (Framingham, MA, USA). Separations were performed using three mobile phases: (A) water with 0.1% acetic acid; (B) ACN: MeOH (9:1, v/v) with 0.1% acetic acid; (C) Isopropanol with 0.1% acetic acid at 40 °C at a flow rate of 0.7 mL/min. The 16-min run used the following gradient: start with 20% B and 1% C; B was increased to 85% between 0.75 and 14 min and C was increased to 15% between 11 and 14 min; and the condition held for 0.5 min prior to column re-equilibration at the starting conditions from 14.8 to 16 min. Data were acquired using Sciex Analyst software (Version 1.7, Framingham, MA, USA) and peak integration used Sciex OS (Version 1.4.0, Framingham, MA, USA).

The high-pH run used a Kinetex® Core-Shell EVO 100 Å C18 column (50 × 2.1 mm, 1.8 µm; Phemomenex, Torrance, CA, USA) on a Shimadzu LCMS-8060 system (Shimadzu, Kyoto, Japan). Separations used mobile phases: (A) 5% ACN with 2 mM ammonium acetate and 0.1% ammonium hydroxide and (B) 95% ACN with 2 mM ammonium acetate and 0.1% ammonium hydroxide at 40 °C at a flow rate of 0.6 mL/min. The gradient started with 1% B; B was increased to 100% from 0.7 to 7.7 min; and 100% B held for 0.75 min prior to re-equilibration at the starting conditions between 8.75 and 11 min. Multiple reaction monitoring (MRM) was utilized in MS/MS acquisition in both the positive and negative electrospray ionization mode with polarity switching. Data were acquired and peaks integrated using LabSolutions (Version 5.97 SP1, Shimadzu, Kyoto, Japan).

2.5.2. Energy Metabolites Analysis

The energy metabolites were analyzed using a hydrophilic interaction liquid chromatography (HILIC) mass spectrometry platform [31]. Briefly, a SeQuant ZIC-cHILIC column (PEEK 100 × 2.1 mm, 3.0 µm particle size; Merck KGaA, Darmstadt, Germany) was used on a Waters UPLC (AcquityTM, Milford, MA, USA) coupled with a Sciex MS (Triple-TOF 5600+, Framingham, MA, USA). The separation method used mobile phases: (A) 90% ACN with 5 mM ammonium acetate at pH 6.8 and (B) 10% ACN with 5 mM ammonium acetate at pH 6.8, at a flow rate of 0.25 mL/min at 30 °C. The gradient method was: 100% A for 2 min; ramping 3–20 min to 60% A; ramping 20–20.1 to 100% A; and re-equilibrated to 35 min with 100% A. The MS data were acquired at a full scan range of 50–900 m/z in the negative ionization mode with curtain gas measured at 39.3 psi, source temperature at 400 °C, and ion source voltage at 4.64 kV by Sciex Analyst (Version 1.7, Framingham, MA, USA), and the peaks were integrated using MultiQuant (Version 3.0.1, Sciex, Framingham, MA, USA).

2.5.3. Amino Acids and Amines Analysis

The amino acids and amines were analyzed by a sheath-liquid Agilent 7100 capillary electrophoresis (CE) system, coupled to an Agilent mass spectrometer (TOF 6230, Waldbronn, Germany), and acquired by MassHunter Data Acquisition (Version B.05.01, Agilent, Santa Clara, CA, USA). Fused-silica capillaries (BGB Analytik, Harderwijk, Netherlands) with a total length of 70 cm and an internal diameter of 50 µm were utilized. The CE separation voltage was 30 kV, and 10% acetic acid in water was used as a background electrolyte (BGE) solution. The sheath-liquid, a mixture of water and isopropanol (50:50, v/v) containing 0.03% acetic acid, was delivered at a flow rate of 3 µL/min by an Agilent 1260 Infinity Isocratic Pump (Waldbronn, Germany). The nebulizer gas was set to 0 psi, the sheath gas flow rate was set at 11 L/min, and the sheath gas temperature was set at 100 °C. The ESI capillary voltage was set at 5500 V. The fragmentor and skimmer voltages were 150 V and 50 V, respectively. MS data were acquired in the positive ion mode between

50 and 1000 m/z with an acquisition rate of 1.5 spectra/s [32]. The amino acids and amines peaks were integrated using MassHunter Quantitative Analysis (Version 05.02, Agilent, Santa Clara, CA, USA).

2.6. Data Analysis

For metabolites for which QC samples had an RSD less than 30%, the response ratios were corrected by QC response ratio and further normalized by the muscle tissue dry weight. For metabolites that can be measured by multiple platforms, i.e., amino acids/amines (which can be measured by the HILIC and CE methods in Sections 2.5.2 and 2.5.3, respectively) and some fatty acids (which can be measured by both low- and high-pH lipid platforms), the method with the smaller QC RSD was utilized (Tables S4 and S5).

For the comparison and evaluation of the developed extraction methods, the extraction recovery and matrix effect were utilized. Extraction recovery was calculated as the ratio of the ISTDs spiked at the start of the extraction procedure and the ISTDs spiked to the injection solvent prior to MS measurement. This value does not reflect the extraction recovery of metabolites from muscle tissue, but the loss of targeted metabolites during the liquid–liquid extraction process. The matrix effect was calculated by Equation (1), where the ratio of ISTDs is extracted from a muscle sample and a blank sample with only extraction solvents:

$$\text{Matrix effect} = \text{ITSDs extracted form muscle samples} \div \text{ISTDs extracted from blank samples} \quad (1)$$

For selection of the optimal extraction method, the percentage of the highest extraction recovery for ISTDs for each extraction method was used and calculated by Equation (2):

$$\text{Percentage (\%)} = \text{The number of highest recovery in ISTDs} \div \text{The number of total ISTDs} \times 100\% \quad (2)$$

For metabolite stability evaluation in mouse muscle tissue, the response ratio was used and obtained by Equation (3):

$$\text{Response ratio} = \text{peak area of the target metabolite} \div \text{peak area of the assigned ISTD} \quad (3)$$

RStudio (Version 1.4.1106) and R (Version 4.0.5) were used for the statistical analysis of the data statistical—all the figures were made by Graphpad Prism (Version 8.1.1, San Diego, CA, USA).

3. Results and Discussion

3.1. Development and Evaluation of the Sample Preparation Methods

Four sample preparation methods, i.e., BD, BMC, BMW, BMMW, were systematically compared and evaluated with respect to extraction recovery and matrix effect for a range of metabolite classes by spiking carbon- or deuterium-labelled metabolites (ISTDs) using pig muscle tissue as a surrogate for mouse muscle during method development.

3.1.1. Extraction of Signaling Lipids

Five classes of lipid metabolites, i.e., oxylipins, lysophospholipids and sphingolipids, free fatty acids, bile acids and steroids, and endocannabinoids, were analyzed in the organic phase for evaluation of the four extraction methods. Figure 1A showed that the extraction recovery of these lipids using BMC (orange), BMW (brown), and BMMW (yellow) were significantly higher than when using the BD (blue) method. This may be due to the utilization of the more non-polar solvents, MTBE and BuOH (relative polarity is 0.124 and 0.586, respectively [33]), for signaling lipids extraction in BMC, BMW, and BMMW than the two less non-polar solvents, i.e., chloroform and MeOH (relative polarity is 0.259 and 0.762, respectively [33]), in BD. The higher non-polar property contributed to a higher partitioning of all signaling lipids in the organic phase in BMC, BMW, and BMMW. Similar results were also reported in [34], which indicated that lipids in a hydrophobically-associated form can more easily be extracted by relatively non-polar solvents, and the polar solvents, i.e., ethanol and methanol, can disrupt the hydrogen bonding or electrostatic forces between membrane-associated lipids and protein.

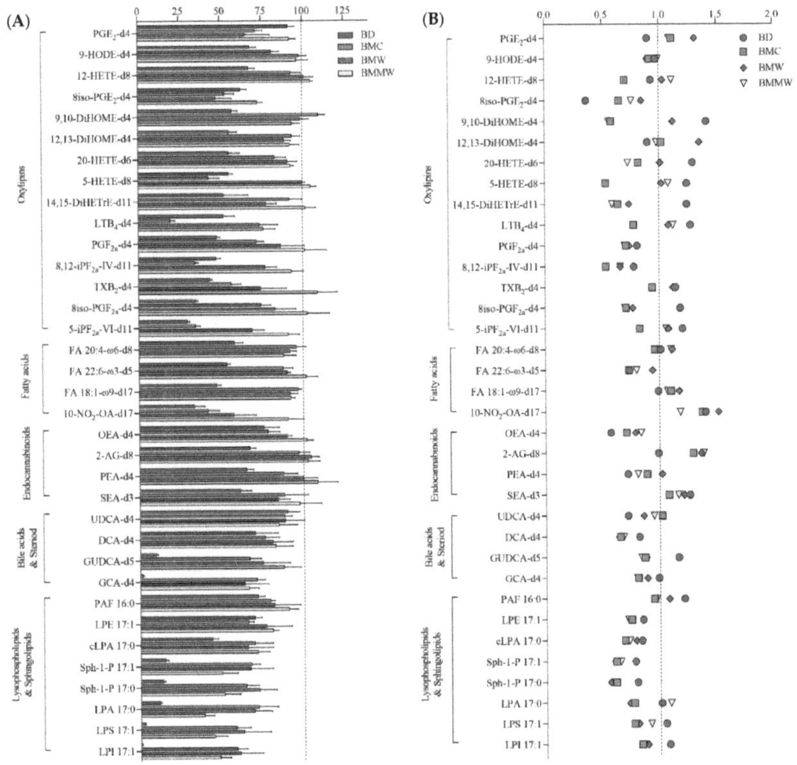

Figure 1. The (**A**) extraction recovery (%) and (**B**) matrix effect of lipids ISTDs by using the four extraction methods: BD, BMC, BMW, and BMMW. Lower recovery values of lysophospholipids and sphingolipids (<91%), and some bile acids, i.e., GCA-d4 (2–71%) and DCA-d4 (70–83%), were observed in all four extraction methods compared to other lipid metabolites. The reason for the lower yield might be because of the less non-polar properties of lysophospholipids and sphingolipids (logP = 2.6–5.4), GCA (logP = 1.4) and DCA (logP = 3.3) than the other classes of lipid metabolites, i.e., fatty acids (logP = 6.0–6.8), endocannabinoids (logP = 5.7–6.7), and oxylipins (logP = 3.1–5.9). The higher recovery of oxylipins reported using the BD method (around 100%) in Alves et al.'s study compared with our BMMW method (>73%) results from the combination of both the organic and aqueous phases for the measurement of these polar lipids [29]. Here, we compared just the organic phase extraction performance for lipid metabolites in the four sample preparation methods.

To determine the matrix effects of the four extraction methods on the targeted lipid measurements, signals of spiked internal standards in samples with and without muscle tissue were investigated (Equation (1)). For most of the signaling lipids, matrix effect values (Figure 1B) ranged between 0.7–1.4, indicating that there is acceptable impact on MS measurements from the muscle tissue matrix for all four extraction methods.

3.1.2. Extraction of Polar Metabolites

As a non-volatile (citric acid/phosphate) buffer was utilized in the published BMC method, the aqueous phase was rendered unsuitable for the intended LC-MS analysis methods. In addition, the exogenous citric acid affected the analysis of one of our target metabolites, citric acid. Therefore, the aqueous phase of the BMC method was not considered for the polar metabolite analysis. Two separation methods for polar metabolites, i.e., HILIC for central energy metabolites, and CE for amino acids and amines, were used to evaluate the extraction of polar metabolites into the aqueous phase for the three extraction methods (BD, BMW, and BMMW). For amino acids and amines, the extraction recoveries in BMW (brown) and BMMW (yellow) were significantly higher than in BD (blue) (Figure 2A). For energy metabolites, the recovery of ATP and UTP in BMW and BMMW was notably better than in BD; however, the recovery of AMP was dramatically lower compared to BD (Figure 2A). This might be the result of one extra 2-min vortex step with chloroform, water, and MeOH at room temperature in BD, which accelerated the hydrolysis of ATP (or ADP) to AMP. Similar results showing ATP hydrolysis at room temperature were also observed in Becker et al.'s study [35]. Bruno et al. preferred the BD method for polar metabolites in mouse muscle over the MeOH/water extraction method but did not evaluate other methods [36]. Given the stability issues, we concluded that the BD method was not the optimal extraction method for the HILIC measurements of energy metabolites from muscle tissues for our study.

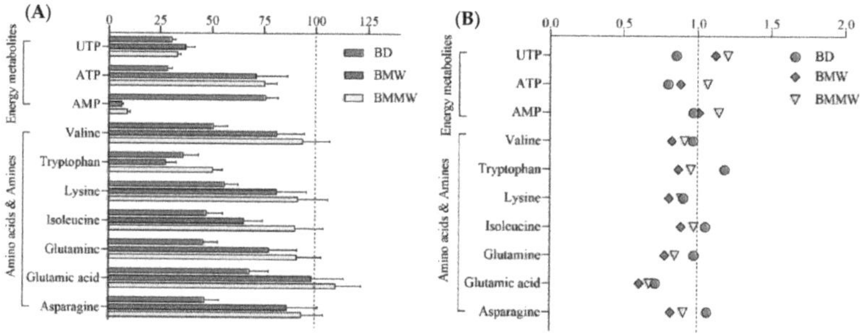

Figure 2. The (**A**) extraction recovery (%) and (**B**) matrix effect of polar ISTDs by using the extraction methods: BD, BMW, and BMMW.

When evaluating the performance of the extraction methods for the CE measurements of amino acids and amines, we note the relatively low recovery obtained for tryptophan (28–50%) as compared to other amino acids and amines. This may be due to its high susceptibility to oxidative degradation [37]—its weakest polar property (logP = −1.1)—and water solubility (1.36 mg/mL) among this class of metabolites (logP ranges from −2.0 to −5.4, water solubility ranges from 80.6 to 210 mg/mL), which contributed to the less passive distribution of tryptophan in the aqueous phase. The weak polar property of UTP (logP = −3.4) may have also contributed to its lower distribution in the aqueous phase compared to ATP (logP = −5.1). The lower extraction recovery of UTP than weakly polar amino acids and amines, i.e., valine (logP = −2.0), may be due to the much higher water solubility of valine (210 mg/mL) than UTP (8.37 mg/mL). Matrix effect values (Figure 2B) were close to one for most of the polar metabolites, demonstrating small impacts from the extraction methods and muscle tissue matrix on MS measurement for the targeted polar metabolites.

3.1.3. Assessment of Sample Preparation Method Yielding Optimal Recovery for Signaling Lipids and Polar Metabolites

The performance of four extraction methods (BD, BMC, BMW, and BMMW) for signaling lipids and three extraction methods (BD, BMW, and BMMW) for polar metabolites were evaluated and compared by calculating the percentage of the highest extraction recovery for each extraction method for different internal controls for each of the two chemical categories (Equation (2)). The BMMW method turned out to give the best recovery, as deduced from reaching the highest percentage of spiked internal standards for both non-polar (63%) and polar (81%) metabolites (Figure 3), thereby demonstrating that this method resulted in the smallest loss of metabolites during the sample preparation procedure in

BMMW for all classes of metabolites of interest. BD was not preferred for mouse muscle extraction not only because of the lower recovery and percentage values, but also due to the rapid hydrolysis observed for ATP (or ADP) to AMP, and the labour required for the reproducible separation of the organic and aqueous phase [24]. Therefore, BMMW was chosen as the extraction method of choice for the targeted non-polar and polar metabolites from small quantities of mouse muscles.

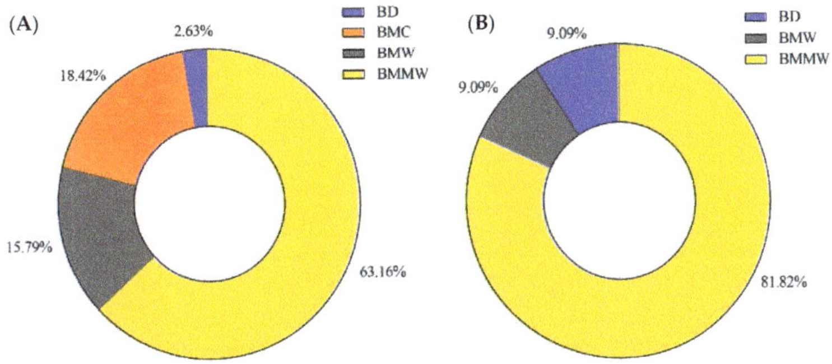

Figure 3. The percentage of the highest extraction recovery each method occupied in (**A**) signaling lipids and (**B**) polar metabolites.

3.2. *Performance of the Optimal Sample Preparation Method in Mouse Muscle Samples*

For the metabolic profiling of mouse muscle, the reported LC-MS and CE-MS detection methods for lipid metabolites [38], energy metabolites [31], and amino acids and amines [32] were utilized. One-hundred-and-nine non-polar and 62 polar targeted metabolites were clearly observed (with a signal to noise ratio > 10) using the LC-MS and CE-MS detection platforms to analyze *Ercc1*$^{\Delta/-}$ mouse muscle tissues (Figure 4). Detailed information of these non-polar (lipid) and polar metabolites for LC-MS and CE-MS analysis is provided in Tables S4 and S5, respectively—in the Supporting Information section. As the sample collection procedure can also influence metabolite stability, the effect of muscle isolation speed on metabolite stability for these targeted metabolites was further investigated.

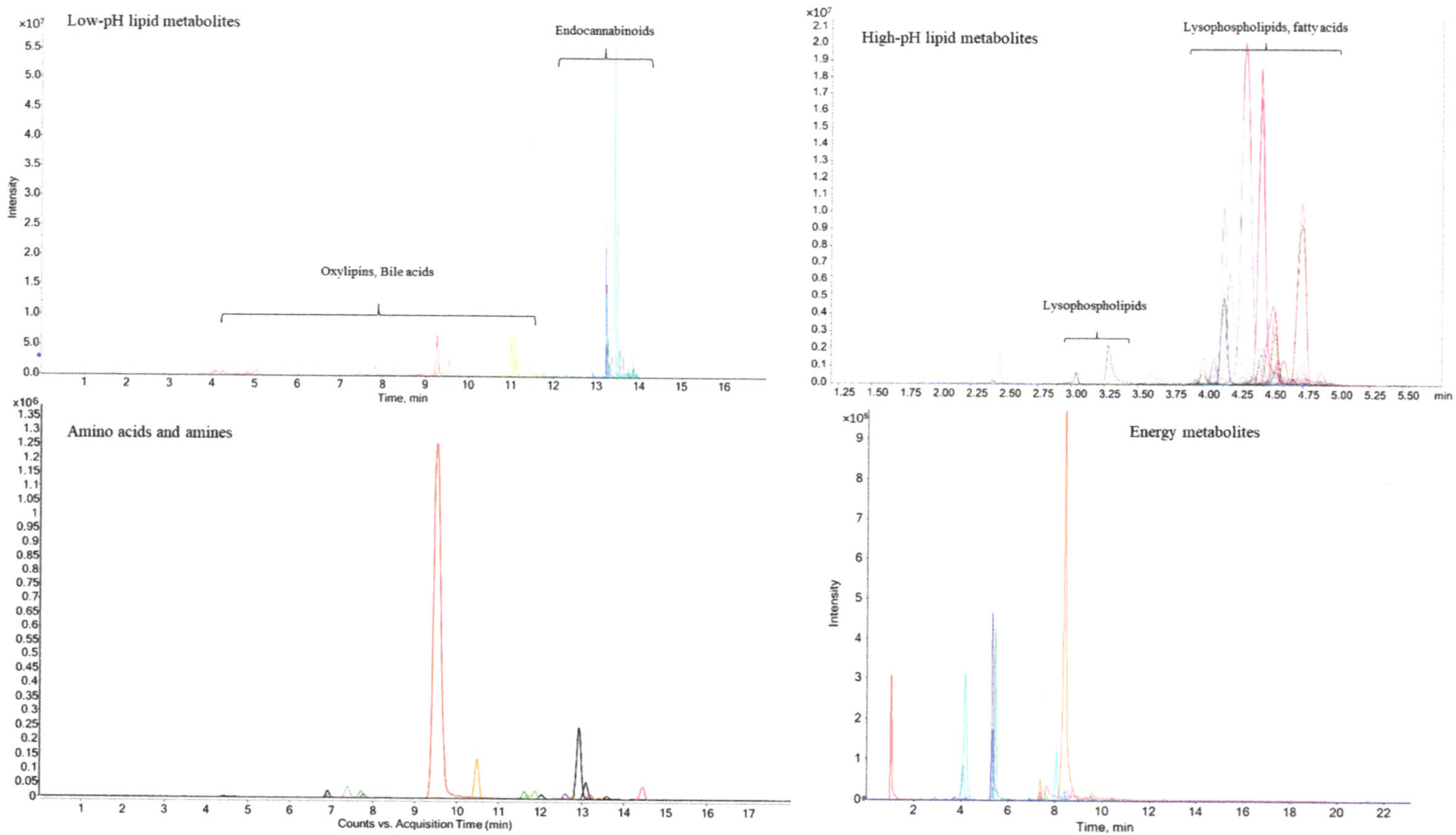

Figure 4. Representative LC/CE-MS chromatograms from different classes of metabolites obtained from *Ercc1*$^{\Delta/-}$ mouse muscle samples.

3.3. The Effects of Sample Isolation Speed on Metabolite Stability

For future metabolomics mechanistic studies of sarcopenia, we must utilize a sensitive sample preparation method and a muscle tissue isolation method that preserves metabolite stability. To deduce the effect of sample collection speed on metabolite stability, the response ratios (Equation (3)) of metabolites in fast and delayed muscle tissue isolation were investigated in three muscle specimens, namely the lower hindlimb muscles gastrocnemius and soleus (Gas + Sol), the extensor digitorum longus + tibialis anterior (EDL + TA), and the upper hindlimb muscle quadriceps (Quadr), which are the most commonly used mouse muscles for molecular analyses. In Gas + Sol, significantly higher unsaturated fatty acids (FA18.1-ω9, FA20.3-ω6, FA20.4-ω6, FA20.5-ω3, FA22.4-ω6) and oxylipins (19-20-DiHDPA, 8-9-DiHETrE) were observed in delayed isolation samples compared to the fast isolation (Table 1). These fatty acids and oxylipins are in the arachidonic acid and eicosapentaenoic acid pathways, which are associated with inflammation and age-related diseases [39], and are oxidation sensitive [40,41]. Fifteen-min at room temperature led to longer oxygen exposure and maybe changed the enzymatic activity in these muscle tissues, which contributed to the oxidation and instability of the unsaturated fatty acids, as well as the generation of their downstream metabolites, i.e., oxylipins [42,43]. The higher lysophospholipids (Table 1), i.e., LPE14.0, LPE16.1, LPE20.4, LPE22.4, LPG16.1, LPI20.4, LPI22.4, and LPI22.6, in 15-min delayed isolation muscle tissues might be due to the hydrolysis of the cellular membrane induced by the longer time of oxidation exposure and oxidative damage [44,45], and/or tissue degeneration. The significantly increased pyruvate in Gas + Sol with delayed isolation (Table 2) may be due to the oxidation of lactate [46]. Creatine phosphate is considered to be the "energy pool" in muscle cells and will be preferentially consumed under the condition of insufficient energy and generate its downstream metabolite, creatine [47]. A higher creatine content in Gas + Sol with the 15-min delayed isolation (Table 2) may be explained by the insufficient energy supply in muscle tissues post-dissection and the consumption of creatine phosphate in the muscle cells before the muscle tissue is isolated and snap frozen [47,48]. The Quadr muscle was much more stable than Gas + Sol with 15-min delayed isolation, as only 3 metabolites, i.e., 7-HDoHE, creatine, and PEA, were significantly affected. More altered metabolites were observed in EDL + TA with 15-min delayed isolation compared to both Gas + Sol and Quadr.

The increase in the number of significantly altered metabolites after delayed isolation in the Gas + Sol muscle, compared to Quadr muscle, may be due to the type-I oxidative muscle (soleus) included in Gas + Sol, as well as the type-II glycolytic muscle of Quadr [49–51]. The oxidative fibers mainly use aerobic respiration to provide ATP, and glycolytic fibers primarily use anaerobic glycolysis as their energy supply [52], which induced more oxidation in Gas + Sol than Quadr. The largest number of significantly altered metabolites was observed in EDL + TA, which may be due to the varied and unsystematic muscle type composition and fiber density in TA [53–55]. Kammoun et. al. found that 57% of type IIB, 3% of hybrid IIAX fibers, and no hybrid IIX/IIB fibers were observed in TA [53]. However, Bloemberg et. al. found mouse white tibialis anterior contained 12.1% hybrid fibers [54]. Lexell et. al. revealed that in TA, the proportion of type-I fibers and fiber density varied significantly but not systematically, and also differed significantly between individuals [55]. Similarly, the fiber types of EDL muscle in *Ercc1*$^{\Delta/-}$ mice are altered in composition compared to normal wild-type controls, having reduced type IIA/IIX and increased type IIB [19]. These variations in TA tissue and/or in mice may have contributed to the observed metabolite alterations in EDL + TA with 15-min delayed isolation at room temperature. The different muscle type proportion may be responsible for the observed differences in the stability of metabolites in the three different kinds of muscle. Given the observed instability of metabolites in muscle tissues with 15-min delayed isolation, fast muscle tissue collection will be preferred for our future sarcopenia study.

Table 1. The effect of sample collection speed on lipid metabolites stability in different muscle types (n = 3).

Analytes	Gas + Sol	Quadr	EDL + TA	Analytes	Gas + Sol	Quadr	EDL + TA
FA16.0	ns	ns	*	LEA	*	ns	*
FA18.0	ns	ns	ns	SEA	ns	ns	ns
FA18.1-ω9	*	ns	*	1-AG & 2-AG	*	ns	ns
FA18.3-ω3	ns	ns	ns	CDCA	ns	ns	ns
FA20.3-ω6	*	ns	ns	GCA	ns	ns	ns
FA20.3-ω9	ns	ns	ns	GCDCA	ns	ns	ns
FA20.4-ω6	*	ns	ns	GDCA	ns	ns	ns
FA20.5-ω3	**	ns	ns	GUDCA	ns	ns	ns
FA22.4-ω6	*	ns	*	cLPA16.1	ns	ns	ns
FA22.5-ω3	ns	ns	ns	cLPA18.0	ns	ns	ns
FA22.5-ω6	ns	ns	ns	cLPA18.1	ns	ns	ns
FA22.6-ω3	ns	ns	*	cLPA18.2	ns	ns	ns
10-HDoHE	ns	ns	ns	LPA14.0	ns	ns	ns
11-HDoHE	ns	ns	ns	LPA16.1	ns	ns	ns
11-HETE	ns	ns	ns	LPA18.0	ns	ns	ns
12-13-DiHOME	ns	ns	ns	LPA18.1	ns	ns	ns
12-HEPE	ns	ns	***	LPA18.2	ns	ns	ns
13-14dihydro-15k-PGD2	ns	ns	ns	LPA20.4	ns	ns	*
13-14dihydro-15k-PGE2	ns	ns	*	LPA22.4	ns	ns	ns
13-14dihydro-PGF2α	ns	ns	ns	LPE14.0	*	ns	ns
13-HODE	ns	ns	ns	LPE16.0	ns	ns	ns
14-15-DiHETrE	ns	ns	*	LPE16.1	*	ns	*
14-HDoHE	ns	ns	ns	LPE18.0	ns	ns	ns
8iso-PGE1	ns	ns	*	LPE18.1	ns	ns	*
8iso-PGF1α	ns	ns	ns	LPE18.2	ns	ns	*
15S-HETrE	ns	ns	ns	LPE18.3	ns	ns	*
17-HDoHE	ns	ns	ns	LPE20.3	ns	ns	**
18-HEPE	ns	ns	***	LPE20.4	*	ns	*
19-20-DiHDPA	*	ns	**	LPE20.5	ns	ns	ns
1a-1b-dihomo-PGF2α	ns	ns	ns	LPE22.4	*	ns	*
20-HETE	ns	ns	*	LPE22.5	ns	ns	ns
5-HETE	ns	ns	ns	LPE22.6	ns	ns	*
5-iPF2α-VI	ns	ns	ns	LPG14.0	ns	ns	*
7-HDoHE	ns	*	ns	LPG16.0	ns	ns	ns
8-12-iso-iPF2α-VI	ns	ns	ns	LPG16.1	*	ns	ns
8-9-DiHETrE	*	ns	*	LPG18.0	ns	ns	ns
8-HDoHE	ns	ns	ns	LPG18.1	ns	ns	ns
8-HETE	ns	ns	ns	LPG18.2	ns	ns	ns
8iso-15R-PGF2α	ns	ns	ns	LPG20.3	ns	ns	*
8iso-PGE2	ns	ns	ns	LPG20.4	ns	ns	ns
8iso-PGF2α	ns	ns	ns	LPG22.4	ns	ns	ns
8S-HETrE	ns	ns	ns	LPI16.1	ns	ns	*
9-10-13-TriHOME	ns	ns	ns	LPI18.0	ns	ns	ns
9-10-DiHOME	ns	ns	ns	LPI18.1	ns	ns	ns
9-HEPE	ns	ns	*	LPI18.2	ns	ns	*
9-HETE	ns	ns	ns	LPI20.4	*	ns	*
9-HODE	ns	ns	ns	LPI22.4	*	ns	*
iPF2α-IV	ns	ns	ns	LPI22.6	*	ns	ns
PGD2	ns	ns	ns	LPS18.1	ns	ns	**
PGD3	ns	ns	ns	LPS18.2	ns	ns	**
PGE2	ns	ns	ns	LPS20.4	ns	ns	***
PGF2α	ns	ns	*	LPS22.4	ns	ns	**
TXB2	ns	ns	ns	LPS22.6	*	ns	*
AEA	*	ns	*	OEA	ns	ns	ns
PEA	ns	*	ns				

Note: ns means no significant difference, * means $p < 0.05$, ** means $p < 0.01$, *** means $p < 0.001$. Orange background color means significantly increased; Blue background color means significantly decreased.

Table 2. The effect of sample collection speed on stability of energy metabolites, amino acids, and amines in different muscle types (n = 3).

Energy Metabolites							
Analytes	**Gas + Sol**	**Quadr**	**EDL + TA**	**Analytes**	**Gas + Sol**	**Quadr**	**EDL + TA**
Acetyl-CoA	ns	ns	**	IMP	ns	ns	ns
Adenosine	**	ns	***	Creatine	*	*	ns
ADP	ns	ns	*	Inosine	ns	ns	ns
AMP	ns	ns	ns	α-Ketoglutarate	*	ns	ns
Ascorbic-acid	ns	ns	ns	6-phosphogluconic-acid	ns	ns	*
ATP	ns	ns	**	Malate	ns	ns	ns
cAMP	ns	ns	*	GTP	ns	ns	**
CDP	ns	ns	ns	Guanosine	ns	ns	ns
cis-Aconitate	ns	ns	ns	Oxiglutathione	ns	ns	ns
CMP	ns	ns	ns	Phosphoenolpyruvate	ns	ns	ns
CTP	ns	ns	ns	Pyruvate	**	ns	*
Cytidine	ns	ns	ns	Succinate	ns	ns	*
Dihydroxyacetone-P	ns	ns	*	UDP	ns	ns	
Fructose-6-P	ns	ns	ns	UMP	ns	ns	*
GABA	*	ns	ns	Uridine	ns	ns	**
GDP	ns	ns	ns	UTP	ns	ns	*
Glucose	ns	ns	ns	Xanthine	*	ns	**
Glucose-1-P	ns	ns	ns	Glycerate-3-P	ns	ns	
Glucose-6-P	ns	ns	ns	GMP	ns	ns	**
Glyceraldehyde-3-P	ns	ns	ns	Hypoxanthine	*	ns	ns
Amino acids and amines							
Alanine	ns	ns	ns	Methionine	ns	ns	*
Arginine	ns	ns	ns	Phenylalanine	*	ns	ns
Asparagine	ns	ns	ns	Proline	ns	ns	ns
Aspartic-acid	ns	ns	*	Serine	ns	ns	ns
Lysine	ns	ns	ns	Spermidine	ns	ns	ns
Creatinine	ns	ns	ns	Tyrosine	ns	ns	ns
Glutamic-acid	ns	ns	ns	Valine	ns	ns	ns
Glutamine	ns	ns	ns	Threonine	ns	ns	ns
Glycine	ns	ns	*	Ornithine	ns	ns	ns
Histidine	ns	ns	ns	4-Hydroxyproline	*	ns	ns
Leucine	ns	ns	ns	Tryptophan	ns	ns	ns

Note: ns means no significant difference, * means $p < 0.05$, ** means $p < 0.01$, *** means $p < 0.001$. Orange background color means significantly increased; Blue background color means significantly decreased.

4. Conclusions

Four extraction methods (BD, BMC, BMW, and BMMW) were compared and evaluated to find the optimal sample preparation method for the simultaneous extraction of targeted non-polar and polar metabolites from a limited amount of muscle tissues. The optimal method, BMMW, had an acceptable matrix effect (close to 1.0) for all metabolites and showed the highest extraction recovery for all types of metabolites, with the best performance of all methods studied for 63% of the signaling lipids and 81% of the polar metabolites. BMMW was used for profiling mouse muscle tissues with quantities as small as 5 mg (dry weight). Our study of sample collection protocols found that fast (<15 min) muscle tissue collection is crucial for metabolite stability. The developed sensitive sample preparation method and fast muscle tissue isolation method will be utilized for future metabolomics mechanistic studies of sarcopenia and animal model studies to evaluate treatments to prevent this syndrome.

Supplementary Materials: The following supporting information can be downloaded at: https://www.mdpi.com/article/10.3390/metabo12080742/s1, Table S1: The information of lipid ISTDs; Table S2: The information of amino acids and amines ISTDs; Table S3: The information of energy metabolites ISTDs; Table S4: Detected lipid metabolites in mouse muscle samples; Table S5: Detected polar metabolites in mouse muscle samples.

Author Contributions: Conceptualization, investigation, formal analysis, writing—original draft, Y.H.; Conceptualization, investigation, writing—review & editing, M.v.M.; Conceptualization, writing—review & editing, W.Y.; Conceptualization, writing—review & editing, L.H.; Conceptualization, writing—review & editing, R.R.; Resources, Y.R.; Resources, writing—review & editing, W.P.V.; Resources, writing—review & editing, J.H.J.H.; Conceptualization, writing—review & editing, A.C.H.; Conceptualization, writing—review & editing, P.W.L.; Supervision, writing—review and editing, funding acquisition T.H. All authors have read and agreed to the published version of the manuscript.

Funding: This research was funded by Netherlands Organisation for Scientific Research (NWO) in the Building Blocks of Life, grant number 737.016.015; the China Scholarship Council (CSC), No. 201706320322. J.H.J.H. was additionally supported by the European Research Council Advanced Grant Dam2Age, NIH grant (PO1 AG017242), the Deutsche Forschungsgemeinschaft—Project-ID 73111208—SFB 829, J.H.J.H. and W.P.V. by ZonMW Memorabel (733050810), and EJP-RD TC-NER RD20-113, and J.H.J.H., W.P.V. and Y.R. by ONCODE (Dutch Cancer Society). This research was part of the Netherlands X-omics Initiative and partially funded by NWO, project 184.034.019.

Institutional Review Board Statement: The animal study protocol was approved by the Institutional Review Board (or Ethics Committee) of the Principles of Laboratory Animal Care and with the guidelines approved by the Dutch Ethical Committee (permit Nos. 139-12-13 and 139-12-18) in full accordance with European legislation.

Informed Consent Statement: Not applicable.

Data Availability Statement: The data presented in this study are accessible through EBI Metabolights repository accession number MTBLS5644 (www.ebi.ac.uk/metabolights/MTBLS5644, accessed on 14 July 2022).

Acknowledgments: We thank Kimberly Smit, Renata Brandt, Sander Barnhoorn and the animal caretakers for general assistance with mouse experiments.

Conflicts of Interest: The authors declare no conflict of interest. The funders had no role in the design of the study; in the collection, analyses, or interpretation of data; in the writing of the manuscript; or in the decision to publish the results.

References

1. Santilli, V.; Bernetti, A.; Mangone, M.; Paoloni, M. Clinical definition of sarcopenia. *Clin. Cases Miner. Bone Metab.* **2014**, *11*, 177–180. [CrossRef] [PubMed]
2. Fielding, R.A.; Vellas, B.; Evans, W.J.; Bhasin, S.; Morley, J.E.; Newman, A.B.; Abellan van Kan, G.; Andrieu, S.; Bauer, J.; Breuille, D.; et al. Sarcopenia: An undiagnosed condition in older adults. Current consensus definition: Prevalence, etiology, and consequences. International working group on sarcopenia. *J. Am. Med. Dir. Assoc.* **2011**, *12*, 249–256. [CrossRef]
3. von Haehling, S.; Morley, J.E.; Anker, S.D. An overview of sarcopenia: Facts and numbers on prevalence and clinical impact. *J. Cachexia Sarcopenia Muscle* **2010**, *1*, 129–133. [CrossRef]
4. Kwak, J.Y.; Kwon, K.-S. Pharmacological Interventions for Treatment of Sarcopenia: Current Status of Drug Development for Sarcopenia. *Ann. Geriatr. Med. Res.* **2019**, *23*, 98–104. [CrossRef]
5. Zhang, A.; Sun, H.; Yan, G.; Wang, P.; Wang, X. Metabolomics for Biomarker Discovery: Moving to the Clinic. *BioMed Res. Int.* **2015**, *2015*, 354671. [CrossRef]
6. Dalle, C.; Ostermann, A.I.; Konrad, T.; Coudy-Gandilhon, C.; Decourt, A.; Barthelemy, J.C.; Roche, F.; Feasson, L.; Mazur, A.; Bechet, D.; et al. Muscle Loss Associated Changes of Oxylipin Signatures During Biological Aging: An Exploratory Study from the PROOF Cohort. *J. Gerontol.* **2019**, *74*, 608–615. [CrossRef] [PubMed]
7. Caligiuri, S.P.B.; Parikh, M.; Stamenkovic, A.; Pierce, G.N.; Aukema, H.M. Dietary modulation of oxylipins in cardiovascular disease and aging. *Am. J. Physiol. Heart Circ. Physiol.* **2017**, *313*, H903–H918. [CrossRef] [PubMed]
8. Mo, C.; Du, Y.; O'Connell, T.M. Applications of Lipidomics to Age-Related Musculoskeletal Disorders. *Curr. Osteoporos. Rep.* **2021**, *19*, 151–157. [CrossRef] [PubMed]
9. Pastoris, O.; Boschi, F.; Verri, M.; Baiardi, P.; Felzani, G.; Vecchiet, J.; Dossena, M.; Catapano, M. The effects of aging on enzyme activities and metabolite concentrations in skeletal muscle from sedentary male and female subjects. *Exp. Gerontol.* **2000**, *35*, 95–104. [CrossRef]
10. Nair, K.S. Aging muscle. *Am. J. Clin. Nutr.* **2005**, *81*, 953–963. [CrossRef] [PubMed]
11. Kirby, B.S.; Crecelius, A.R.; Voyles, W.F.; Dinenno, F.A. Impaired Skeletal Muscle Blood Flow Control With Advancing Age in Humans Attenuated ATP Release and Local Vasodilation During Erythrocyte Deoxygenation. *Circ. Res.* **2012**, *111*, 220–230. [CrossRef]
12. Fujita, S.; Volpi, E. Amino acids and muscle loss with aging. *J. Nutr.* **2006**, *136*, 277s–280s. [CrossRef]

13. Sheffield-Moore, M.; Paddon-Jones, D.; Urban, R.J. Amino acid supplementation and skeletal muscle metabolism in ageing populations. *Horm. Res.* **2006**, *66*, 93–97. [CrossRef]
14. Vermeij, W.P.; Hoeijmakers, J.H.; Pothof, J. Genome Integrity in Aging: Human Syndromes, Mouse Models, and Therapeutic Options. *Annu. Rev. Pharmacol. Toxicol.* **2016**, *56*, 427–445. [CrossRef]
15. Dolle, M.E.; Kuiper, R.V.; Roodbergen, M.; Robinson, J.; Vlugt, S.; Wijnhoven, S.W.; Beems, R.B.; Fonteyne, L.; With, P.; Pluijm, I.; et al. Broad segmental progeroid changes in short-lived *Ercc1*$^{-/\Delta 7}$ mice. *Pathobiol. Aging Age Relat. Dis.* **2011**, *1*, 1–14. [CrossRef] [PubMed]
16. Weeda, G.; Donker, I.; Wit, J.; Morreau, H.; Janssens, R.; Vissers, C.J.; Nigg, A.; Steeg, H.; Bootsma, D.; Hoeijmakers, J.H. Disruption of mouse *ERCC1* results in a novel repair syndrome with growth failure, nuclear abnormalities and senescence. *Curr. Biol.* **1997**, *7*, 427–439. [CrossRef]
17. Graaf, E.L.; Vermeij, W.P.; Waard, M.C.; Rijksen, Y.; Pluijm, I.; Hoogenraad, C.C.; Hoeijmakers, J.H.; Altelaar, A.F.; Heck, A.J. Spatio-temporal analysis of molecular determinants of neuronal degeneration in the aging mouse cerebellum. *Mol. Cell Proteomics* **2013**, *12*, 1350–1362. [CrossRef] [PubMed]
18. Yousefzadeh, M.J.; Zhao, J.; Bukata, C.; Wade, E.A.; McGowan, S.J.; Angelini, L.A.; Bank, M.P.; Gurkar, A.U.; McGuckian, C.A.; Calubag, M.F.; et al. Tissue specificity of senescent cell accumulation during physiologic and accelerated aging of mice. *Aging Cell* **2020**, *19*, e13094. [CrossRef]
19. Alyodawi, K.; Vermeij, W.P.; Omairi, S.; Kretz, O.; Hopkinson, M.; Solagna, F.; Joch, B.; Brandt, R.M.C.; Barnhoorn, S.; Vliet, N.; et al. Compression of morbidity in a progeroid mouse model through the attenuation of myostatin/activin signalling. *J. Cachexia Sarcopenia Muscle* **2019**, *10*, 662–686. [CrossRef]
20. Vermeij, W.P.; Dollé, M.E.T.; Reiling, E.; Jaarsma, D.; Payan-Gomez, C.; Bombardieri, C.R.; Wu, H.; Roks, A.J.M.; Botter, S.M.; Eerden, B.C.; et al. Restricted diet delays accelerated ageing and genomic stress in DNA-repair-deficient mice. *Nature* **2016**, *537*, 427–431. [CrossRef]
21. Niedernhofer, L.J.; Garinis, G.A.; Raams, A.; Lalai, A.S.; Robinson, A.R.; Appeldoorn, E.; Odijk, H.; Oostendorp, R.; Ahmad, A.; Leeuwen, W. A new progeroid syndrome reveals that genotoxic stress suppresses the somatotroph axis. *Nature* **2006**, *444*, 1038–1043. [CrossRef]
22. Marteijn, J.A.; Lans, H.; Vermeulen, W.; Hoeijmakers, J.H. Understanding nucleotide excision repair and its roles in cancer and ageing. *Nat. Rev. Mol. Cell Biol.* **2014**, *15*, 465–481. [CrossRef]
23. Lans, H.; Hoeijmakers, J.H.; Vermeulen, W.; Marteijn, J.A. The DNA damage response to transcription stress. *Nat. Rev. Mol. Cell Biol.* **2019**, *20*, 766–784. [CrossRef]
24. Ulmer, C.Z.; Jones, C.M.; Yost, R.A.; Garrett, T.J.; Bowden, J.A. Optimization of Folch, Bligh-Dyer, and Matyash sample-to-extraction solvent ratios for human plasma-based lipidomics studies. *Anal. Chim. Acta* **2018**, *1037*, 351–357. [CrossRef]
25. Vvedenskaya, O.; Wang, Y.; Ackerman, J.M.; Knittelfelder, O.; Shevchenko, A. Analytical challenges in human plasma lipidomics: A winding path towards the truth. *TrAC Trends Anal. Chem.* **2019**, *120*, 115277. [CrossRef]
26. Medina, J.; Velpen, V.; Teav, T.; Guitton, Y.; Gallart-Ayala, H.; Ivanisevic, J. Single-Step Extraction Coupled with Targeted HILIC-MS/MS Approach for Comprehensive Analysis of Human Plasma Lipidome and Polar Metabolome. *Metabolites* **2020**, *10*, 495. [CrossRef]
27. Löfgren, L.; Ståhlman, M.; Forsberg, G.-B.; Saarinen, S.; Nilsson, R.; Hansson, G.I. The BUME method: A novel automated chloroform-free 96-well total lipid extraction method for blood plasma. *J. Lipid Res.* **2012**, *53*, 1690–1700. [CrossRef]
28. Di Zazzo, A.; Yang, W.; Coassin, M.; Micera, A.; Antonini, M.; Piccinni, F.; De Piano, M.; Kohler, I.; Harms, A.C.; Hankemeier, T.; et al. Signaling lipids as diagnostic biomarkers for ocular surface cicatrizing conjunctivitis. *J. Mol. Med.* **2020**, *98*, 751–760. [CrossRef] [PubMed]
29. Alves, R.D.A.M.; Dane, A.D.; Harms, A.; Strassburg, K.; Seifar, R.M.; Verdijk, L.B.; Kersten, S.; Berger, R.; Hankemeier, T.; Vreeken, R.J. Global profiling of the muscle metabolome: Method optimization, validation and application to determine exercise-induced metabolic effects. *Metabolomics* **2014**, *11*, 271–285. [CrossRef]
30. Shinin, V.; Gayraud-Morel, B.; Tajbakhsh, S. Template DNA-strand co-segregation and asymmetric cell division in skeletal muscle stem cells. *Methods Mol. Biol.* **2009**, *482*, 295–317.
31. Buijink, M.R.; van Weeghel, M.; Gulersonmez, M.C.; Harms, A.C.; Rohling, J.H.T.; Meijer, J.H.; Hankemeier, T.; Michel, S. The influence of neuronal electrical activity on the mammalian central clock metabolome. *Metabolomics* **2018**, *14*, 122. [CrossRef]
32. Drouin, N.; Pezzatti, J.; Gagnebin, Y.; González-Ruiz, V.; Schappler, J.; Rudaz, S. Effective mobility as a robust criterion for compound annotation and identification in metabolomics: Toward a mobility-based library. *Anal. Chim Acta* **2018**, *1032*, 178–187. [CrossRef]
33. Reichardt, C.; Welton, T. *Solvents and Solvent Effects in Organic Chemistry*; John Wiley & Sons: Hoboken, NJ, USA, 2011.
34. Kates, M. Chapter 3 Lipid extraction procedures. In *Laboratory Techniques in Biochemistry and Molecular Biology*; Elsevier: Chicago, IL, USA, 1972; pp. 347–353.
35. Becker, E.W. The roles of ATP in the dynamics of the actin filaments of the cytoskeleton. *Biol. Chem.* **2006**, *387*, 401–406. [CrossRef]
36. Bruno, C.; Patin, F.; Bocca, C.; Nadal-Desbarats, L.; Bonnier, F.; Reynier, P.; Emond, P.; Vourc'h, P.; Joseph-Delafont, K.; Corcia, P.; et al. The combination of four analytical methods to explore skeletal muscle metabolomics: Better coverage of metabolic pathways or a marketing argument? *J. Pharm. Biomed.* **2018**, *148*, 273–279. [CrossRef]

37. Bellmaine, S.; Schnellbaecher, A.; Zimmer, A. Reactivity and degradation products of tryptophan in solution and proteins. *Free. Radic. Biol. Med.* **2020**, *160*, 696–718. [CrossRef]
38. Schoeman, J.C.; Harms, A.C.; van Weeghel, M.; Berger, R.; Vreeken, R.J.; Hankemeier, T. Development and application of a UHPLC-MS/MS metabolomics based comprehensive systemic and tissue-specific screening method for inflammatory, oxidative and nitrosative stress. *Anal. Bioanal. Chem.* **2018**, *410*, 2551–2568. [CrossRef]
39. Das, U.N. Ageing: Is there a role for arachidonic acid and other bioactive lipids? A review. *J. Adv. Res.* **2018**, *11*, 67–79. [CrossRef]
40. Sardenne, F.; Puccinelli, E.; Vagner, M.; Pecquerie, L.; Bideau, A.; Le Grand, F.; Soudant, P. Post-mortem storage conditions and cooking methods affect long-chain omega-3 fatty acid content in Atlantic mackerel (*Scomber scombrus*). *Food Chem.* **2021**, *359*, 129828. [CrossRef]
41. Hsieh, R.J.; Kinsella, J.E. Oxidation of polyunsaturated fatty acids: Mechanisms, products, and inhibition with emphasis on fish. *Adv. Food Nutr Res.* **1989**, *33*, 233–341.
42. Kaur, N.; Chugh, V.; Gupta, A.K. Essential fatty acids as functional components of foods—A review. *J. Food Sci. Technol.* **2014**, *51*, 2289–2303. [CrossRef]
43. Suganuma, H.; Collins, C.T.; McPhee, A.J.; Leemaqz, S.; Liu, G.; Andersen, C.C.; Bonney, D.; Gibson, R.A. Effect of parenteral lipid emulsion on preterm infant PUFAs and their downstream metabolites. *Prostaglandins Leukot. Essent. Fat. Acids* **2021**, *164*, 102217. [CrossRef]
44. El-Bacha, T.; Torres, A.G. Phospholipids: Physiology. *Encycl. Food Health* **2016**, 352–359. [CrossRef]
45. Kondakova, I.V.; Borunov, E.V.; Antipova, E.V. Activation of phospholipase hydrolysis in the process of oxidative cell damage. *Biull. Eksp. Biol. Med.* **1991**, *112*, 285–287. [CrossRef]
46. Brooks, G.A.; Dubouchaud, H.; Brown, M.; Sicurello, J.P.; Butz, C.E. Role of mitochondrial lactate dehydrogenase and lactate oxidation in the intracellular lactate shuttle. *Proc. Natl. Acad. Sci. USA* **1999**, *96*, 1129–1134. [CrossRef]
47. Rhoades, R.A.; Bell, D.R. *Medical Phisiology: Principles for Clinical Medicine*, 4th ed.; Lippincott Williams & Wilkins: Philadelphia, PA, USA, 2012; pp. 156–157.
48. Vasudevan, D.M.; Sreekumari, S.; Vaidyanathan, K. *Textbook of Biochemistry for Medical Students*, 4th ed.; Jaypee Brothers Medical Publishers: New Delhi, India, 2019.
49. Jacobs, R.A.; Diaz, V.; Soldini, L.; Haider, T.; Thomassen, M.; Nordsborg, N.B.; Gassmann, M.; Lundby, C. Fast-Twitch Glycolytic Skeletal Muscle Is Predisposed to Age-Induced Impairments in Mitochondrial Function. *J. Gerontol. Biol. Sci. Med. Sci.* **2013**, *68*, 1010–1022. [CrossRef]
50. Shu, J.T.; Xiao, Q.; Shan, Y.J.; Zhang, M.; Tu, Y.J.; Ji, G.G.; Sheng, Z.W.; Zou, J.M. Oxidative and glycolytic skeletal muscles show marked differences in gene expression profile in Chinese Qingyuan partridge chickens. *PLoS ONE* **2017**, *12*, e0183118.
51. Soukup, T.; Zacharová, G.; Smerdu, V. Fibre type composition of soleus and extensor digitorum longus muscles in normal female inbred Lewis rats. *Acta Histochem.* **2002**, *104*, 399–405. [CrossRef]
52. Biga, L.M.; Dawson, S.; Harwell, A.; Hopkins, R.; Kaufmann, J.; LeMaster, M.; Matern, P.; Morrison-Graham, K.; Quick, D.; Runyeon, J. *Anatomy & Physiology*; OpenStax; Oregon State University: Corvallis, OR, USA, 2020.
53. Kammoun, M.; Cassar-Malek, I.; Meunier, B.; Picard, B. A simplified immunohistochemical classification of skeletal muscle fibres in mouse. *Eur. J. Histochem.* **2014**, *58*, 2254. [CrossRef]
54. Bloemberg, D.; Quadrilatero, J. Rapid determination of myosin heavy chain expression in rat, mouse, and human skeletal muscle using multicolor immunofluorescence analysis. *PLoS ONE* **2012**, *7*, e35273. [CrossRef]
55. Lexell, J.; Jarvis, J.C.; Currie, J.; Downham, D.Y.; Salmons, S. Fibre type composition of rabbit tibialis anterior and extensor digitorum longus muscles. *J. Anat.* **1994**, *185*, 95–101. [PubMed]

metabolites

MDPI

Article

The Human Meconium Metabolome and Its Evolution during the First Days of Life

Nihel Bekhti [1], Florence Castelli [1], Alain Paris [2], Blanche Guillon [1], Christophe Junot [1], Clémence Moiron [3], François Fenaille [1,*] and Karine Adel-Patient [1,*]

[1] Département Médicaments et Technologies pour la Santé (DMTS), MetaboHUB, CEA, INRAE, Université Paris-Saclay, 91191 Gif-sur-Yvette, France; bekhtinihel@gmail.com (N.B.); florence.castelli@cea.fr (F.C.); blanche.guillon@cea.fr (B.G.); christophe.junot@cea.fr (C.J.)

[2] UMR7245 MNHN-CNRS, Muséum National d'Histoire Naturelle, 75005 Paris, France; alain.paris@mnhn.fr

[3] Clinique Sainte Thérèse, 75017 Paris, France; pharma.stherese@almaviva-sante.com

* Correspondence: francois.fenaille@cea.fr (F.F.); karine.adel-patient@cea.fr (K.A.-P.)

Abstract: Meconium represents the first newborn stools, formed from the second month of gestation and excreted in the first days after birth. As an accumulative and inert matrix, it accumulates most of the molecules transferred through the placenta from the mother to the fetus during the last 6 months of pregnancy, and those resulting from the metabolic activities of the fetus. To date, only few studies dealing with meconium metabolomics have been published. In this study, we aimed to provide a comprehensive view of the meconium metabolic composition using 33 samples collected longitudinally from 11 healthy newborns and to analyze its evolution during the first 3 days of life. First, a robust and efficient methodology for metabolite extraction was implemented. Data acquisition was performed using liquid chromatography coupled to high-resolution mass spectrometry (LC-HRMS), using two complementary LC-HRMS conditions. Data preprocessing and treatment were performed using the Workflow4Metabolomics platform and the metabolite annotation was performed using our in-house database by matching accurate masses, retention times, and MS/MS spectra to those of pure standards. We successfully identified up to 229 metabolites at a high confidence level in human meconium, belonging to diverse chemical classes and from different origins. A progressive evolution of the metabolic profile was statistically evidenced, with sugars, amino acids, and some bacteria-derived metabolites being among the most impacted identified compounds. Our implemented analytical workflow allows a unique and comprehensive description of the meconium metabolome, which is related to factors, such as maternal diet and environment.

Keywords: meconium metabolome; untargeted metabolomics; LC-HRMS; day-to-day variations

Citation: Bekhti, N.; Castelli, F.; Paris, A.; Guillon, B.; Junot, C.; Moiron, C.; Fenaille, F.; Adel-Patient, K. The Human Meconium Metabolome and Its Evolution during the First Days of Life. *Metabolites* **2022**, *12*, 414. https://doi.org/10.3390/metabo12050414

Academic Editor: Joana Pinto

Received: 25 March 2022
Accepted: 27 April 2022
Published: 5 May 2022

1. Introduction

Meconium, i.e., the first stools of the neonate, starts accumulating in the fetal intestine from the 12th week of gestation and is excreted within the first 24–79 h post-partum [1,2]. Meconium represents an accumulative matrix with a low metabolic activity. It thus provides the longest historical record of fetal exposure but also contains the essential nutrients to shape the future primordial microbiota. Meconium is composed of ~80% water, and, in a decreasing order of abundance, lipids, proteins, and metabolites; it also contains intestinal epithelial cells, neonatal hairs, and minerals [3]. The different substances found in meconium are either produced by the fetus itself or result from trans-placental transfer. The latter substances notably include metabolites derived from the mother's endogenous and microbiota metabolisms, and from various maternal exogenous factors (diet, medication, and environmental contaminants).

Targeted analyses of meconium have largely been performed to evidence fetal exposure to specific xenobiotics [4,5]. As a representative example, Ostrea et al. compared the pesticide content detected in hair, umbilical cord blood, and meconium collected from

598 infants, and evidenced higher levels of some xenobiotics in meconium due to its accumulative nature [6]. Other mass spectrometry (MS)-based targeted analyses of meconium evidenced signatures of fetal exposure to alcohol [7], tobacco [8], and drugs [9]. Targeted analyses using liquid chromatography coupled to tandem mass spectrometry (LC-MS/MS) and subsequent quantification of 33 bile acids also showed an association between primary and secondary bile acid concentrations in meconium and gestational age [10].

Few studies have dealt with untargeted metabolomics analyses of meconium, and essentially provided lists of metabolites that discriminate different mother/child health outcomes. For instance, using nuclear magnetic resonance (NMR) analysis, Peng et al. identified nine meconium metabolites allowing the diagnosis of gestational diabetes mellitus (GDM) [11]. Thanks to a set of 113 metabolites detected by LC coupled to high-resolution MS (LC-HRMS) in meconium samples, Chen et al. then described associations between GDM and alterations in taurine, pyrimidine, and bile acid metabolic pathways [12]. On the other side, NMR-based metabolomics described 16 water-soluble metabolites (e.g., amino acids, organic acids, and ketone bodies) while sterols (e.g., cholesterol, squalene) and fatty acids were the major lipid classes detected in the organic fraction. Both the concentrations of nine water-soluble metabolites, on the one hand, and the fatty acid concentrations and their unsaturation index, on the other hand, significantly increased with postpartum time [13], reflecting breastfeeding initiation [14]. An analysis of 70 fecal samples from 21 newborns, collected between day 1 and day 30 after birth, allowed identification and/or quantification of 33 metabolites, the concentrations of which changed during the first days of life and which correlated with intestinal bacterial species appearance [15]. In a recent study, Bittinger et al. performed a multi-omics analysis of meconium/fecal samples collected between day 1 and day 7 after birth. Metabolomic analysis was performed by LC-HRMS using a single liquid chromatography condition. In total, 45 metabolites were identified to show different profiles in samples collected after 16 h post-partum, where more bacteria were detected [16]. Wandro et al. performed a non-targeted gas chromatography (GC)-MS analysis of samples collected from day 7 after birth and annotated a total of 224 endogenous metabolites, including amino acids, bile acids, fatty acids, nucleotides, and sugars [17]. The largest mapping of early feces metabolome was very recently provided by Petersen et al. Within this study, a non-targeted LC-HRMS analysis of 100 meconium samples was performed by Metabolon, Inc (Morrisville, NC, USA), which reported the detection of 714 compounds belonging to different metabolic pathways, including predominately complex lipid species (e.g., lysophospholipids, sphingomyelins), fatty acids, amino acids, xenobiotics, vitamins, and cofactors [18].

In the present study, we aimed to provide a comprehensive description of the meconium metabolome, and provide a list of metabolites identified at a high confidence level. We thus implemented an MS-based metabolomics workflow involving two untargeted distinct and complementary LC-HRMS platforms combined with annotation based on an in-house chemical database comprising more than 1200 pure authentic standards analyzed under identical conditions [19–21] and confirmed by MS^2 analysis. Under these conditions, 229 metabolites were confidently annotated. We then analyzed global and annotated metabolome evolution during the first three days of life.

2. Results

2.1. *Optimization of the Workflow for Metabolome Analysis of Meconium*

To obtain the most precise view of the human meconium metabolome, we devised an original sample preparation protocol for obtaining robust metabolic fingerprints. Thus, we first performed preliminary experiments to optimize the different steps of the sample preparation thanks to a pool of meconium samples. The steps considered and the optimized final conditions are provided in Figure 1. As a first step, manual homogenization of freshly collected samples was performed prior to aliquoting to avoid any topographical position bias [22]. Then, freeze-drying was performed prior to metabolic extraction to allow future standardization from the dry weight while avoiding potential biases linked to sample-to-sample variation in the water content. We observed that 74 to 78% of the initial meconium weight was lost upon freeze-drying, which is consistent with the reported median value of water content in human feces (~75%) [23]. In line with previous observations [16], methanol proved to be the most efficient solvent for metabolites extraction. Thus, 10 mg of freeze-dried meconium were suspended in 750 µL of methanol/water mixture (4:1, *v*/*v*). Meconium dissociation and homogenization were performed using a Precellys apparatus (Bertin Technologies, Montigny-le-Bretonneux, France) and tubes preloaded with ceramic beads, which was the most efficient and reproducible dissociation method of those we tested (e.g., compared to a sonication bath or probe). Metabolomic analyses of the resulting extracts were performed using two complementary LC-HRMS platforms involving either a reversed-phase column with MS detection in the positive ionization mode (C18-ESI$^+$) or a hydrophilic interaction liquid chromatography column with detection in the negative ionization mode (HILIC-ESI$^-$), allowing the analysis of hydrophobic and polar metabolites, respectively. It is important to mention that we have recently demonstrated the efficiency and robustness of such a protocol to analyze the metabolic profiles of adults feces [24]. The reproducibility of the sample preparation was assessed by analyzing 5 analytical replicates that showed an average coefficient of variation (CV) below 15% for all the metabolite features (see below). Of note, this study focused on the analysis of small-molecular-weight metabolites, the detection of complex lipids (such as lysophospholipids or sphingomyelins) or long-chain fatty acids would imply the use of a dedicated lipidomics platform.

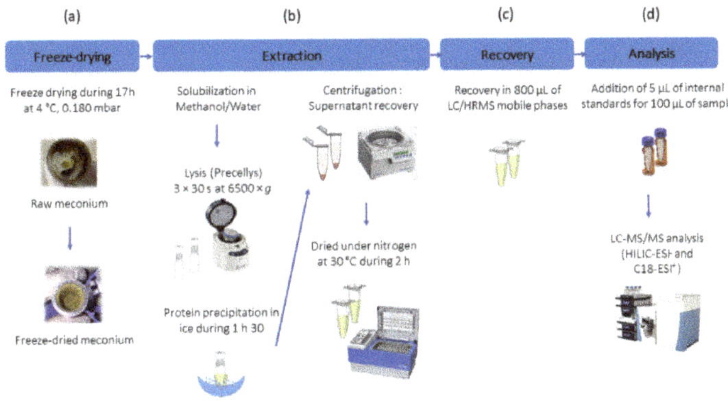

Figure 1. Optimized protocol for meconium preparation prior to LC-HRMS analysis. Four major steps were considered, and the corresponding optimized conditions are provided: (**a**) freeze-drying parameters, (**b**) metabolites extraction method, (**c**) recovery of the pellet containing the metabolites, and (**d**) LC-HRMS analytical conditions.

2.2. *Characterization of the Human Meconium Metabolome*

2.2.1. Comprehensive View of the Meconium Metabolome

The LC-HRMS-based metabolomic methods and their associated data processing workflow (see Material and Methods) were applied to analyze 33 meconium samples collected at different time points from 11 newborns (Table S1). The number and time of excretion varied greatly from one newborn to another, with the collection time ranging from 1 to 79 h after birth. From the analyzed samples, we extracted 9274 and 12,397 features in the HILIC-ESI$^-$ and C18-ESI$^+$ analytical conditions, respectively. Of these, 6843 and 8555 features were found to be analytically relevant, i.e., satisfying our 3 quality evaluation criteria (biological to blank samples intensity ratio, CV between QCs, and correlation within the diluted QC series; see the details in the Material and Methods section). A given metabolite is commonly detected under ESI conditions as a multiplicity of molecular species (i.e., monoisotopic peak, adducts, dimers, and/or fragments), resulting in an overestimation of the real number of unique metabolites in the meconium samples [25].

Overall, 224 metabolic features from the C18-ESI$^+$ conditions and 149 from the HILIC-ESI$^-$ conditions matched with accurate RT and m/z values of a pure standard included in our chemical reference database. Additional MS/MS (MS^2) analyses were performed to confirm these putative annotations, and we finally identified 197 and 89 metabolites for the HILIC-ESI$^-$ and C18-ESI$^+$ conditions, respectively. Only 57 metabolites were identified by both methods, further demonstrating the complementarity of the 2 LC-HRMS platforms. When excluding the overlapping metabolites, 229 unique metabolites were finally identified in meconium with at least 2 orthogonal parameters (retention time, m/z or/and MS/MS spectra) as proposed by the Metabolomics Standards Initiative (MSI) [26]. The corresponding list of metabolites is provided in Table S2.

2.2.2. A Map of the Human Meconium Metabolome

The chemical families of the 229 identified metabolites were assigned using the human metabolome database (HMDB) [27]. The meconium showed a rich metabolic composition covering several molecular families, as summarized in Figure 2a. We observed a major representation of amino acids, peptides, and analogues (32%); carbohydrates and carbohydrate conjugates (15%); and nucleosides, nucleotides, and analogues (13%), with these 3 families covering 60% of the whole meconium metabolome. Within the remaining 40%, we detected mainly organic acids and derivatives (11%) and fatty acids conjugates (7%). The high representation of amino acid, peptide, and analogue families is consistent with previous studies [15,16], and this may also be linked to the high representation of amino acids and derivatives in our chemical library.

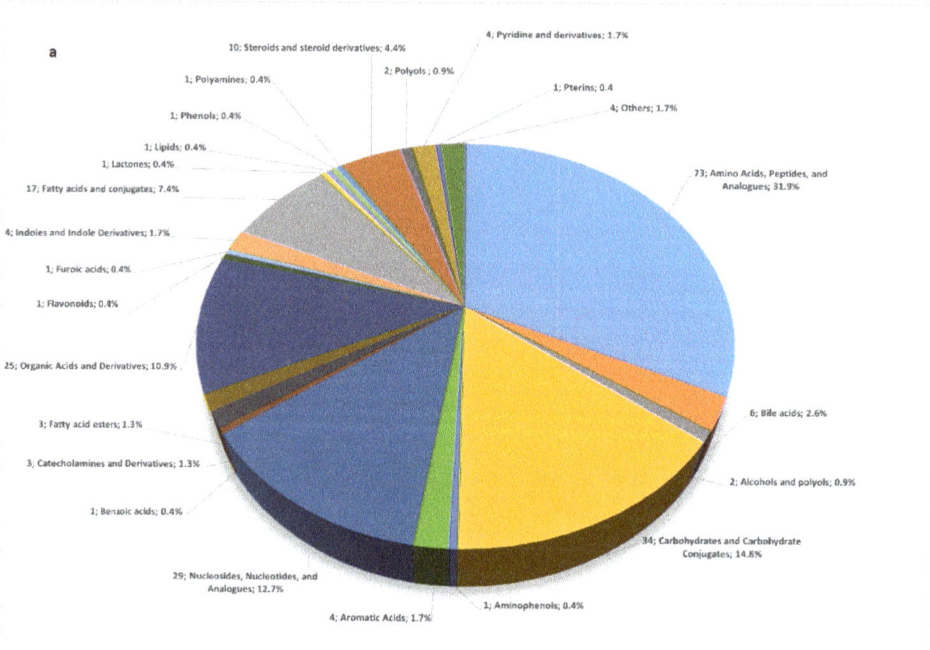

Figure 2. *Cont.*

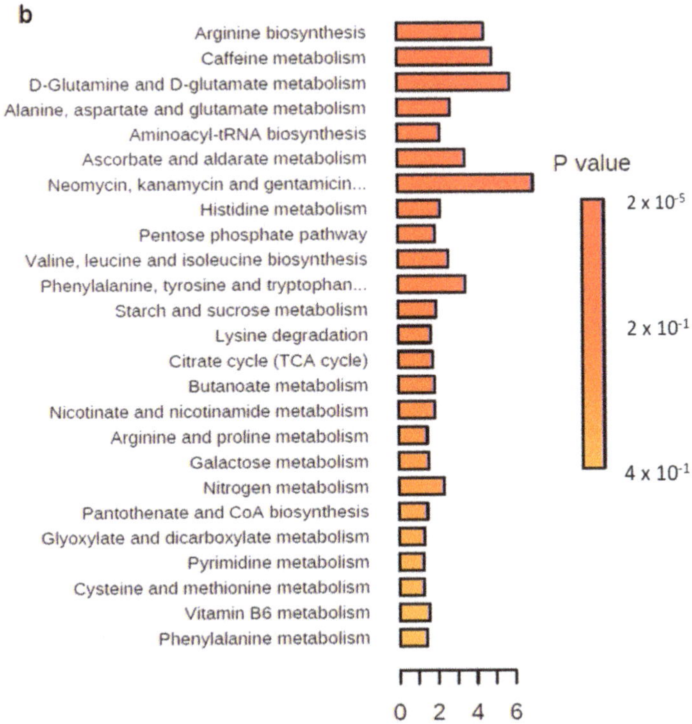

Figure 2. (**a**) Main chemical families represented in meconium and their numbers and percentages (number; name of the chemical family; percentage). (**b**) Enriched metabolic pathways identified through interrogation of the KEGG human metabolic pathways using Metaboanalyst 5.0 tools [28]. The enrichment ratio on the x-axis represents the number of metabolites assigned per metabolic pathway out of the total number of metabolites belonging to the pathway studied.

Within our meconium samples, some microbial metabolites were identified, such as muramic acid, 2-isopropylmalic acid, secondary bile acids (e.g., lithocholic acid (LCA)), and short-chain fatty acids (SCFAs, e.g., isobutyric acid). Other compounds that may result from tryptophan metabolization by bacteria were also detected, particularly indole derivatives (e.g., indoleacetic acid, indoleacrylic acid, indolelactic acid, indole-3-carboxylic acid, and indoxyl sulfate) [29].

The 229 identified metabolites were then repositioned into metabolic pathways [28]. A total of 52 enriched metabolic pathways were identified, and the most enriched ones are shown in Figure 2b. They include biosynthesis and degradation of amino acids pathways (arginine, glutamine, histidine, valine, leucine, phenylalanine, tryptophan, . . . ; amino-acyl-tRNA biosynthesis), and pathways related to the metabolism of caffeine (paraxanthine, theobromine, . . .) or antibiotics (neomycin, kanamycin, . . .).

2.3. Analysis of the Whole Dataset Evidenced Rapid Evolution of the Meconium Metabolome during the First Days of Life

2.3.1. Preliminary Analysis of the Whole Dataset through Non-Supervised PCA and Hierarchical Clustering

A first global analysis of all the analytically relevant features obtained under the C18-ESI$^+$ (8555 features) and the HILIC-ESI$^-$ (6843 features) analytical conditions was performed using a non-supervised multivariate analysis (principal component analysis, PCA) to obtain a first rough picture of the samples and the data distribution. The first 2

PCA components explained 42% (C18-ESI$^+$, Figure 3A) and 45% (HILIC-ESI$^-$, Figure 3B) of the total variance. Almost the same outliers were identified in both conditions, in particular the NB08.270 and NB06.2520 samples. These two samples were not excreted by the same neonate and the other samples collected from the same neonates were not identified as outliers. They do not correspond to a specific gender, and do not reflect a particular excretion time (270 and 2520 min). Moreover, these samples were not close in the analytical sequence, excluding a possible analytical bias during the data acquisition. Thus, there is no obvious reason for explaining the particular behavior of these two samples. They may either have a true but unexplained different metabolic composition or were partially contaminated during their collection from the diaper. As only a few samples were available for this study, we decided to keep all the samples.

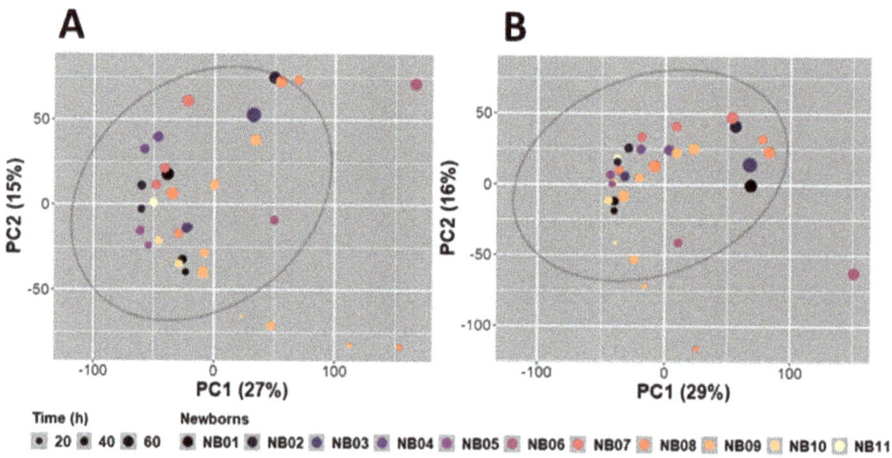

Figure 3. PCA scores plot (PC1 vs. PC2) built using all relevant features obtained in C18-ESI$^+$ ((**A**) 8555 features) and HILIC-ESI$^-$ ((**B**) 6843 features) MS detection conditions. The colors used represent the different newborns (NB01 to NB11) the meconium samples were collected from. The size of each point traduces the collection time-points, expressed in hours (h). Data were log10-transformed and mean-centered before PCA. Ellipses represent the confidence intervals of the scores projected on factorial plans at a probability $p = 0.975$.

Unsupervised PCA highlighted similar structuration of the whole dataset in both analytical conditions (Figure 3). A time-dependent distribution of samples was evidenced, with a non-linear and progressive change in the overall metabolome composition of the meconium (Figure 3). The proximity of samples with close sampling times, and, to a lesser extent, excreted by the same newborn were highlighted using non-supervised hierarchical clustering (not shown). We observed a group of features with a low intensity in the early collection times that increased over time while another group of features showed an opposite trend.

2.3.2. Canonical and Regression Analyses Confirmed a Major Impact of Time on the Meconium Metabolome

To more deeply characterize the inter and intra metabolic variability within the newborns, canonical analyses were performed between the C18-ESI$^+$ and HILIC-ESI$^-$ whole datasets, on the one hand, and the dummy matrix built from repeatedly collected newborns combined with the true time scale at which feces were sampled, on the other hand. Both canonical analyses showed that the first principal components (PC1) represented 38.4% and 38.8% of the total variance and revealed nearly continuous variation in the metabolome for both datasets with rather limited intra-newborn variance (Figure 4). Interestingly, PC2s, which represent 17.2% and 24.0% of the total variance, respectively, displayed a particular behavior for 4 newborns (NB10 for the C18-ESI$^+$ dataset, and NB08, NB09, and NB11 newborns for the HILIC-ESI$^-$ dataset).

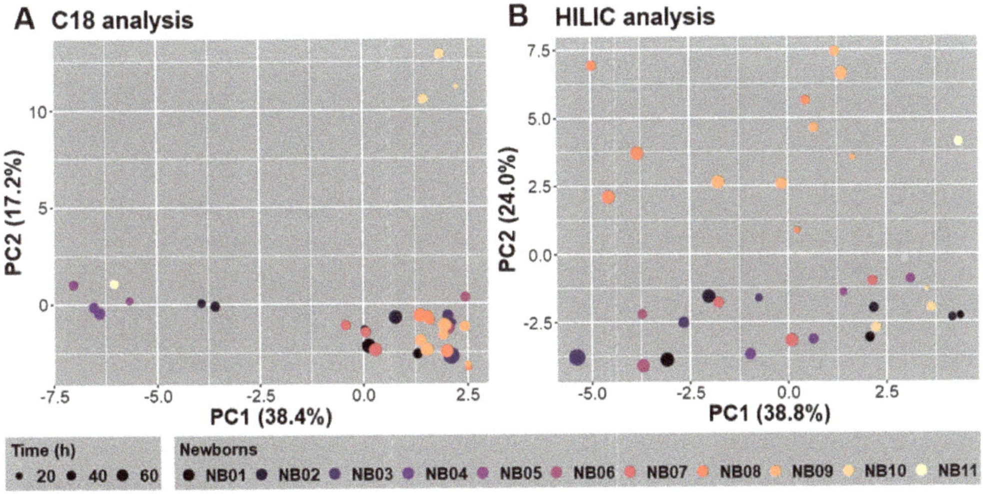

Figure 4. Canonical analyses between all relevant features obtained in C18-ESI$^+$ ((**A**) 8555 features) and HILIC-ESI$^-$ ((**B**) 6843 features) LC-HRMS conditions, and a dummy matrix summarizing feces samples collected for the different newborns (NBs) completed with the time date (expressed in hours, h) of sample collection.

The statistical models used here, which considered the meconium sampling time, highlighted a significant correlation between PC1 scores and time for both analytical conditions (regressions not shown). Accordingly, sparse PLS (sPLS) regressions were significantly established between the 500 most informative features selected from either the C18-ESI$^+$ or HILIC-ESI$^-$ datasets, on the one side, and the time variable, on the other side.

Among the 500 features retained for sPLS regression analysis from either the C18-ESI$^+$ or HILIC-ESI$^-$ dataset, the 20 variables that correlated most with the collection time were sorted according to the absolute value of the correlation with the Comp[1] variable in the sPLS regression. In parallel, thanks to the sPLS regression model used, VIP values were also calculated (Table S3). Whatever the dataset considered, the absolute values of the correlation of the first 20 features explaining the regressions were above 0.862 when their VIP values were higher than 6.50 for the C18-ESI$^+$ dataset and 2.50 for the HILIC-ESI$^-$ dataset. Unfortunately, among these 20 features, none was assigned to metabolites present in our in-house database. Among the 500 features selected, only 2 annotated metabolites were found for the C18-ESI$^+$ dataset (glycyl-leucine, xylulose), whereas 9 were found for the HILIC-ESI$^-$ dataset (ribose phosphate, N-acetylglycine, etc.) (Table S3); however, all were sorted in the less correlated feature sets with a rank above 112.

These global regressions analyzed by sPLS were then reinforced by a mixed model analysis between the Comp[1] scores and the time variable considering the newborn factor as a random factor. For both datasets, every newborn-specific linear regression displayed a positive slope (Figure 5A,C), except for NB05 in the HILIC-ESI$^-$ dataset (Figure 5C). Moreover, the quality of the statistical models was estimated by a quantile-quantile diagram (or QQ plot) of the estimated residues (Figure 5B,D). All samples, except those collected for NB09, were predicted very well by such modeling.

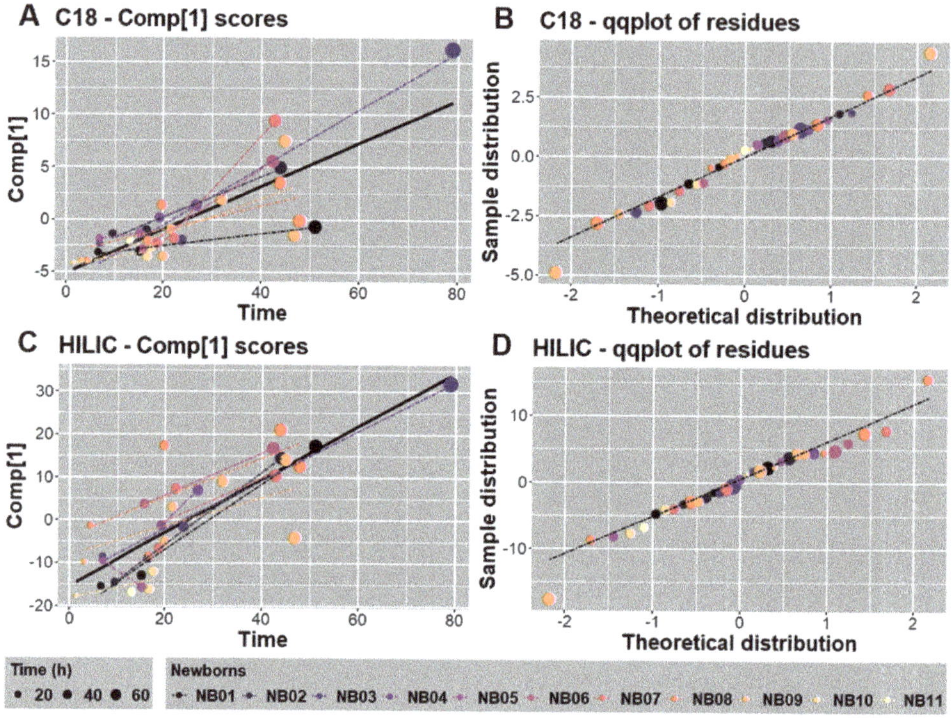

Figure 5. Modeling of the variation in the Comp[1] scores with time (h) for the C18-ESI$^+$ (**A**) or HILIC-ESI$^-$ (**C**) analytical conditions. Comp[1] scores were calculated according to a sPLS regression between the metabolomic datasets and the time at which fecal matrices were collected. The global regressions were drawn according to the thick continuous black line. For every newborn, a mixed model linear regression was applied to model the regression between Comp[1] scores and time and individual regressions are plotted as dashed lines. The distributions of residues calculated after regression were compared to the theoretical one for metabolites detected under the C18-ESI$^+$ (**B**) or HILIC-ESI$^-$ (**D**) analytical conditions. The Comp[1] scores were weakly predicted by the mixed model linear regression only for 1 or 2 samples from NB09, both collected after 40 h, in the C18-ESI$^+$ (**B**) or HILIC-ESI$^-$ (**D**) analytical conditions, respectively.

2.3.3. Modeling of the Whole Datasets and Metabolic Changes Associated with Time-Dependent Variation

Global PLS regression thus illustrated the association between the metabolomic datasets and collection time. Better modeling of the link between these two variables, i.e., Comp[1] and time, was examined considering a log-transformation of the time variable and curvilinear modeling based on polynomials expressed in log(time) with degrees 4 and 3 for the C18-ESI$^+$ and HILIC-ESI$^-$ datasets, respectively (Figure 6 and Table S4). For both datasets, the model prediction was superimposed on the loess modeling of the data well (Figure 6). In addition, the statistical parameters summarizing this curvilinear modeling were conveniently optimized, as shown by the high significance of the coefficients assigned to polynomials equal to or above degree 2 (Table S4). These 2 curves were adjusted with strongly significant alpha risks of 5.2×10^{-15} and 2.2×10^{-16}, respectively. Interestingly, if we analyze the parallel variations of these 2 curves by discarding the simultaneous variation of log(time) and keeping the higher degree of polynomials, that is, 4 and 3, respectively, we can roughly estimate the allometric variation between C18-ESI$^+$ Comp[1] and HILIC-ESI$^-$ Comp[1] with an exponentiation coefficient equal to 0.75. When we more precisely considered the log-transformation of C18-ESI$^+$ and HILIC-ESI$^-$ data predicted using polynomials, as indicated in Figure 6, we obtained the following allometric equation:

$$\text{Predicted-C18 ESI}^+ = (\text{Predicted-HILIC-ESI}^-)^{0.7285} - 0.1034 \quad (1)$$

with adjusted $R^2 = 0.9668$, $F_{1,29} = 874$, and p-value $< 2 \times 10^{-16}$ (Supplementary Figure S1), that is, with an exponentiation coefficient of 0.7285, which is very close to 0.75, the roughly estimated exponent indicated above.

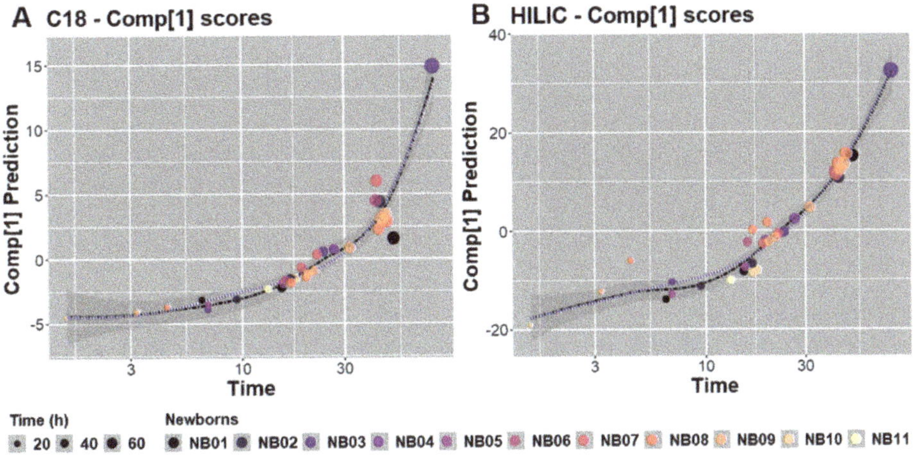

Figure 6. Curvilinear regressions modeled with polynomials of degrees 4 and 3 for the prediction of Comp[1] scores upon log(time) for metabolites analyzed under C18-ESI$^+$ (**A**) and HILIC-ESI$^-$ (**B**) conditions, respectively. As dashed black lines, the loess-supported regressions are shown with a confidence interval of 5% shown in dark grey, and the dotted blue lines display the polynomial regressions.

2.4. Identification of Metabolites Showing Time-Dependent Variations: Complementarity of Non-Supervised and Supervised Analyses

Based on all these results, and to interpret more deeply and directly the observed differences and/or similarities between the samples, we finally performed non-supervised and supervised analyses solely using the 229 annotated metabolites and considering the 33 samples independently of their origin (i.e., not taking into account the fact that stool samples were repeatedly collected for some newborns). We first performed PCA, which globally reproduced the same structuring observed for the whole dataset (Supplementary Figure S2A vs. Figure 3). This suggests that the set of annotated metabolites is quite representative of the whole datasets. Non-supervised hierarchical clustering of the 33 samples based on the 229 annotated metabolites identified 2 clusters of samples (clusters 1 and 2) (Figure S2B). The first cluster of samples is composed of 23 meconium samples, of which 20 were excreted before 24 h, while the second cluster is composed of 10 samples, 8 of which were excreted after 24 h. This spontaneous classification suggests a 24-h cutoff regarding the global metabolic composition in our cohort of meconium samples. The intensities of some metabolites increased (mainly enriched in the cluster > 24 h) while others decreased (mainly enriched in the cluster < 24 h) over time (see below). As shown by non-supervised analysis of the whole dataset, evolution of the meconium composition is thus mainly driven by the time post-partum, independently of the newborn. The distinction between samples excreted in the first 24 h from others was already detectable in the PCA score plot (Figure S2A).

According to these results, we finally performed univariate and multivariate supervised analyses on the 229 annotated features, with samples classified into 2 groups, i.e., excreted before or after 24 h. First, we performed univariate Wilcoxon testing, using a false discovery rate (FDR) correction for multiple comparisons. Around 70% of the annotated metabolites showed comparable intensities before and after 24 h, whereas 67 out the 229'features (29%) showed significantly different intensities between the 2 groups (adjusted *p*-values < 0.05). Among these metabolites, 50 showed significantly higher intensities in late samples (Table S2, red color), including N-acetylglycine, L-threonic acid, or methylguanine, but also some potential microbiota-derived metabolites (e.g., indole derivatives, isobutyric acid). The 17 remaining metabolites displayed significantly more intense signals in samples collected before 24 h (Table S2, blue colors), such as glycerol phosphate, xanthopterin, and N-acetylneuraminic acid. Metabolite up- or downregulation were in agreement with the correlation (negative or positive) calculated in sPLS regression (Table S3). We finally performed supervised multivariate analysis on the 229 annotated features, (i.e., PLS-DA modeling), considering 2 groups of samples (before or after 24 h). A model was successfully built, with a good predictive value (pR2Y < 0.05, pQ2 < 0.05). We then identified 36 metabolites with variable importance in projection (VIP) values above 1.5, i.e., the metabolites that participate most in the model building and then discriminate the most between the 2 groups of samples (Table S2, yellow color). Within these discriminant metabolites, some with the highest VIP values were already contributing highly in the t1-component of the non-supervised PCA (Figure S2C; e.g., glyceric acid, N-acetyl- glutamine, N-acetyl- aspartic acid, threitol, methyl-succinic acid), and all these metabolites were identified as being impacted by the collection time in the univariate analyses.

3. Discussion

The goal of this study was to provide the most precise view of the meconium metabolic composition, with an emphasis on small-molecular-weight metabolites and excluding complex lipids (such as lysophospholipids or sphingomyelins) or other long-chain fatty acids whose detection would imply the use of a dedicated lipidomics platform. In that context, we first implemented a sample preparation approach to robustly extract metabolites from meconium. This protocol was then applied to 33 samples collected from 11 healthy newborns during their first 3 days of life. Metabolite profiling was then performed using two complementary untargeted LC-HRMS approaches, i.e., reversed phase and HILIC chromatography coupled to a Q-Exactive mass spectrometer. Over 17,000 metabolite features comprising redundancy were identified in meconium samples. On the highest MSI level 1, 216 metabolites were identified thanks to our in-house database and using pure authentic standards with an accurate mass, retention time, and MS/MS matching. An additional set of 13 metabolites were annotated with the same procedure but at level 2 since some isomers could not be distinguished. Overall, 229 metabolites were annotated in meconium, which compares favorably with the 222 unique metabolites counted by Aristizabal-Henao et al. using 3 distinct MS-based platforms (2 LC-HRMS and 1 GC-MS) [30]. Using the same analytical approach and database, we identified more than 400 unique metabolites in adult stools [31], highlighting the comparatively lower metabolite richness of meconium. Of note, only 74 metabolites from our dataset were also described by Petersen et al. within their list of 714 detected metabolites and complex lipids (level of confidence not provided) [18].

Non-supervised analysis and modeling of the whole datasets evidenced that time post-partum drastically affected the meconium composition, which evolves rapidly independently of the newborn. Within 79 h post-partum, quantitative and qualitative changes in the metabolome core found in the HILIC-ESI$^-$ dataset were more pronounced than in the C18-ESI$^+$ dataset, although these changes appeared highly dynamically coordinated. Interestingly, most of the metabolites highly impacted by time were not annotated, i.e., not present in our in-house database. This probably reflects that they mostly correspond to metabolites derived from microbiota metabolism.

Non-supervised analysis, i.e., PCA and hierarchical clustering, of the 229 annotated metabolites further suggested a 24-h cutoff for the metabolic composition of our samples. Supervised univariate statistical analysis, performed on samples collected before or after

24 h, revealed that 67 metabolites were impacted by the collection time (29% of the annotated metabolites). In total, 36 metabolites of those 67 also had a PLS-DA-derived VIP score > 1.5 (Table S2). These 67 compounds represented different chemical families, i.e., amino acids, carbohydrates, and organic acids. The corresponding enriched pathways included the pentose phosphate pathway, and glycerolipid, taurine and hypotaurine, and ascorbate and aldarate metabolism. Interestingly, 50 metabolites were accumulated in late samples (Table S2, red color), in line with different studies [13,15,16]. These increased metabolites included N-acetylglycine, threonic acid, and methylguanine, but also taurine and phenylalanine as already described by an NMR analysis of meconium collected between days 1 and 3 by Righetti et al. [2]. Interestingly, some potential microbiota-derived metabolites were also found within this group of accumulating metabolites (indole derivatives, isobutyric acid), in line with progressive microbiota establishment and activity. On the other hand, an additional set of 17 metabolites displayed significantly decreased signals in samples collected after 24 h (Table S2, blue colors), such as glycerol phosphate, xanthopterin, and N-acetylneuraminic acid.

Different studies have reported similar trends to those we observed, which were associated with multiple processes starting after birth: the establishment of the child's intestinal microbiota and its associated metabolic functions [15,32], which lead to the consumption of in utero accumulated metabolites to benefit new bacteria-derived metabolites, the initiation of breastfeeding or infant formula intake [13], and also the intensification of the neonate endogenous digestive and metabolic functions.

Altogether, we implemented an analytical workflow and provided a unique and comprehensive description of the meconium metabolome. We evidenced its rapid change over the first days of life. Core metabolome accumulating in utero is related to factors, such as the maternal diet and environment.

4. Material and Methods

4.1. Study Subjects

Meconium samples were collected in the maternity ward at Sainte-Thérèse Clinic (Paris, France; February 2019). In total, 33 meconium samples were included in the study, obtained within the first 79 h post-partum from 11 anonymized newborns, 2 of which were girls. All babies were born at term by vaginal delivery.

Meconium was recovered by scraping stained diaper with a sterile disposable spatula, taking care to not touch the nappy surface. The diapers were stored at +4 °C until sample collection, which was performed within 12 h (median 4 h). Samples were then immediately stored at −20 °C until transport to the laboratory where they were stored at −80 °C. The meconium samples were collected until a change in color and/or texture was noticed, reflecting the appearance of the first stools. To limit experimental bias related to contamination by chemicals, we provided the same diapers to all participants. The newborn code (NBx), gender, and time of sample collection in minutes postpartum (and the equivalent in hours) are provided in Table S1.

4.2. Meconium Sample Preparation

Frozen meconium was further freeze-dried using a Triad™ Labconco (Missouri, USA) freeze dryer with temperatures fixed at 4 °C for the tray and −83 °C for the trap; the vacuum was fixed at 0.180 mbar. Freeze-dried samples were homogenized, aliquoted, and stored at −80 °C until analysis. To precipitate proteins, 10 mg of freeze-dried meconium were suspended in 750 µL of methanol/H_2O (4:1, *v*/*v*). The samples were then homogenized using a Precellys 24® (Bertin Technologies, Montigny-le-Bretonneux, France) and CK14 ceramic beads (6500× *g*; 4 °C; 3 × 30 s), and then incubated on ice for 1.5 h. After centrifugation (20,000× *g*; 4 °C; 15 min), supernatants (containing metabolites) were recovered and dried under a stream of nitrogen at 30 °C using a TurboVap® concentration workstation (Biotage, France). Samples were stored dried at −80 °C until further analysis.

The pellets were resuspended in a volume of 800 μL of ammonium carbonate (10 mM pH 10.5)/acetonitrile (ACN) (40:60, *v*/*v*) or H_2O + 0.1% formic acid (FA)/ACN + 0.1% FA (95:5, *v*/*v*) for chromatographic separation using HILIC and C18 columns, respectively. Quality control samples (QC) were prepared by pooling equivalent volumes of all samples. Dilution series of QC samples were prepared (1/2, 1/4 and 1/8) to allow data filtration. In total, 100 μL of each biological, QC, and diluted QC samples were spiked with 5 μL of a standard mixture (Table S5).

4.3. Metabolic Profiling

Metabolic profiling experiments were performed by LC-HRMS following optimized protocols routinely used in our laboratory [20,21]. LC-HRMS was performed on an Ultimate 3000 chromatographic system coupled to a Q-Exactive mass spectrometer (both from Thermo Fisher Scientific, Courtaboeuf, France) fitted with an electrospray (ESI) source operating in the positive (ESI^+) and negative (ESI^-) ionization modes. LC was performed using two types of columns to obtain a more comprehensive description of the metabolic landscape: C18 (Hypersil GOLD C18 column, 1.9 μm, 2.1 × 150 mm, Thermo Fisher Scientific; ESI^+) and ZIC-pHILIC (Hydrophilic Interaction Liquid Chromatography; Sequant ZICpHILIC column, 5 μm, 2.1 × 150 mm, Merck, Darmstadt, Germany; ESI^-). Diluted QC samples were analyzed in triplicates at the beginning of the sequence while non-diluted QC samples were introduced every 5 biological samples for data normalization/standardization purposes.

Raw data (.raw files) were manually inspected using the Qual-browser module of Xcalibur (version 4.1, Thermo Fisher Scientific) and then converted to .mzXML format using MSconvert (ProteoWizard). Peak extraction, peak picking, alignment, and integration were performed using the Workflow4Metabolomics (W4M) platform [32]. Data were filtered based on three criteria: (i) ratio of chromatographic peak areas obtained for+ biological to blank samples > 3, (ii) coefficient of variation (CV) of metabolites in the QC samples < 30%, and (iii) correlation between QC dilution factors and areas of chromatographic peaks > 70%. The output files were used for metabolite annotation and further statistical analyses. Metabolite annotation was performed thanks to an in-house chemical database by matching accurate measured masses and chromatographic retention times to those of more than 1200 pure authentic standards analyzed under identical conditions [19–21]. Retention time (RT) tolerances accepted were ±15 and ±90 s for the C18 and ZIC-pHILIC columns, respectively. The mass to charge (m/z) tolerance was 10 ppm for both the positive and negative ionization modes. Each annotated peak was manually checked on the Qual-browser module of Xcalibur by considering the peak shape, isotope pattern, and presence of the considered peak in at least 6 successive MS scans. To limit the presence of irrelevant peaks, an intensity cut-off of 10,000 and 30,000 was applied for the C18-ESI^+ and HILIC-ESI^- conditions, respectively.

MS/MS analysis was then conducted on the relevant signals to confirm their annotation. In total, 4 normalized collision energies (NCEs) were used to obtain the optimal MS/MS spectra (10, 20, 40, and 80%). Of note, some isomeric metabolites could not be resolved using the LC-HRMS approach. Thus, some chromatographic peaks could correspond to more than one metabolite (e.g., hexoses).

Statistical analysis of log-transformed data was conducted using the W4M platform version 3.4.4 [32]. Univariate Wilcoxon tests were conducted to compare data, and adjusted p-values were calculated taking into account multiple testing (false discovery rate, FDR). Adjusted p-values < 0.05 were considered significant. Multivariate principal component analysis (PCA) and partial least squares-discriminant analysis (PLS-DA) were carried out using log-transformed data. Hierarchical classification of the samples and features (centered and reduced data) was also carried out, and represented in the form of a "heatmap". Complementary statistical analyses, such as sparse partial least squares (sPLS) regression and canonical correlation analysis, were performed on R 3.6.4 (R Core Team 2019 [33,34]). Mixed model analysis was performed using the R package lme4 [35]. Metabolic pathways enrichment analysis was conducted using the KEGG database and the "pathway enrichment" tool of MetaboAnalyst 5.0 [28].

Supplementary Materials: The following supporting information can be downloaded at: https://www.mdpi.com/article/10.3390/metabo12050414/s1, Figure S1: Linear variation of log(Predicted-C18-ESI$^+$) data according to log(Predicted-HILIC-ESI$^-$) data established from the respective modelled Comp[1]; Figure S2: A. PCA performed on the dataset built with the 229 metabolites identified in meconium, B. Non-supervised clustering of samples (in columns, n = 33 samples) based on the 229 identified metabolites (in rows); C-D loadings on t1 and t2, respectively, of the 25 first features from the 229 annotated metabolites; Table S1: Description of meconium samples collected; Table S2: List of metabolites annotated in meconium samples; Table S3: Assigned and non-assigned features most correlated to the sparse PLS regressions Comp[1] from C18-ESI$^+$ or HILIC-ESI$^-$ datasets; Table S4: Summary of statistical modeling of the time-dependence variation of the metabolome distribution in C18-ESI$^+$ and HILIC-ESI$^-$ datasets. Table S5: List of external standards used for meconium analysis.

Author Contributions: Conceptualization, C.J., F.F. and K.A.-P.; methodology, B.G., F.C. and N.B.; validation, N.B.; formal analysis, A.P. and N.B.; investigation, A.P., F.F., K.A.-P. and N.B.; resources, C.M.; data curation, B.G., F.C., F.F. and N.B.; writing—original draft preparation, K.A.-P., F.F., A.P. and N.B.; writing—correction, review and editing, C.J., F.F. and K.A.-P.; visualization, N.B.; supervision, F.F. and K.A.-P.; project administration, F.F. and K.A.-P.; funding acquisition, C.J., F.F. and K.A.-P. All authors have read and agreed to the published version of the manuscript.

Funding: This research received no external funding. Nihel Bekhti benefited from a PhD allocation from CEA.

Institutional Review Board Statement: The study was conducted according to the guidelines of the Declaration of Helsinki, and an internal agreement was obtained from the clinic scientific committee. Ethical review was waived for this study, due to fact meconium collection is non-invasive as it is directly recovered by scraping stained diaper.

Informed Consent Statement: Informed signed consents were obtained from all parents and all samples were anonymized.

Data Availability Statement: Raw data have been uploaded to MassIVE with the accession number MSV000089260.

Acknowledgments: We would like to thank the parents involved in the study, and the staff from the Clinique Sainte-Thérèse for their involvement in the study.

Conflicts of Interest: The authors declare no conflict of interest.

References

1. Meyer-Monath, M.; Chatellier, C.; Rouget, F.; Morel, I.; Lestremau, F. Development of a Multi-Residue Method in a Fetal Matrix: Analysis of Meconium. *Anal. Bioanal. Chem.* **2014**, *406*, 7785–7797. [CrossRef] [PubMed]
2. Berton, T.; Mayhoub, F.; Chardon, K.; Duca, R.-C.; Lestremau, F.; Bach, V.; Tack, K. Development of an Analytical Strategy Based on LC–MS/MS for the Measurement of Different Classes of Pesticides and Theirs Metabolites in Meconium: Application and Characterisation of Foetal Exposure in France. *Environ. Res.* **2014**, *132*, 311–320. [CrossRef] [PubMed]
3. Unsworth, J.; Vause, S. Meconium in Labour. *Obstet. Gynaecol. Reprod. Med.* **2010**, *20*, 289–294. [CrossRef]
4. Ostrea, E.M. Testing for Exposure to Illicit Drugs and Other Agents in the Neonate: A Review of Laboratory Methods and the Role of Meconium Analysis. *Curr. Probl. Pediatr.* **1999**, *29*, 41–56. [CrossRef]
5. Ostrea, E.M. Meconium Analysis to Assess Fetal Exposure to Nicotine by Active and Passive Maternal Smoking. *J. Pediatr.* **1994**, *124*, 471–476. [CrossRef]
6. Ostrea, E.M.; Bielawski, D.M.; Posecion, N.C.; Corrion, M.; Villanueva-Uy, E.; Bernardo, R.C.; Jin, Y.; Janisse, J.J.; Ager, J.W. Combined Analysis of Prenatal (Maternal Hair and Blood) and Neonatal (Infant Hair, Cord Blood and Meconium) Matrices to Detect Fetal Exposure to Environmental Pesticides. *Environ. Res.* **2009**, *109*, 116–122. [CrossRef]
7. Bearer, C.F.; Lee, S.; Salvator, A.E.; Minnes, S.; Swick, A.; Yamashita, T.; Singer, L.T. Ethyl Linoleate in Meconium: A Biomarker for Prenatal Ethanol Exposure. *Alcohol. Clin. Exp. Res.* **1999**, *23*, 487–493. [CrossRef]
8. Lamy, S.; Hennart, B.; Houivet, E.; Dulaurent, S.; Delavenne, H.; Benichou, J.; Allorge, D.; Marret, S.; Thibaut, F. Assessment of Tobacco, Alcohol and Cannabinoid Metabolites in 645 Meconium Samples of Newborns Compared to Maternal Self-Reports. *J. Psychiatr. Res.* **2017**, *90*, 86–93. [CrossRef]

9. Ristimaa, J.; Gergov, M.; Pelander, A.; Halmesmäki, E.; Ojanperä, I. Broad-Spectrum Drug Screening of Meconium by Liquid Chromatography with Tandem Mass Spectrometry and Time-of-Flight Mass Spectrometry. *Anal. Bioanal. Chem.* **2010**, *398*, 925–935. [CrossRef]
10. Naritaka, N.; Suzuki, M.; Sato, H.; Takei, H.; Murai, T.; Kurosawa, T.; Iida, T.; Nittono, H.; Shimizu, T. Profile of Bile Acids in Fetal Gallbladder and Meconium Using Liquid Chromatography-Tandem Mass Spectrometry. *Clin. Chim. Acta* **2015**, *446*, 76–81. [CrossRef]
11. Peng, S.; Zhang, J.; Liu, L.; Zhang, X.; Huang, Q.; Alamdar, A.; Tian, M.; Shen, H. Newborn Meconium and Urinary Metabolome Response to Maternal Gestational Diabetes Mellitus: A Preliminary Case–Control Study. *J. Proteome Res.* **2015**, *14*, 1799–1809. [CrossRef] [PubMed]
12. Chen, T.; Qin, Y.; Chen, M.; Zhang, Y.; Wang, X.; Dong, T.; Chen, G.; Sun, X.; Lu, T.; White, R.A.; et al. Gestational Diabetes Mellitus Is Associated with the Neonatal Gut Microbiota and Metabolome. *BMC Med.* **2021**, *19*, 120. [CrossRef] [PubMed]
13. Righetti, C.; Peroni, D.G.; Pietrobelli, A.; Zancanaro, C. Proton Nuclear Magnetic Resonance Analysis of Meconium Composition in Newborns. *J. Pediatric Gastroenterol. Nutr.* **2003**, *36*, 498–501. [CrossRef]
14. Jensen, R.G.; Ferris, A.M.; Lammi-Keefe, C.J.; Henderson, R.A. Lipids of Bovine and Human Milks: A Comparison. *J. Dairy Sci.* **1990**, *73*, 223–240. [CrossRef]
15. Del Chierico, F.; Vernocchi, P.; Petrucca, A.; Paci, P.; Fuentes, S.; Praticò, G.; Capuani, G.; Masotti, A.; Reddel, S.; Russo, A.; et al. Phylogenetic and Metabolic Tracking of Gut Microbiota during Perinatal Development. *PLoS ONE* **2015**, *10*, e0137347. [CrossRef] [PubMed]
16. Bittinger, K.; Zhao, C.; Li, Y.; Ford, E.; Friedman, E.S.; Ni, J.; Kulkarni, C.V.; Cai, J.; Tian, Y.; Liu, Q.; et al. Bacterial Colonization Reprograms the Neonatal Gut Metabolome. *Nat. Microbiol.* **2020**, *5*, 838–847. [CrossRef]
17. Wandro, S.; Osborne, S.; Enriquez, C.; Bixby, C.; Arrieta, A.; Whiteson, K. The Microbiome and Metabolome of Preterm Infant Stool Are Personalized and Not Driven by Health Outcomes, Including Necrotizing Enterocolitis and Late-Onset Sepsis. *mSphere* **2018**, *3*, e00104-18. [CrossRef]
18. Petersen, C.; Dai, D.L.Y.; Boutin, R.C.T.; Sbihi, H.; Sears, M.R.; Moraes, T.J.; Becker, A.B.; Azad, M.B.; Mandhane, P.J.; Subbarao, P.; et al. A Rich Meconium Metabolome in Human Infants Is Associated with Early-Life Gut Microbiota Composition and Reduced Allergic Sensitization. *Cell Rep. Med.* **2021**, *2*, 100260. [CrossRef]
19. Boudah, S.; Olivier, M.-F.; Aros-Calt, S.; Oliveira, L.; Fenaille, F.; Tabet, J.-C.; Junot, C. Annotation of the Human Serum Metabolome by Coupling Three Liquid Chromatography Methods to High-Resolution Mass Spectrometry. *J. Chromatogr. B* **2014**, *966*, 34–47. [CrossRef]
20. Roux, A.; Lison, D.; Junot, C.; Heilier, J.-F. Applications of Liquid Chromatography Coupled to Mass Spectrometry-Based Metabolomics in Clinical Chemistry and Toxicology: A Review. *Clin. Biochem.* **2011**, *44*, 119–135. [CrossRef]
21. Aros-Calt, S.; Muller, B.H.; Boudah, S.; Ducruix, C.; Gervasi, G.; Junot, C.; Fenaille, F. Annotation of the Staphylococcus Aureus Metabolome Using Liquid Chromatography Coupled to High-Resolution Mass Spectrometry and Application to the Study of Methicillin Resistance. *J. Proteome Res.* **2015**, *14*, 4863–4875. [CrossRef] [PubMed]
22. Gratton, J.; Phetcharaburanin, J.; Mullish, B.H.; Williams, H.R.T.; Thursz, M.; Nicholson, J.K.; Holmes, E.; Marchesi, J.R.; Li, J.V. Optimized Sample Handling Strategy for Metabolic Profiling of Human Feces. *Anal. Chem.* **2016**, *88*, 4661–4668. [CrossRef] [PubMed]
23. Karu, N.; Deng, L.; Slae, M.; Guo, A.C.; Sajed, T.; Huynh, H.; Wine, E.; Wishart, D.S. A Review on Human Fecal Metabolomics: Methods, Applications and the Human Fecal Metabolome Database. *Anal. Chim. Acta* **2018**, *1030*, 1–24. [CrossRef] [PubMed]
24. Tabone, M.; Bressa, C.; García-Merino, J.A.; Moreno-Pérez, D.; Van, E.C.; Castelli, F.A.; Fenaille, F.; Larrosa, M. The Effect of Acute Moderate-Intensity Exercise on the Serum and Fecal Metabolomes and the Gut Microbiota of Cross-Country Endurance Athletes. *Sci. Rep.* **2021**, *11*. [CrossRef] [PubMed]
25. Sindelar, M.; Patti, G.J. Chemical Discovery in the Era of Metabolomics. *J. Am. Chem. Soc.* **2020**, *142*, 9097–9105. [CrossRef]
26. Sumner, L.W.; Amberg, A.; Barrett, D.; Beale, M.H.; Beger, R.; Daykin, C.A.; Fan, T.W.-M.; Fiehn, O.; Goodacre, R.; Griffin, J.L.; et al. Proposed Minimum Reporting Standards for Chemical Analysis. *Metabolomics* **2007**, *3*, 211–221. [CrossRef]
27. Wishart, D.S.; Feunang, Y.D.; Marcu, A.; Guo, A.C.; Liang, K.; Vázquez-Fresno, R.; Sajed, T.; Johnson, D.; Li, C.; Karu, N.; et al. HMDB 4.0: The Human Metabolome Database for 2018. *Nucleic Acids Res.* **2018**, *46*, D608–D617. [CrossRef]
28. Brydges, C.; Fiehn, O.; Mayberg, H.; Kaddurah-Daouk, R. Indoxyl sulfate, a gut microbiome-derived uremic toxin, is associated with psychic anxiety and its functional magnetic resonance imaging-based neurologic signature. *Sci. Rep.* **2021**, *11*, 21011. Available online: https://www.nature.com/articles/s41598-021-99845-1 (accessed on 20 December 2021). [CrossRef]
29. Chong, J.; Soufan, O.; Li, C.; Caraus, I.; Li, S.; Bourque, G.; Wishart, D.S.; Xia, J. MetaboAnalyst 4.0: Towards More Transparent and Integrative Metabolomics Analysis. *Nucleic Acids Res.* **2018**, *46*, W486–W494. [CrossRef]
30. Aristizabal-Henao, J.J.; Lemas, D.J.; Griffin, E.K.; Costa, K.A.; Camacho, C.; Bowden, J.A. Metabolomic Profiling of Biological Reference Materials Using a Multiplatform High-Resolution Mass Spectrometric Approach. *J. Am. Soc. Mass Spectrom.* **2021**, *32*, 2481–2489. [CrossRef]
31. Tabone, M.; García-Merino, J.A.; Bressa, C.; Rocha Guzman, N.E.; Herrera Rocha, K.; Chu Van, E.; Castelli, F.A.; Fenaille, F.; Larrosa, M. Chronic Consumption of Cocoa Rich in Procyanidins Has a Marginal Impact on Gut Microbiota and on Serum and Fecal Metabolomes in Male Endurance Athletes. *J. Agric. Food Chem.* **2022**, *70*(6), 1878–1889. [CrossRef] [PubMed]

32. Giacomoni, F.; Le Corguillé, G.; Monsoor, M.; Landi, M.; Pericard, P.; Pétéra, M.; Duperier, C.; Tremblay-Franco, M.; Martin, J.-F.; Jacob, D.; et al. Workflow4Metabolomics: A Collaborative Research Infrastructure for Computational Metabolomics. *Bioinformatics* **2015**, *31*, 1493–1495. [CrossRef] [PubMed]
33. Bunn, A.; Korpela, M. *An Introduction to DplR*; Washington University: Seattle, WA, USA, 2020.
34. Rohart, F.; Gautier, B.; Singh, A.; Cao, K.-A.L. MixOmics: An R Package for 'omics Feature Selection and Multiple Data Integration. *PLOS Comput. Biol.* **2017**, *13*. [CrossRef] [PubMed]
35. Bates, D.; Mächler, M.; Bolker, B.; Walker, S. Fitting Linear Mixed-Effects Models Using Lme4. *arXiv* **2014**, arXiv:1406.5823.

metabolites

MDPI

Protocol

Quantitative Profiling of Bile Acids in Feces of Humans and Rodents by Ultra-High-Performance Liquid Chromatography–Quadrupole Time-of-Flight Mass Spectrometry

Xiaoxu Zhang [1], Xiaoxue Liu [2], Jiufang Yang [3], Fazheng Ren [1] and Yixuan Li [1,*]

1 Key Laboratory of Precision Nutrition and Food Quality, Department of Nutrition and Health, China Agricultural University, Beijing 100193, China; xiaoxuzhang0220@cau.edu.cn (X.Z.); renfazheng@263.net (F.R.)
2 College of Food Science and Nutritional Engineering, China Agricultural University, Beijing 100083, China; lxiaoxue0828@163.com
3 Institut Necker Enfants Malades, Université Paris Cité, F-75015 Paris, France; jiufang.yang@inserm.fr
* Correspondence: liyixuan@cau.edu.cn

Abstract: A simple, sensitive, and reliable quantification and identification method was developed and validated for simultaneous analysis of 58 bile acids (BAs) in human and rodent (mouse and rat) fecal samples. The method involves an extraction step with a 5% ammonium–ethanol aqueous solution; the BAs were quantified by high-resolution mass spectrometry (ultra-high-performance liquid chromatography coupled with quadrupole-time-of-flight mass spectrometry, UPLC–Q-TOF). The recoveries were 80.05–120.83%, with coefficient variations (CVs) of 0.01–9.82% for three biological species. The limits of detection (LODs) were in the range of 0.01–0.24 μg/kg, and the limits of quantification (LOQs) ranged from 0.03 to 0.81 μg/kg. In addition, the analytical method was used to identify and quantify BAs in end-stage renal disease (ESRD) patients, C57BL/6 mice, and Sprague-Dawley (SD) rats. The fecal BA profile and analysis of BA indices in these samples provide valuable information for further BA metabolic disorder research.

Keywords: bile acids; UPLC–Q-TOF; wet feces; sulfation; isomerization; BA indices

Citation: Zhang, X.; Liu, X.; Yang, J.; Ren, F.; Li, Y. Quantitative Profiling of Bile Acids in Feces of Humans and Rodents by Ultra-High-Performance Liquid Chromatography–Quadrupole Time-of-Flight Mass Spectrometry. *Metabolites* **2022**, *12*, 633. https://doi.org/10.3390/metabo12070633

Academic Editors: Paula Guedes de Pinho and Joana Pinto

Received: 30 May 2022
Accepted: 8 July 2022
Published: 11 July 2022

1. Introduction

Bile acids (BAs) are biosynthesized in hepatocytes from cholesterol, which present an important biological function in humans and animals, such as digestion and absorption of lipids and other fat-soluble components [1]. BAs synthesized in the liver are called primary BA (PBA). PBAs' composition is quite different in different biological species. In humans, cholic acid (CA) and chenodeoxycholic acid (CDCA) are major PBAs; in addition, muricholic acid (MCA) is also a PBA in rodents PBAs are synthesized by the classical and alternative pathways. The classical pathway is initiated via 7α-hydroxylation of cholesterol under 7α-hydroxylase (CYP7A1) action and then 12-α hydroxylation of the intermediates by sterol 12-α hydroxylase (CYP8B1), followed by side-chain oxidation by sterol 27 hydroxylase (CYP27A1). The alternative pathway begins with the hydroxylation of the cholesterol side chain by CYP27A1, followed by 7-α hydroxylation of the oxysterol intermediates by oxysterol 7-α hydroxylase (CYP7B1). In rodents, most CDCA is immediately converted into MCA [2]. Then PBAs are amidated with glycine and taurine in the liver rapidly, then flow into the duodenum [1]. The conjugated BAs mainly incur deconjugation, 7α-dehydroxylation, oxidation, and epimerization reactions in the colon by several bacteria to produce secondary BAs (SBA) [3–5]. Other reactions also undergo liver and intestinal metabolisms, such as sulfation, glycosyl esterification, and glycosylation [6,7]. Almost 95% of the BAs are reabsorbed and recycled during the enterohepatic circulation, and only 5% are excreted in the feces [8].

Fecal BAs act as important biomarkers and signaling molecules in several studies because of the complex interplay between BAs and gut microbiota [9,10]. Gut microbiota could modify the BA pool composition and size, especially disordered gut microbiome producing abundant of conjugated BAs and BA structurally similar metabolites, which could indirectly indicate disease states [10–12]. In turn, some BAs might have antimicrobial activities, such as changing the pH of the intestinal microenvironment or disrupting intestinal microbial membranes [13]. Since BAs are regarded as a communication bridge between the gut microbiome and the various organs, such as liver and brain (gut–liver–brain axis), more and more studies focus on the fecal detection of BAs to find the relationship between diseases and intestinal microorganism [10,14–17]. More than that, BAs have a correlation with aging [12,18,19]. Sato et al. found some particular SBA, including iso-, 3-oxo-, allo-, 3-oxoallo-, and isoallo-lithocholic acid in centenarians' feces, which generated by enriched gut microbes [13]. Based on these, developing a robust, accurate, and high throughput BA analysis method in feces is extremely important. However, the multiple variations of BA structure and their similar chemical properties have presented challenges in their separation and detection.

Different from the serum or plasma matrix, the fecal matrix is more complex because of the presence of proteins, lipids, salts, and others. These impurities lead to difficulty in fecal BAs detection. Moreover, the feces condition, wet or dry, also affects the extraction efficiency. Recently, Shafaei et al. compared the extraction efficiency of 12 BAs from wet and dry feces [8]. They found that the recoveries of all the target BAs were quite lower in dried fecal samples than in wet samples. Especially, the glycine conjugated BA recoveries were below 30%. Therefore, an appropriate BA preparation process is quite necessary for accurate qualitative and quantitative analysis, especially when LC-MS is used as the detector. The common preparation processes include BA extraction, purification, and dilution for high BA concentration. A small amount of ammonium hydroxide or sodium hydroxide was often added to the extraction solution to attenuate the binding of proteins to BAs, thus improving fecal BAs extraction efficiency [6]. Solid-phase extraction (SPE) and liquid–liquid extraction (LLE) are two optional purification and concentration methods [6,20]. An efficient preparation procedure could decrease matrix effects and improve sensitivity in BAs analysis. Therefore, for different experimental substrates, pretreatment methods still need to be optimized.

Many studies have reported BA detection technologies in the last few decades, including thin-layer chromatography (TLC), high-performance liquid chromatography with UV detection (HPLC-UV), gas chromatography with flame ionization (GC-FID), gas chromatography mass spectrometry (GC-MS), liquid chromatography mass spectrometry (LC-MS), ultrahigh-performance liquid chromatography-tandem mass spectrometry (UHPLC-MS/MS), enzyme-linked immunosorbent assay (ELISA), and nuclear magnetic resonance spectroscopy (NMR) [20–22]. Recently, high-resolution mass spectrometers (HRMS), such as Orbitrap or TOF, have been increasingly used for BA identification, characterization, and quantifications. These HRMS offer high mass resolution (>100,000 fwhm) and high mass accuracy (<5 ppm). They could improve the BA isomers' separation, sensitivity, and specificity [23,24]. Importantly, once full-scan mass spectra are obtained during sample acquisition, valuable information about other BA metabolites, metabolite modifications, or degradation products could be available for further data analysis.

Hence, our aim was to develop a simple, robust, and reliable bioanalytical method for wide structural coverage of BA analytes (Mono-OH, Di-OH, Tri-OH, and oxo-, nor-, iso-) measured quantitatively in feces from clinical and preclinical study samples (two rodents, rats and mouse) with UPLC–Q-TOF. Our research group was studying the abnormal lipid metabolism in end-stage renal disease patients (ESRD) who lost renal function. These patients are always accompanied by lipid metabolic disturbance and intestinal microbiota disorders. Thus, the feces of ESRD patients were chosen as the clinical sample to validate our BA method. The feces of rodents (C57BL/6 mice and Sprague- Dawley rats) were also used to validate the method. Finally, the BA indices were compared in these biological species, especially the SBA composition analysis, to provide basic data for future gut/intestinal-X axis research.

2. Result and Discussion

Along with the discovery of the important role of gut microbes in diseases and health, more and more studies have focused on the composition and changes of microbial metabolites, especially for BAs. The composition of microbes and the regulation of the micro-environment on microbial metabolism would change the structure of BAs, causing them to isomerize and form variable isomers, which play an important regulatory role in humans and animals. Many studies have developed and profiled the BAs in plasma and urine of multiple biological species, such as humans, monkeys, rabbits, rats, and mice [20,25–28]. However, it is crucial to profile various BAs in feces because the fecal BA could evaluate intestinal microorganism status directly [6,29,30]. Therefore, our goal was to develop a robust, high throughput method for comprehensive analysis of fecal BA in human and preclinical animals (mouse and rat).

2.1. BA Extraction Methods Comparison

Shafaei et al. found poor BA recovery present in dried fecal material, while wet samples could provide better efficiency and repeatability [8]. Hence, our extraction approach optimization was carried out on wet feces.

In order to develop a simple, time-saving, and high BA species coverage extraction method for different fecal biological samples, three different procedures were compared, including ethanol extraction (S1), reversed-phase SPE with high pH (S2), and high pH ethanol (S3) extraction. The pooled sample was used to evaluate the above protocols. As shown in Figure 1, S2 protocol NaOH-SPE gave the highest concentrations for unconjugated BAs but a much lower content of conjugated and sulfated BAs than the other two protocols ($p < 0.05$). In this protocol, one-hour preincubation for the sample before extraction may be responsible for this low conjugated content, especially the taurine-conjugated BAs. The dehydrogenase and desulfatase enzymes of the fecal microorganism may hydrolyze conjugated and sulfated BAs during this preincubation period [31]. Ethanol extraction (S1) also gave a lower yield for glycine and taurine conjugated BAs than the high pH ethanol protocol (S3). The alkaline condition could benefit from breaking the bonds between conjugated BA and fecal protein [32]. To summarize, the S3 protocol (5% ammonium–ethanol) was chosen in further experiments as a routine extraction protocol for feces.

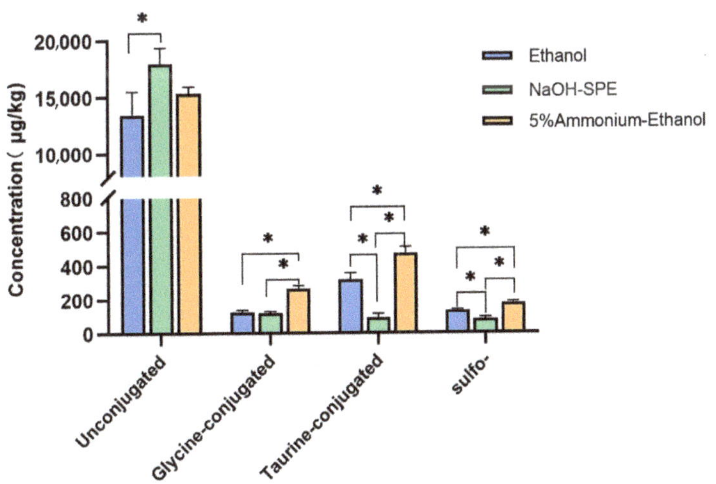

Figure 1. Extraction conditions comparison of bile acids profiles. Differences considered significant ($p < 0.05$) are denoted by *.

2.2. *Chromatography Separation Optimization*

For high-resolution mass spectrometry, effective chromatography separation reduces matrix effects and improves the accuracy of identification and quantification. Therefore, the mobile phase constitution was optimized. Ammonium-based buffer and formic acid are common additives in aqueous solvents (mobile phase A) for negative ionization mode [33]. Additionally, the acidic condition was beneficial for the separation of BA structural isomer. For example, GUDCA, GHDCA, GCDCA, and GDCA have the same molecular formula, $C_{26}H_{43}NO_5$, but different in -OH position (shown in Figure 2). The separations of their addition ion forms $[M\text{-}H]^-$ (m/z 448.3063) were significantly affected by ammonium acetate and formic acid additive amount. GUDCA and GHDCA were not separated under individual ammonium acetate conditions (Figure 2a). When 0.05% formic acid was added in the mobile phase simultaneously, isobaric BAs can be differentiated (Figure 2b). Furthermore, the analysis times of these analytes could save 1–2 min when formic acid addiction volume was 0.01% instead of 0.05% (Figure 2c). Therefore, in terms of peak shapes, analysis times, and solvent saving, the best mobile phase compromise was 2 mM ammonium acetate and 0.01% formic acid in H_2O. This is because weakly acidic mobile phase conditions contribute to the deprotonation of the analytes [34]. Information about the observed ions and retention times (RTs) of all analyzed compounds is given in Table 1. The extracted ion chromatograms (EICs) of all the unconjugated, glycine-conjugated, taurine-conjugated BAs, sulfated Bas, and deuterium-labeled BAs are shown in Figure 3.

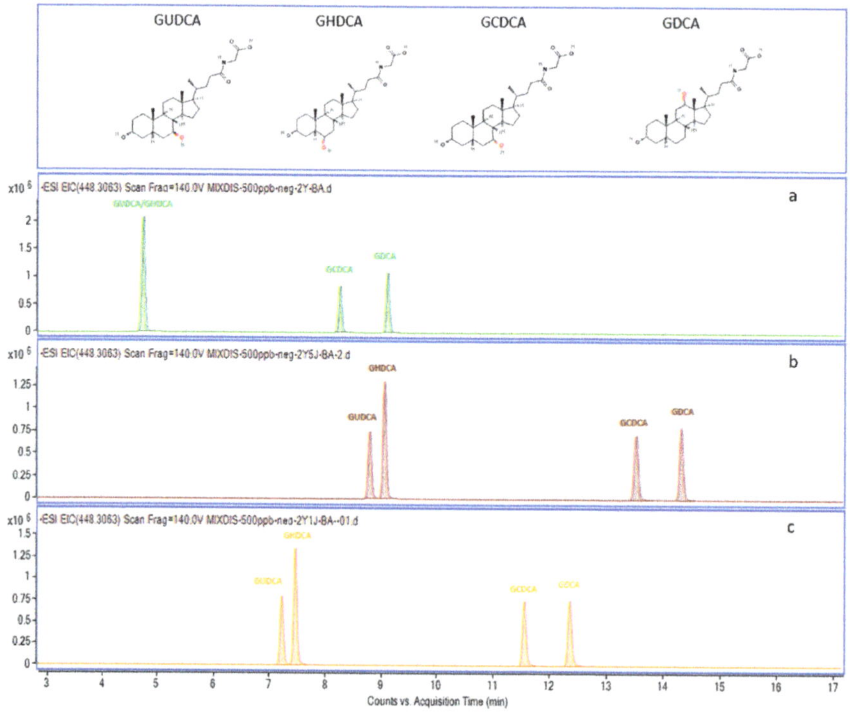

Figure 2. Mobile phase optimization for isobaric compounds chromatography separation. The GUDCA, GHDCA, GCDCA, and GDCA have the same molecular formula, $C_{26}H_{43}NO_5$ ([M-H]- m/z 448.3063), but different -OH positions. (**a**) 2 mM ammonium acetate in aqueous solvent; (**b**) 2mM ammonium acetate and 0.05% formic acid in aqueous solvent; (**c**) 2 mM ammonium acetate and 0.01% formic acid in aqueous solvent.

Table 1. Bile acid analyte structural, quantitative, and qualitative information.

Compounds	Formula	RT [1]	Transition	Adduct Ion	IS [2]
Unconjugated					
Nordeoxycholic acid (norDCA)	$C_{23}H_{38}O_4$	15.63	377.2697	[M − H]−	CA-d4
Dehydrocholic acid (DHCA)	$C_{24}H_{34}O_5$	8.10	401.2333	[M − H]−	CA-d4
7,12-Diketolithocholic acid (7,12-diketoLCA)	$C_{24}H_{36}O_5$	7.40	449.2545	[M + COOH]−	CA-d4
6,7-Diketolithocholic acid (6,7-diketoLCA)	$C_{24}H_{36}O_5$	15.56	403.2490	[M − H]−	CA-d4
3-Ketolithocholic acid (3-ketoLCA)	$C_{24}H_{38}O_3$	23.75	373.2748	[M − H]−	CA-d4
7-Ketolithocholic acid (7-ketoLCA)	$C_{24}H_{38}O_4$	15.35	435.2752	[M + COOH]−	CA-d4
12-Ketolithocholic acid (12-ketoLCA)	$C_{24}H_{38}O_4$	9.68	405.2646	[M − H]−	CA-d4
7-Ketodeoxycholic acid (7-DHCA)	$C_{24}H_{38}O_5$	12.29	405.2646	[M − H]−	CA-d4
3-Ketodeoxycholic acid (3-DHCA)	$C_{24}H_{38}O_5$	15.90	389.2679	[M + COOH]−	CA-d4
Apocholic acid (apoCA)	$C_{24}H_{38}O_4$	15.75	435.2752	[M − H]−	CA-d4
Isolithocholic acid (isoLCA)	$C_{24}H_{40}O_3$	21.23	375.2905	[M − H]−	CA-d4
Allolithocholic acid (isoalloLCA)	$C_{24}H_{40}O_3$	21.00	375.2905	[M − H]−	CA-d4
Allocholic acid (alloCA)	$C_{24}H_{40}O_5$	12.44	407.2803	[M − H]−	CA-d4
Ursocholic acid (UCA)	$C_{24}H_{40}O_5$	6.92	407.2803	[M − H]−	CA-d4
Lithocholic acid (LCA)	$C_{24}H_{40}O_3$	22.91	421.2959	[M + COOH]−	LCA-d4
Ursodeoxycholic acid (UDCA)	$C_{24}H_{40}O_4$	13.41	437.2909	[M + COOH]−	UDCA-d4
Chenodeoxycholic acid (CDCA)	$C_{24}H_{40}O_4$	17.58	437.2909	[M + COOH]−	CDCA-d4
Isodeoxycholic acid (isoDCA)	$C_{24}H_{40}O_4$	21.25	391.2854	[M − H]−	DCA-d4
murideoxycholic acid (MDCA)	$C_{24}H_{40}O_4$	11.85	437.2909	[M + COOH]−	CA-d4
Deoxycholic acid (DCA)	$C_{24}H_{40}O_4$	18.10	391.2854	[M − H]−	CA-d4
Hyodeoxycholic acid (HDCA)	$C_{24}H_{40}O_4$	13.78	437.2909	[M + COOH]−	CA-d4
α-Muricholic acid (α-MCA)	$C_{24}H_{40}O_5$	9.39	453.2860	[M + COOH]−	CA-d4
β-Muricholic acid (β-MCA)	$C_{24}H_{40}O_5$	10.04	453.2858	[M + COOH]−	CA-d4
γ-Muricholic acid (γ-MCA)	$C_{24}H_{40}O_5$	11.41	453.2858	[M + COOH]−	CA-d4
ω-muricholic acid (ω-MCA)	$C_{24}H_{40}O_5$	8.70	453.2858	[M + COOH]−	CA-d4
Cholic acid (CA)	$C_{24}H_{40}O_5$	12.83	407.2803	[M − H]−	CA-d4
Glycine-conjugated					
Glycodehydrocholic acid (GDHCA)	$C_{26}H_{37}NO_6$	3.34	458.2548	[M − H]−	GCA-d4
Glycolithocholic acid (GLCA)	$C_{26}H_{43}NO_4$	16.81	432.3119	[M − H]−	GLCA-d4
Glycodeoxycholic acid (GDCA)	$C_{26}H_{43}NO_5$	12.25	448.3063	[M − H]−	GDCA-d4
Glycochenodeoxycholic acid (GCDCA)	$C_{26}H_{43}NO_5$	11.44	448.3063	[M − H]−	GCDCA-d4
Glycoursodeoxycholic acid (GUDCA)	$C_{26}H_{43}NO_5$	7.09	448.3068	[M − H]−	GUDCA-d4
Glycohyocholic acid (GHCA)	$C_{26}H_{43}NO_6$	5.78	464.3018	[M − H]−	GCA-d4
Glycohyodeoxycholic acid (GHDCA)	$C_{26}H_{43}NO_5$	7.34	448.3068	[M − H]−	GCA-d4
Glycocholic acid (GCA)	$C_{26}H_{43}NO_6$	7.36	464.3018	[M − H]−	GCA-d4
Taurine-conjugated					
Taurolithocholic acid (TLCA)	$C_{26}H_{45}NO_5S$	14.24	482.2946	[M − H]−	TCA-d4
Taurochenodeoxycholic acid (TCDCA)	$C_{26}H_{45}NO_6S$	9.23	498.2895	[M − H]−	TCDCA-d5
Taurodeoxycholic acid (TDCA)	$C_{26}H_{45}NO_6S$	10.05	498.2895	[M − H]−	TCDCA-d5
Tauroursodeoxycholic acid (TUDCA)	$C_{26}H_{45}NO_6S$	5.48	498.2895	[M − H]−	TCDCA-d5
Taurohyodeoxycholic acid (THDCA)	$C_{26}H_{45}NO_6S$	5.58	498.2895	[M − H]−	TCDCA-d5
Tauro α-Muricholic acid (T-α-MCA)	$C_{26}H_{45}NO_7S$	3.19	514.2844	[M − H]−	TCA-d4
Tauro β-Muricholic acid (T-β-MCA)	$C_{26}H_{45}NO_7S$	4.34	514.2844	[M − H]−	TCA-d4
Taurohyocholic Acid (THCA)	$C_{26}H_{45}NO_7S$	5.92	514.2844	[M − H]−	TCA-d4
Taurocholic acid (TCA)	$C_{26}H_{45}NO_7S$	3.04	514.2844	[M − H]−	TCA-d4
Sulfo-					
Lithocholic Acid-3-Sulfate (LCA-3S)	$C_{24}H_{40}O_6S$	15.27	455.2473	[M − H]−	CDCA-3S-d4
Ursodeocycholic Acid-3-Sulfate (UDCA-3S)	$C_{24}H_{40}O_7S$	7.40	471.2422	[M − H]−	CDCA-3S-d4
Chenodeoxycholic Acid-3-Sulfate (CDCA-3S)	$C_{24}H_{40}O_7S$	11.32	471.2422	[M − H]−	CDCA-3S-d4
Deoxycholic Acid-3-Sulfate (DCA-3S)	$C_{24}H_{40}O_7S$	11.63	471.2422	[M − H]−	CDCA-3S-d4
Cholic Acid-3-Sulfate (CA-3S)	$C_{24}H_{40}O_8S$	7.72	487.2371	[M − H]−	CDCA-3S-d4
Glycine-Sulfo-					
Glycolithocholic Acid-3-Sulfate (GLCA-3S)	$C_{26}H_{43}NO_7S$	9.98	512.2687	[M − H]−	GDCA-3S-d4
Glycoursodeoxycholic Acid-3-Sulfate (GUDCA-3S)	$C_{26}H_{43}NO_8S$	3.42	528.2637	[M − H]−	GDCA-3S-d4

Table 1. *Cont.*

Compounds	Formula	RT [1]	Transition	Adduct Ion	IS [2]
Glycochenodeoxychlolic Acid-3-Sulfate (GCDCA-3S)	$C_{26}H_{43}NO_8S$	6.50	528.2637	[M − H]−	GDCA-3S-d4
Glycodeoxycholic Acid-3-Sulfate (GDCA-3S)	$C_{26}H_{43}NO_8S$	6.90	528.2637	[M − H]−	GDCA-3S-d4
Glycocholic Acid-3-Sulfate (GCA-3S)	$C_{26}H_{43}NO_9S$	3.91	544.2586	[M − H]−	GDCA-3S-d4
Taurine-Sulfo-					
Taurolithocholic Acid-3-Sulfate (TLCA-3S)	$C_{26}H_{45}NO_8S_2$	6.98	562.2514	[M − H]−	TUDCA-3S-d4
Tauroursodeoxycholic Acid-3-Sulfate (TUDCA-3S)	$C_{26}H_{45}NO_9S_2$	2.06	578.2463	[M − H]−	TUDCA-3S-d4
Taurochenodeoxycholic Acid-3-Sulfate (TCDCA-3S)	$C_{26}H_{45}NO_9S_2$	4.19	578.2463	[M − H]−	TUDCA-3S-d4
Taurodeoxycholic Acid-3-Sulfate (TDCA-3S)	$C_{26}H_{45}NO_9S_2$	4.57	578.2463	[M − H]−	TUDCA-3S-d4
Taurocholic Acid-3-Sulfate (TCA-3S)	$C_{26}H_{45}NO_{10}S_2$	2.46	594.2412	[M − H]−	TUDCA-3S-d4
Internal standard					
Cholic acid-d4 (CA-d4)	$C_{24}H_{36}D_4O_5$	12.86	411.3054	[M − H]−	
Lithocholic Acid-d4 (LCA-d4)	$C_{24}H_{36}D_4O_3$	22.96	379.3156	[M − H]−	
Ursodeoxycholic Acid-d4 (UDCA-d4)	$C_{24}H_{36}D_4O_4$	13.45	395.3105	[M − H]−	
Chenodeoxycholic Acid-d4 (CDCA-d4)	$C_{24}H_{36}D_4O_4$	17.54	395.3105	[M − H]−	
Deoxycholic Acid-d4 (DCA-d4)	$C_{24}H_{36}D_4O_4$	18.13	395.3105	[M − H]−	
Glycolithocholic Acid-d4 (GLCA-d4)	$C_{26}H_{39}D_4NO_4$	16.86	436.3370	[M − H]−	
Glycoursodeoxycholic Acid-d4 (GUDCA-d4)	$C_{26}H_{39}D_4NO_5$	7.12	452.3320	[M − H]−	
Glycodeoxycholic Acid-d4 (GDCA-d4)	$C_{26}H_{39}D_4NO_5$	6.90	452.3320	[M − H]−	
Glycochenodeoxycholic Acid-d4 (GCDCA-d4)	$C_{26}H_{39}D_4NO_5$	12.26	452.3320	[M − H]−	
Glycocholic Acid-d4 (GCA-d4)	$C_{26}H_{39}D_4NO_6$	7.43	468.3269	[M − H]−	
Taurocholic Acid-d4 (TCA-d4)	$C_{26}H_{41}D_4NO_7S$	5.96	518.3095	[M − H]−	
Taurochenodeoxycholic Acid-d5 (TCDCA-d5)	$C_{26}H_{40}D_5NO_6S$	9.26	503.3209	[M − H]−	
Chenodeoxycholic Acid-3-Suflate-d4 (CDCA-3S-d4)	$C_{24}H_{36}D_4O_7S$	11.38	475.2673	[M − H]−	
Glycodeoxycholic Acid-3-Sulfate-d4 (GDCA-3S-d4)	$C_{26}H_{39}D_4NO_8S$	11.42	532.2888	[M − H]−	
Tauroursodeoxycholic Acid-3-Sulfate-d4 (TUDCA-3S-d4)	$C_{26}H_{41}D_4NO_9S_2$	2.25	582.2714	[M − H]−	

[1] RT—retention time. [2] IS:—internal standard.

2.3. Method Validation

The method was then validated following the recommendations for bioanalytical method validation [35] and confirmed to be selective and specific. For 58 BA analytes, the LOD ranged from 0.01 to 0.24 μg/kg (corresponding 0.15–3.26 nM) and the LOQ from 0.03 to 0.81 μg/kg (corresponding 0.44–9.77 nM) (Table S1). Our quantification limits were lower than previously reported values, 10–12.5 nM (methanol extraction in wet feces and detection by LC-MS/MS) [8] and 12.6–73.2 nM (NaOH-SPE extraction in dried feces and detection by LC-MS/MS) [32].

All calibration curves covered quantities from 0.5 to 1000 μg/L (corresponding 1–2000 nM) were linear, with squared correlation coefficients (r^2) ranging from 0.995 to 0.999 without extra weighting analysis in the calibration curves algorithm. In other HRMS-based methods, nonlinear correlations have been observed in the quantitative analysis [23]. The possible reason is the limited dynamic range of the detector. Therefore, different regression algorithms (linear and quadratic with $1/x$ or $1/x^2$ weightings) were used in calibration curves calculation.

The accuracy of 58 bile acids at low, medium, and high spiked concentrations were 92.39–115.55%, 89.14–111.04%, and 94.88–106.25%, respectively. The precisions of intra-day and inter-day, expressed as variation coefficients %, were less than 10% (0.04–9.78% and 0.12–9.98%, respectively). Results are shown in Table S1.

The recovery percentages of three biological fecal matrices were evaluated at different concentration levels. Table S2 presented the recovery of BA in each biological sample. All the recovery values ranged from 80 to 120%.

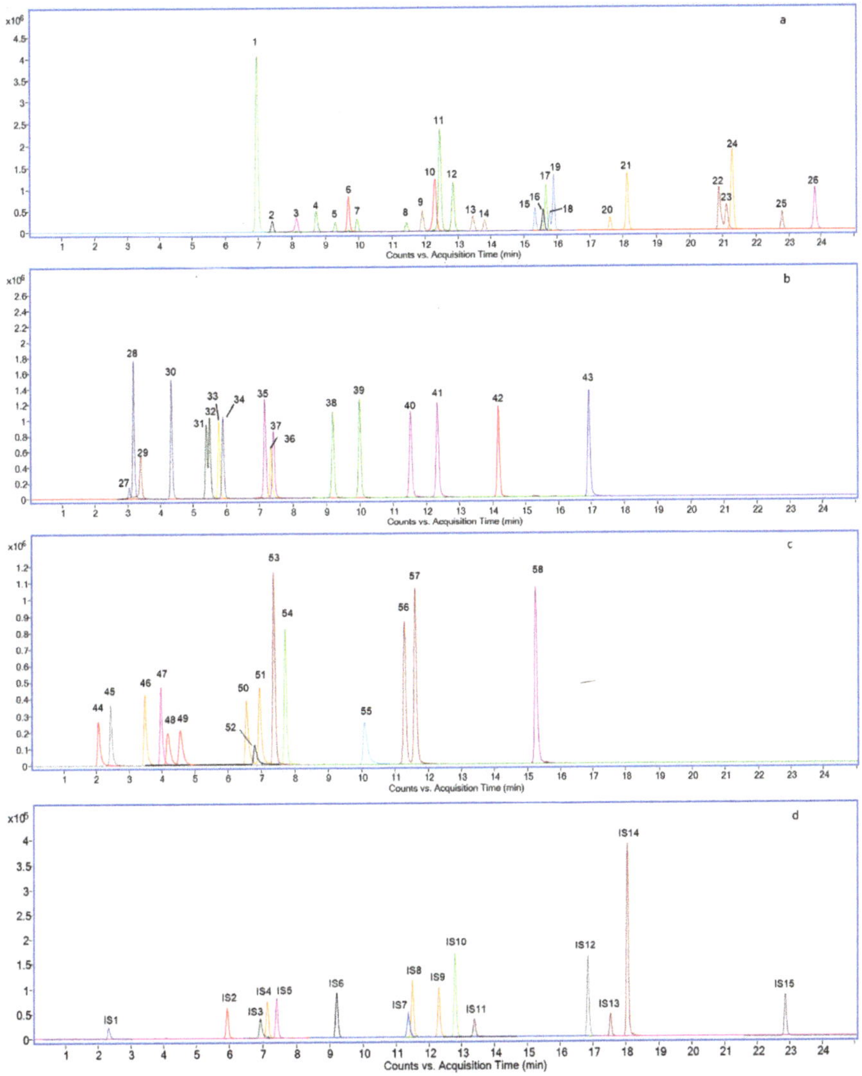

Figure 3. Extracted ion chromatograms of 26 unconjugated (**a**), 8 glycine-conjugated and 9 taurine-conjugated BAs (**b**), 5 taurine-sulfated, 5 glycine-sulfated, and 5 sulfated BAs (**c**), and 15 deuterium (D)-labeled BAs as Internal Standards (**d**). (1) UCA (2) 7,12-diketoLCA (3) DHCA (4) ω-MCA (5) α-MCA (6) 7-DHCA (7) β-MCA (8) γ-MCA (9) MDCA (10) 3-DHCA (11) alloCA (12) CA (13) UDCA (14) HDCA (15) 7-ketoLCA (16) 6,7-diketoLCA (17) norDCA (18) apoCA (19)12-ketoLCA (20) CDCA (21) DCA (22) isoalloLCA (23) isoDCA (24) isoLCA (25) LCA (26) 3-ketoLCA (27) T-α-MCA (28) T-β-MCA (29) GDHCA (30) THCA (31) TUDCA (32) THDCA (33) GHCA (34) TCA (35) GHDCA (36) GCA (37) GUDCA (38) TCDCA (39) TDCA (40) GDCA (41) GCDCA (42) TLCA (43) GLCA (44) TUDCA-3S (45) TCA-3S (46) GUDCA-3S (47) GCA-3S (48) TCDCA-3S (49) TDCA-3S (50) GCDCA-3S (51) GDCA-3S (52) TLCA-3S (53) UDCA-3S (54) CA-3S (55) GLCA-3S (56) CDCA-3S (57) DCA-3S (58) LCA-3S (IS1) TUDCA-3S-d4 (IS2) TCA-d4 (IS3) GDCA-3S-d4 (IS4) GUDCA-d4 (IS5) GCA-d4 (IS6) TCDCA-d5 (IS7) CDCA-3S-d4 (IS8) GDCA-d4 (IS9) GCDCA-d4 (IS10) CA-d4 (IS11) UDCA-d4 (IS12) GLCA-d4 (IS13) CDCA-d4 (IS14) DCA-d4 (IS15).

The matrix effect percentages were evaluated at different concentration levels of each biological sample; detailed results are listed in Table S3. Matrix effect values were considered negligible, being in a range of 80–120% for all analytes in rat, mouse, and human fecal matrix samples. An appropriate extraction method, high-resolution detection instrument, and calibration of the internal standard could effectively reduce the matrix effect [24]. Moreover, the usage of 15 deuterium-labeled standards (1 for each of the 15 analyzed BA species) also compensated for the matrix effects. All these results were acceptable according to the FDA guidelines [35].

2.4. BA Profiling in Humans, Rats, and Mice

The composition of individual BAs and the concentration proration of each BA in all their forms (unconjugated, amino acid conjugated (glycine or taurine), sulfo-conjugated, and double conjugated) are shown in Figure 4.

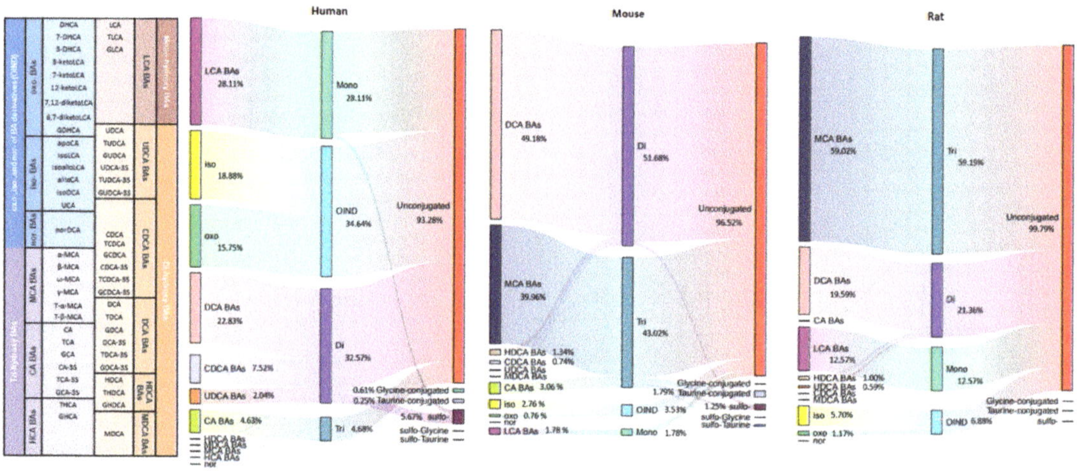

Figure 4. Sankey diagram of different bile acid concentration distribution in humans, rats, and mice.

Unconjugated BAs were the most abundant BAs in humans, rats, and mice, all above 90%, with the corresponding BA indices of 93.28%, 96.52%, and 99.79%. Among the conjugated BAs, the highest sulfation of BAs was observed in humans (5.67%), while sulfation of BAs in mice and rats was 1.25% and 0.1%, respectively. Thakare et al. reported humans present better sulfation capability than rats and mice [28]. The higher percent sulfation capacity of humans is also consistent with plasma and urine matrix [25,36]. BA sulfation could increase BA hydrophilicity and promotes excretion in feces and urine, so it is an important mechanism for BA detoxification [30]. Glycine-conjugated BAs were predominant in humans, but only a tiny amount in rodent species, while the BA indices of taurine-conjugated were on the contrary. This tendency was consistent with previous reports [28]. Double conjugated BAs were also detected, but at very low levels, only 0.18% in humans, 0.41% in mice, and 0.03% in rats. For these double conjugated BAs, little data were reported in feces, which may be due to the limit of the extraction and the detection methods.

The number of hydroxyl groups affects the hydrophobic of BAs. The hydrophobicity was increased in the order of tri-OH BAs (CA, MCA, and HCA), di-OH BAs (CDCA and DCA), and mono-OH BA (LCA). The mono-OH BA (LCA) indices were more dominant in humans (28.11%) than in rodent species (mice 12.57% and rats 1.78%). The percentage of di-OH BAs was highest in rats (51.68%), followed by humans (32.57%), and was lowest in mice (21.36%). The percentage of tri-OH BAs was highest in mice (59.19%) and followed by rats (43.02%), and was lowest in humans (4.68%). Compared with humans, rodents possess a bigger hydrophilic BA pool because of the abundant presence of MCAs. It is well known that MCAs are scarcely reported in humans, but García-Cañaveras et al. and Li et al. detected MCAs in healthy and ESRD human serum, respectively [27,37]. Interestingly, for the first time, we detected ω-MCA and γ-MCA in ESRD patient fecal samples. The average concentration of total MCA was 0.7 μg/kg, and the percentage of these two MCA form BAs was 0.06%.

BAs have many derivatives, including iso-, oxo-, and nor- (OIND BAs). OIND BAs were highest in humans (34.64%), followed by mice (3.53%) and rats (6.88%). This result is opposite to the plasma, in which much higher OIND BAs were found in rats than in humans [25]. This might be due to the fact that more OIND BAs were absorbed by rats as compared to humans. Among these OIND BAs, the percentage of the form of iso- BAs was higher than oxo among human and rodent species (human iso- 18.88%, oxo- 15.75%, nor 0.01%; mouse iso- 2.76%, oxo- 0.76%, nor 0.01%; rat iso- 5.70%, oxo- 1.17%, nor 0.01%). However, higher oxo-BAs in plasma were reported [25]. It indicated that oxo-BA was easier to be reabsorbed into the blood, while iso-BA was excreted in feces.

2.5. BA Indices in Humans, Rats, and Mice

The BA composition ratios are called BA indices, which could comprehensively assess the composition of BA pools, including primary to secondary BA ratio, CA/CDCA ratio, 12α-OH/non-12α-OH ratio, DCA/DCA+CA ratio, and LCA/LCA+CDCA ratio [28]. The BA indices are able to describe the metabolic transformation and biological function of BAs [25,28]. The BA indices vary among biological species, and the changes in BA indices could be an early warning of disease [29]. Therefore, in order to provide reference data for future research, the BA indices of human and rodent species (mice and rats) were present (Figure 5).

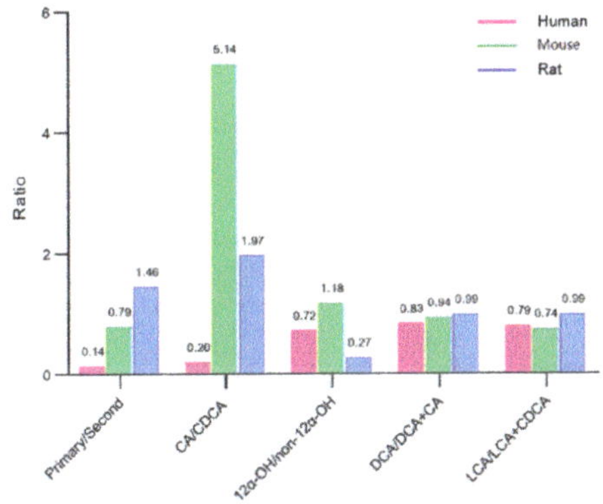

Figure 5. Bile acid indices in humans, rats, and mice.

The ratio of primary to secondary BAs was calculated as the ratio of the sum of the concentrations of CDCA, CA, MCA, and HCA to the sum of the concentrations of DCA, LCA, UDCA, HDCA, and MDCA in all their forms [36]. This rate was lowest in humans (0.14), followed by mice (0.79) and rats (1.46). Rhishikesh et al. found the values of PBA/SBA in humans, rats, and mice were 1.1, 2.3, and 4.5 in plasma matrix while 0.8, 1.9, and 16.5 in urine matrix [28]. These results indicated that human gut microbiota is able to produce more abundant SBA than rodent species.

The ratio of total CA/total CDCA was quite lower in humans (0.20) than in rodents (mice 5.14 and rats 1.97). This ratio reflects the BA synthesis pathway preference (classical or alternative pathways) [36]. In the plasma and urine samples, this rate was also lower in humans than in mice and rats [28]. Thus, it could be speculated that BA synthesis is the preferred classical pathway in humans.

The ratio of 12α-OH/non-12α-OH was calculated as the ratio of the sum of the concentrations of DCA and CA to the sum of the concentrations of CDCA, HDCA, MDCA, LCA, UDCA, HCA, and MCA in all their forms. This ratio was highest in rats (1.18), followed by humans (0.72) and mice (0.27). This tendency was also in line with plasma and urine matrix. The extent of 12α-OH/non-12α-OH BA ratio has been linked to the deficiency of 12α-hydroxylase (CYP8B1) activity [38]. Moreover, several studies have reported that this ratio could reflect host metabolic status [2].

The 7α-dehydroxylase converts CA in DCA and CDCA in LCA, so the ratio of DCA/DCA+CA and the ratio of LCA/LCA+CDCA both reflected the 7α-dehydroxylase activity [9]. These two ratios were quite similar in human and rodent species, ranging from 0.74–0.99. Thus, it could be speculated that 7α-dehydroxylase activity was equal in these species.

3. Materials and Methods

3.1. Materials & Methods

LC-MS grade acetonitrile, ammonium acetate, and formic acid were obtained from Merck (Darmstadt, Germany). Water was obtained from the Merck Millipore Ultra-pure water purification system at 18.2 MΩ/cm (Merck Millipore, Darmstadt, Germany). SPE C18 columns (30 mg/300 cc) were purchased from Waters (Milford, CT, USA). The authentic compounds of 26 unconjugated, 8 glycine conjugated, 9 taurine conjugated, and 15 sulfo-BAs were ordered from either Toronto Research Chemicals (Toronto, ON, Canada), BePure (Bejing, China), or zzstandard® (Shanghai, China). Fifteen deuterium D4-labeled BAs were purchased from BePure and were used as isotope-labeled internal standard (IS) for quantitation. The details on these authentic compounds are provided in Table 1.

3.2. Standard Solutions and Calibration Curves

Stock solutions (1 mg/mL each) of individual BAs were prepared by dissolving the respective compounds separately in methanol. These stock solutions were further diluted with methanol to give final concentrations of 0.01–50 mg/L. These standard solutions were used to determine the limits of detection (LODs) and the limits of quantitation (LOQs). A mixed-standard solution containing 100 μg/L of each of the 15 D4-labeled BAs was prepared in methanol and was used as the IS solution. For the preparation of the calibration curves, each working standard solution was mixed with an equal volume of the IS solution. The 58 standard stock solutions were then pooled together to obtain a 5 mg/L solution, further diluted in methanol to obtain 10 levels in the calibration curve ranging from 0.5–1000 μg/L.

3.3. Sample Preparation

Wet feces were thoroughly homogenized after reception and then stored at −80 °C. The fecal moisture content percentage of ESRD patients and rodent species were 70 ± 10% and 50 ± 5%, respectively. The sample was treated according to 3 extraction protocols (S1–S3).

3.3.1. S1 Extraction with Ethanol

Briefly, 15 mg fecal sample was homogenized with 1 mL cold ethanol (containing Mix ISs) in 2 mL tubes filled with 3–4 mm glass beads. Homogenization was performed using an automated Precellys 24 Tissue Homogenizer (Bertin Technologies, Bretonneux, France) at middle speed for 30 s. The mixtures were centrifuged at 1350× *g* for 10 min at 4 °C. Supernatant was collected and dried under a nitrogen stream at 24 °C. The residue was dissolved in 150 μL initial mobile phase.

3.3.2. S2 Extraction with 0.1 mol/L NaOH Followed by SPE

First, 1 mL of NaOH (0.1 mol/L) was added to 15 mg feces, Vortex shaken 30 s, and incubated for 1 h at 60 °C. Then, 2 mL water was added, homogenized 30 s, and centrifuged at 1350× *g* for 10 min at 4 °C. The supernatant was collected and purified with an SPE cartridge. The 30 mg SPE cartridge was pre-conditioned with 5 mL methanol and 5 mL water, then loaded with supernatant of extract solution and rinsed successively with 20 mL water, 10 mL hexane, and other 20 mL water. The BAs were then eluted with 5 mL methanol. The eluted fraction was collected and dried under a nitrogen stream at 24 °C. The residue was dissolved in a 150 μL initial mobile phase.

3.3.3. S3 Extraction with 5% Ammonium-Ethanol

Briefly, 15 mg fecal sample was homogenized with 1 mL 5% Ammonium-Ethanol (contain Mix ISs) in 2 mL tubes filled with 3–4 mm glass beads. Homogenization was performed using an automated Precellys 24 Tissue Homogenizer (Bertin Technologies, Bretonneux, France) at middle speed for 30 s. The mixtures were centrifuged at 1350× *g* for 10 min at 4 °C. Another 1 mL extraction solution added, repeat the extraction process. Supernatant from the two extraction steps were pooled and dried under a nitrogen stream at 24 °C. The residue was dissolved in 150 μL initial mobile phase.

Reconstituted solution was diluted according to the endogenous BA content before LC injection. The results obtained from the analysis, expressed as μg/L of extract, were converted to μg/kg of dry feces by applying the following formula:

$$C = C_0 \times (V/m) \times n \times (1 - M_F) \quad (1)$$

where

C represents the concentration expressed as μg/kg;
C_0 represents the concentration expressed as μg/L;
V represents the extraction volume (in L);
m represents the weight of wet feces (kg) subjected to extraction;
n represents the dilution ratio;
M_F represents the fecal moisture content (%).

3.4. *UPLC–Q-TOF Analysis*

BAs analysis was performed by using an Agilent 1290 II UPLC system coupled with a G6545 quadrupole-time-of-flight mass spectrometer (UPLC–Q-TOF) from Agilent Technologies (Santa Clara, CA, USA), equipped with an Agilent Jet Stream electrospray (AJS ESI) source.

Separation of BAs was carried out using a using a BEH C18 (2.1 mm × 100 mm, 1.7 μm) UPLC column and a C18 guard column (2.1 × 10 mm, 1.7 μm), both from Waters Inc. (Milford, CT, USA). The column was kept at 30 °C. The mobile phase was consisted of 0.01% formic acid and 2 mM ammonium acetate in water (A) and acetonitrile (B). A linear gradient elution program was applied as follows: 0 min 25% B, 12.0 min 60% B, 26.0 min 75% B, 28.0 min 100% B, and hold on 2.0 min for equilibration. The flow rate was 0.3 mL/min, and the injection volume was 5 μL.

An Agilent Jet Stream electrospray ionization (ESI) interface was used, and its parameters were set as follows: dry gas temperature, 325 °C; dry gas flow, 7 L/min; nebulizer pressure, 35 psig; sheath gas temperature, 350 °C; and sheath gas flow 12 L/min, fragmentor voltage of 140 V and a capillary voltage of 3000 V. The detector operated in a low mass range (1700 *m/z*) and a 2 GHz extended dynamic range. In addition, the centroid mode was used for data collection and storage. Mass accuracy during the analysis was ensured by direct infusion into the source of a reference solution containing TFANH4 (112.9855 *m/z*) and HP-921 (1033.9881 *m/z*). This instrument gave a resolution greater than 10,000 full widths at half maximum (FWHM) at 112.985587 *m/z* and greater than 30,000 FWHM at 1633.949786 *m/z*. The analysis was carried out in the MS mode. The MS scan range was 100–1200 *m/z*, and the acquisition rate was two spectra per second. Data acquisition was accomplished using Agilent MassHunter Workstation Data Acquisition Version B.10.01 software (Santa Clara, CA, USA).

3.5. Standard Solutions and Calibration Curves

3.5.1. Linearity, LOD, and LOQ

Linearity was determined by analysis of calibration curves for all commercially available standards of BAs. The method was validated using a ten-point calibration curve of 0.5–1000 μg/L.

The LOQ was defined as the lowest concentration at which the peak response was ten times that of the noise (10 S/N), and the LOD was the extrapolated concentration with a signal-to-noise ratio of three (3 S/N).

3.5.2. Recovery and Matrix Effect

A relative blank matrix was used in method validation. In order to obtain the relative blank matrix, the fecal sample was extracted by extract solution, certificated, and vaporized solvent. The recovery and matrix effect were assessed at different spiked concentrations with six replicates. The spiked concentration of low, medium, and high were 2.5, 25, and 50 μg/kg, respectively. Moreover, according to the real endogenous BA content in different biological species, some higher spiked concentrations for individual abundant BA were also carried out, with detailed spiked concentrations shown in Table S1. The recovery was evaluated by comparing the peak areas of the analytes before extraction to the corresponding peak area in samples after extraction. The recovery rates must be within $100 \pm 20\%$. The matrix effect was determined by comparing the peak area of post-extraction spiked BA and the standard solution of the same concentration.

3.5.3. Precision and Accuracy

The precisions were evaluated as the intra- and inter-day coefficient of variation (CV, %) for BA analyses with a low spiked concentration in the pooled sample. The intra- and inter-day variations were determined using 5 replicates of spiked samples on the same day and on 5 different days. The precision CV was calculated from the ratio of the relative standard deviation to the mean of the measured analyte concentration. The accuracies were evaluated using solutions with spiked samples with low, medium, and high three certain concentrations, and the accuracy was calculated from the measured/theoretical concentrations$\times$100%. The acceptable range of precision and accuracy for the maximal variation is within$\pm$20% according to the FDA guidelines [35].

3.5.4. Clinical and Preclinical Samples Collection

This methodology application in ESRD patients was an ancillary study to our ESRD patient lipid metabolism trial. Therefore, twenty ESRD patients (10 males and 10 females, aged from 35 to 69) were randomly selected from total of 284 original samples. These patients underwent regular hemodialysis at the hemodialysis center in Beijing, China. Fresh fecal samples of patients were collected from all bowel motions before hemodialysis in hospital. Then individual samples were homogenized immediately and frozen at −80 °C.

Six C57BL/6 mice (male, 8 weeks of age) were purchased from the Experimental Animal Center of the First Affiliated Hospital of Tianjin University of Traditional Chinese Medicine. Six Sprague-Dawley (SD) rats (male, 8 weeks age) were purchased by Beijing HFK Bioscience Co. Ltd. (Beijing, China). Mice and rats were housed at animal facility with free access to a normal diet and water at 22 ± 2 °C, the relative humidity of 45 ± 15%, and a 12 h light/dark cycle. All experiments were approved by the Animal Ethical and Welfare Committee. The discharged feces of each rodent were collected into a centrifuge tube and frozen at −80 °C.

3.5.5. Data Processing and Statistical Study

For LC-MS/MS data, MassHunter Quantitative Analysis vB.10.01 (Agilent Technologies, Inc., Santa Clara, CA, USA) was used for quantification. If the analyte concentration was below the LOQ, the value of LOQ/2 was used for statistical calculation. Tukey HSD all-pairwise comparisons test was used for multiple comparisons of data. A difference of $p < 0.05$ was considered significant. Statistical analysis and graphs were performed in GraphPad Prism v8.0.1 (GraphPad Software, San Diego, CA, USA). Sankey diagram was used in this study to visualize the BA types distribution in each biological species under the R Graph Gallery (https://www.data-to-viz.com/graph/sankey.html, accessed on 1 May 2022). Different BA types (classified by BA structure, detailed in Table 1) are represented by rectangles. Their links are represented with arcs that have a width proportional to the sub-category of the BA types.

4. Conclusions

In this study, we developed and validated a simple, effective and sensitive UPLC–Q-TOF method that simultaneously performs quantitative and qualitative analysis of 58 BAs, including unconjugated, amino acid conjugated (glycine or taurine), sulfo-conjugated, and double conjugated, as well as iso-, nor-, and oxo- BA metabolites in feces. All the methodology results were acceptable according to the FDA guidelines. This method could be applied in the global profiling of BA in humans, rats, and mice. In general, a higher proportion of sulfated BAs and mono-OH BA (LCA) was present in humans rather than in rodents. OIND BAs were also abundant in humans, especially iso- and oxo- BA. In terms of BA indices, PBA/SBA ratio and total CA/total CDCA ratio were fairly low in humans, while the DCA/DCA + CA ratio and LCA/LCA + CDCA ratio were equal in humans and rodents.

Supplementary Materials: The following supporting information can be downloaded at: https://www.mdpi.com/article/10.3390/metabo12070633/s1, Table S1: Sensitivity, linearity, accuracy, and precision of bile acids by ultra-high-performance liquid chromatography coupled with quadrupole-time-of-flight mass spectrometer analysis; Table S2: The recovery (%) of spiked concentration (μg/kg) for bile acids in each biological sample; Table S3: The matrix effect (%) of spiked concentration (μg/kg) for bile acids in each biological sample.

Author Contributions: Conceptualization, X.Z. and F.R.; methodology, X.Z. and X.L.; validation, X.Z. and Y.L.; formal analysis, X.Z. and J.Y.; investigation, X.Z., and X.L.; resources, F.R. and Y.L.; data curation, X.Z. and X.L.; writing—original draft preparation, X.Z.; writing—review and editing, J.Y.; visualization, X.Z. and Y.L.; supervision, X.Z. and F.R.; project administration, Y.L. All authors have read and agreed to the published version of the manuscript.

Funding: This research received no external funding.

Institutional Review Board Statement: The study was conducted in accordance with the Declaration of Helsinki and approved on 20 October 2020 by the Ethics Committee of China Agricultural University (No. CAUPCKD-02). The animal study protocol was approved by the Ethics Committee of China Agricultural University (No. IRMDWLL-2017095).

Informed Consent Statement: Informed consent was obtained from all subjects involved in the study.

Data Availability Statement: The data presented in this study are available in the article and Supplementary Materials.

Acknowledgments: We would like to thank the parents involved in the study.

Conflicts of Interest: The authors declare no conflict of interest.

References

1. Di Ciaula, A.; Garruti, G.; Lunardi Baccetto, R.; Molina-Molina, E.; Bonfrate, L.; Wang, D.Q.H.; Portincasa, P. Bile Acid Physiology. *Ann. Hepatol.* **2017**, *16*, S4–S14. [CrossRef] [PubMed]
2. Jia, W.; Wei, M.; Rajani, C.; Zheng, X. Targeting the alternative bile acid synthetic pathway for metabolic diseases. *Protein Cell* **2021**, *12*, 411–425. [CrossRef] [PubMed]
3. Schoeler, M.; Caesar, R. Dietary lipids, gut microbiota and lipid metabolism. *Rev. Endocr. Metab. Disord.* **2019**, *20*, 461–472. [CrossRef] [PubMed]
4. Vital, M.; Rud, T.; Rath, S.; Pieper, D.H.; Schlüter, D. Diversity of Bacteria Exhibiting Bile Acid-inducible 7α-dehydroxylation Genes in the Human Gut. *Comput. Struct. Biotechnol. J.* **2019**, *17*, 1016–1019. [CrossRef]

5. Funabashi, M.; Grove, T.L.; Wang, M.; Varma, Y.; McFadden, M.E.; Brown, L.C.; Guo, C.; Higginbottom, S.; Almo, S.C.; Fischbach, M.A. A metabolic pathway for bile acid dehydroxylation by the gut microbiome. *Nature* **2020**, *582*, 566–570. [CrossRef]
6. Marion, S.; Desharnais, L.; Studer, N.; Dong, Y.; Notter, M.D.; Poudel, S.; Menin, L.; Janowczyk, A.; Hettich, R.L.; Hapfelmeier, S.; et al. Biogeography of microbial bile acid transformations along the murine gut. *J. Lipid Res.* **2020**, *61*, 1450–1463. [CrossRef]
7. Alnouti, Y. Bile Acid Sulfation: A Pathway of Bile Acid Elimination and Detoxification. *Toxicol. Sci.* **2009**, *108*, 225–246. [CrossRef]
8. Shafaei, A.; Rees, J.; Christophersen, C.T.; Devine, A.; Broadhurst, D.; Boyce, M.C. Extraction and quantitative determination of bile acids in feces. *Anal. Chim. Acta* **2021**, *1150*, 338224. [CrossRef]
9. Fiorucci, S.; Biagioli, M.; Zampella, A.; Distrutti, E. Bile Acids Activated Receptors Regulate Innate Immunity. *Front. Immunol.* **2018**, *9*, 1853. [CrossRef]
10. Thomas, C.; Pellicciari, R.; Pruzanski, M.; Auwerx, J.; Schoonjans, K. Targeting bile-acid signalling for metabolic diseases. *Nat. Rev. Drug Discov.* **2008**, *7*, 678–693. [CrossRef]
11. Chiang, J.Y.L.; Ferrell, J.M. Bile Acids as Metabolic Regulators and Nutrient Sensors. *Annu. Rev. Nutr.* **2019**, *39*, 175–200. [CrossRef] [PubMed]
12. Perino, A.; Demagny, H.; Velazquez-Villegas, L.; Schoonjans, K. Molecular Physiology of Bile Acid Signaling in Health, Disease, and Aging. *Physiol Rev.* **2021**, *101*, 683–731. [CrossRef] [PubMed]
13. Sato, Y.; Atarashi, K.; Plichta, D.R.; Arai, Y.; Sasajima, S.; Kearney, S.M.; Suda, W.; Takeshita, K.; Sasaki, T.; Okamoto, S.; et al. Novel bile acid biosynthetic pathways are enriched in the microbiome of centenarians. *Nature* **2021**, *599*, 458–464. [CrossRef] [PubMed]
14. Wang, Y.; Gao, X.; Zhang, X.; Xiao, Y.; Huang, J.; Yu, D.; Li, X.; Hu, H.; Ge, T.; Li, D.; et al. Gut Microbiota Dysbiosis Is Associated with Altered Bile Acid Metabolism in Infantile Cholestasis. *mSystems* **2019**, *4*, e00463-19. [CrossRef]
15. Vasavan, T.; Ferraro, E.; Ibrahim, E.; Dixon, P.; Gorelik, J.; Williamson, C. Heart and bile acids—Clinical consequences of altered bile acid metabolism. *Biochim. Biophys. Acta Mol. Basis Dis.* **2018**, *1864*, 1345–1355. [CrossRef]
16. Mahmoudian Dehkordi, S.; Arnold, M.; Nho, K.; Ahmad, S.; Jia, W.; Xie, G.; Louie, G.; Kueider-Paisley, A.; Moseley, M.A.; Thompson, J.W.; et al. Altered bile acid profile associates with cognitive impairment in Alzheimer's disease—An emerging role for gut microbiome. *Alzheimer's Dement.* **2018**, *15*, 76–92. [CrossRef]
17. Hang, S.; Paik, D.; Yao, L.; Kim, E.; Trinath, J.; Lu, J.; Ha, S.; Nelson, B.N.; Kelly, S.P.; Wu, L.; et al. Bile acid metabolites control TH17 and Treg cell differentiation. *Nature* **2019**, *576*, 143–148. [CrossRef]
18. Sivamaruthi, B.S.; Fern, L.A.; Rashidah Pg Hj Ismail, D.S.N.; Chaiyasut, C. The influence of probiotics on bile acids in diseases and aging. *Biomed Pharm.* **2020**, *128*, 110310. [CrossRef]
19. Gao, T.; Feridooni, H.A.; Howlett, S.E.; Pelis, R.M. Influence of age on intestinal bile acid transport in C57BL/6 mice. *Pharm. Res. Perspect.* **2017**, *5*, e00287. [CrossRef]
20. Han, J.; Liu, Y.; Wang, R.; Yang, J.; Ling, V.; Borchers, C.H. Metabolic Profiling of Bile Acids in Human and Mouse Blood by LC–MS/MS in Combination with Phospholipid-Depletion Solid-Phase Extraction. *Anal. Chem.* **2015**, *87*, 1127–1136. [CrossRef]
21. Sharma, K.R. Review on bile acid analysis. *Int. J. Pharm. Biomed. Sci.* **2012**, *3*, 28–34.
22. Liu, Y.; Rong, Z.; Xiang, D.; Zhang, C.; Liu, D. Detection technologies and metabolic profiling of bile acids: A comprehensive review. *Lipids Health Dis.* **2018**, *17*, 121. [CrossRef] [PubMed]
23. Lee, G.; Lee, H.; Hong, J.; Lee, S.H.; Jung, B.H. Quantitative profiling of bile acids in rat bile using ultrahigh-performance liquid chromatography–orbitrap mass spectrometry: Alteration of the bile acid composition with aging. *J. Chromatogr. B* **2016**, *1031*, 37–49. [CrossRef] [PubMed]
24. Haag, M.; Hofmann, U.; Mürdter, T.E.; Heinkele, G.; Leuthold, P.; Blank, A.; Haefeli, W.E.; Alexandrov, A.; Urban, S.; Schwab, M. Quantitative bile acid profiling by liquid chromatography quadrupole time-of-flight mass spectrometry: Monitoring hepatitis B therapy by a novel Na(+)-taurocholate cotransporting polypeptide inhibitor. *Anal. Bioanal. Chem.* **2015**, *407*, 6815–6825. [CrossRef] [PubMed]
25. Sangaraju, D.; Shi, Y.; Van Parys, M.; Ray, A.; Walker, A.; Caminiti, R.; Milanowski, D.; Jaochico, A.; Dean, B.; Liang, X. Robust and Comprehensive Targeted Metabolomics Method for Quantification of 50 Different Primary, Secondary, and Sulfated Bile Acids in Multiple Biological Species (Human, Monkey, Rabbit, Dog, and Rat) and Matrices (Plasma and Urine) Using Liquid Chromatography High Resolution Mass Spectrometry (LC-HRMS) Analysis. *J. Am. Soc. Mass Spectrom.* **2021**, *32*, 2033–2049. [CrossRef] [PubMed]
26. Choucair, I.; Nemet, I.; Li, L.; Cole, M.A.; Skye, S.M.; Kirsop, J.D.; Fischbach, M.A.; Gogonea, V.; Brown, J.M.; Tang, W.H.W.; et al. Quantification of bile acids: A mass spectrometry platform for studying gut microbe connection to metabolic diseases. *J. Lipid Res.* **2020**, *61*, 159–177. [CrossRef] [PubMed]
27. Li, R.; Zeng, L.; Xie, S.; Chen, J.; Yu, Y.; Zhong, L. Targeted metabolomics study of serum bile acid profile in patients with end-stage renal disease undergoing hemodialysis. *PeerJ* **2019**, *7*, e7145. [CrossRef]
28. Thakare, R.; Alamoudi, J.A.; Gautam, N.; Rodrigues, A.D.; Alnouti, Y. Species differences in bile acids I. Plasma and urine bile acid composition. *J. Appl. Toxicol.* **2018**, *38*, 1323–1335. [CrossRef]
29. Jia, W.; Xie, G.; Jia, W. Bile acid–microbiota crosstalk in gastrointestinal inflammation and carcinogenesis. *Nat. Rev. Gastroenterol. Hepatol.* **2018**, *15*, 111–128. [CrossRef]
30. Jones, B.; Begley, M.; Hill, C.; Gahan, C.; Marchesi, J. Functional and comparative metagenomic analysis of bile salt hydrolase activity in the human gut microbiome. *Proc. Natl. Acad. Sci. USA* **2008**, *105*, 13580–13585. [CrossRef]

31. Perwaiz, S.; Mignault, D.; Tuchweber, B.; Yousef, I.M. Rapid and improved method for the determination of bile acids in human feces using MS. *Lipids* **2002**, *37*, 1093–1100. [CrossRef] [PubMed]
32. Humbert, L.; Maubert, M.; Wolf, C.; Duboc, H.; Mahé, M.; Farabos, D.; Seksik, P.; Mallet, J.-M.; Trugnan, G.; Masliah, J.; et al. Bile acid profiling in human biological samples: Comparison of extraction procedures and application to normal and cholestatic patients. *J. Chromatogr. B Anal. Technol. Biomed. Life Sci.* **2012**, *899*, 135–145. [CrossRef] [PubMed]
33. Yin, S.; Su, M.; Xie, G.; Li, X.; Wei, R.; Liu, C.; Lan, K.; Jia, W. Factors affecting separation and detection of bile acids by liquid chromatography coupled with mass spectrometry in negative mode. *Anal. Bioanal. Chem.* **2017**, *409*, 5533–5545. [CrossRef] [PubMed]
34. Griffiths, W.J.; Sjovall, J. Bile acids: Analysis in biological fluids and tissues. *J. Lipid Res.* **2010**, *51*, 23–41. [CrossRef]
35. FDA. *Guidance for Industry, Bioanalytical Method Validation*; U.S. Department of Health and Human Services, Center for Drug Evaluation and Research (CDER), Center for Veterinary Medicine (CV): Silver Spring, MD, USA, 2001.
36. Thakare, R.; Alamoudi, J.; Gautam, N.; Rodrigues, A.; Alnouti, Y. Species differences in bile acids II. Bile acid metabolism. *J. Appl. Toxicol.* **2018**, *38*, 1590–1600. [CrossRef] [PubMed]
37. García-Cañaveras, J.C.; Donato, M.T.; Castell, J.V.; Lahoz, A. Targeted profiling of circulating and hepatic bile acids in human, mouse, and rat using a UPLC-MRM-MS-validated method. *J. Lipid Res.* **2012**, *53*, 2231–2241. [CrossRef]
38. Haeusler, R.; Pratt-Hyatt, M.; Welch, C.; Klaassen, C.; Accili, D. Impaired Generation of 12-Hydroxylated Bile Acids Links Hepatic Insulin Signaling with Dyslipidemia. *Cell Metab.* **2011**, *15*, 65–74. [CrossRef]

metabolites

MDPI

Article

An Update on Sphingolipidomics: Is Something Still Missing? Some Considerations on the Analysis of Complex Sphingolipids and Free-Sphingoid Bases in Plasma and Red Blood Cells

Camillo Morano [1], Aida Zulueta [2], Anna Caretti [3], Gabriella Roda [1], Rita Paroni [3] and Michele Dei Cas [3,*]

1 Department of Pharmaceutical Sciences, Università degli Studi di Milano, 20133 Milan, Italy; camillo.morano@unimi.it (C.M.); gabriella.roda@unimi.it (G.R.)
2 Neurorehabilitation Department, IRCCS Istituti Clinici Scientifici Maugeri di Milano, 20138 Milan, Italy; aida.zuluetamorales@icsmaugeri.it
3 Department of Health Sciences, Università degli Studi di Milano, 20142 Milan, Italy; anna.caretti@unimi.it (A.C.); rita.paroni@unimi.it (R.P.)
* Correspondence: michele.deicas@unimi.it

Abstract: The main concerns in targeted "*sphingolipidomics*" are the extraction and proper handling of biological samples to avoid interferences and achieve a quantitative yield well representing all the sphingolipids in the matrix. Our work aimed to compare different pre-analytical procedures and to evaluate a derivatization step for sphingoid bases quantification, to avoid interferences and improve sensitivity. We tested four protocols for the extraction of sphingolipids from human plasma, at different temperatures and durations, and two derivatization procedures for the conversion of sphingoid bases into phenylthiourea derivatives. Different columns and LC-MS/MS chromatographic conditions were also tested. The protocol that worked better for sphingolipids analysis involved a single-phase extraction in methanol/chloroform mixture (2:1, *v*/*v*) for 1 h at 38 °C, followed by a 2 h alkaline methanolysis at 38 °C, for the suppression of phospholipids signals. The derivatization of sphingoid bases promotes the sensibility of non-phosphorylated species but we proved that it is not superior to a careful choice of the appropriate column and a full-length elution gradient. Our procedure was eventually validated by analyzing plasma and erythrocyte samples of 20 volunteers. While both extraction and methanolysis are pivotal steps, our final consideration is to analyze sphingolipids and sphingoid bases under different chromatographic conditions, minding the interferences.

Keywords: sphingolipids; sphingolipidomics; sphingoid bases; lipidomics; mass spectrometry

Citation: Morano, C.; Zulueta, A.; Caretti, A.; Roda, G.; Paroni, R.; Dei Cas, M. An Update on Sphingolipidomics: Is Something Still Missing? Some Considerations on the Analysis of Complex Sphingolipids and Free-Sphingoid Bases in Plasma and Red Blood Cells. *Metabolites* **2022**, *12*, 450. https://doi.org/10.3390/metabo12050450

Academic Editor: Joana Pinto

Received: 1 May 2022
Accepted: 13 May 2022
Published: 17 May 2022

1. Introduction

Sphingolipids are a ubiquitous class of lipids, whose structure always comprises a long-chain base, usually sphingosine (Sph) or sphinganine. Their name derives from the mythological figure of the sphynx, because of their enigmatic nature [1]. Sphingolipids are commonly divided into two major classes: ceramides (Cer) and complex sphingolipids. Cer are "de novo" biologically synthesized by attaching a fatty acid to the amine group of dihydrosphingosine (dhSph) through an amidic bond, and are mostly found in the outer leaflet of the plasma membrane. Cer are then catabolized to Sph and sphingosine-1P (S1P) which will exit the pathway by degradation to palmitoyl aldehyde and phosphoethanolamine. Complex sphingolipids, on the other hand, comprise many different subclasses, such as sphingomyelins (SM), made up of a polar head such as choline or serine, and glycosphingolipids, which are, in turn, classified according to the number of sugar residues attached to the carbon chain [2,3]. Other than their role in the formation and modulation of biological membranes, sphingolipids, especially Cer, the "central hub" of sphingolipids metabolism, and S1P, are believed to be responsible for many different signaling functions in the organism such as apoptosis, inflammation, cell proliferation, and differentiation.

Due to the various roles that sphingolipids have, any alteration of their metabolism could be part of pathological mechanisms or, sometimes, could be the reason for the diseases themselves [4–8]. The analysis of the whole set of sphingolipids in a biological system is referred to as "*sphingolipidomics*", and is now standardly carried out through liquid chromatography coupled to mass spectrometry to characterize and differentiate simultaneously the numerous species of sphingolipids belonging to the different subclasses [9,10]. However, due to the high variety of chemical structures, one of the main issues remains the extraction and proper handling of samples to achieve a yield that could well represent the actual concentrations of all sphingolipids in the system under analysis as already postulated in untargeted lipidomics [11]. In fact, on one hand, Cer and complex sphingolipids can be easily characterized using a solvent extraction followed by alkaline methanolysis [12,13], which remains the method of choice for sample handling; on the other, free sphingoid bases are hard to extract and their analysis can be quite challenging. Indeed, while there is a growing general interest in achieving a common protocol for the analysis of free sphingoid bases such as Sph and S1P, as they appear to be, as mentioned, important biological mediators, they are fairly difficult to be detected using LC-MS/MS. The reason for this is double-fold: (1) short liquid chromatography does not allow one to properly separate free sphingoid bases from any interferent in the system; and (2) their chemical nature makes it difficult to obtain proper ionization of the compounds.

Our work aimed to compare different methods of sample handling and extraction for sphingolipids. Furthermore, we evaluated whether a derivatization step by phenylisothiocyanate (PITC) could improve the detection and analysis of free sphingoid bases.

2. Results and Discussion

2.1. Set Up of the Extraction Procedure

We tested four different extraction protocols (Materials and Methods) to evaluate whether different conditions could deeply affect the recovery of sphingolipids. As displayed in Figure 1A, the more complex classes of sphingolipids do not seem to be impacted using the four different procedures, except SM, which appears to be underestimated using the first two protocols. Alkaline methanolysis is useful to disrupt the ester bond in phospholipids while maintaining the amide linkage unaltered, which is characteristic of sphingolipids. Especially using low-resolution triple quadrupole, the need for distinguishing or chromatographically separating phosphatidylcholine (PC) and SM is factual since they can co-elute and/or overlap in mass transition (e.g., SM 38:3 m/z 771.6115 > 184 and PC O-36:2 m/z 772.6209 > 184; SM 42:4 m/z 809.6494 > 184 and PC 38:4 m/z 810.6004 > 184) competing irremediably in their quantification [14]. In every condition (Figure 1B) here reported, incubation at 38 °C from 1 h to 12 h can effectively reduce the plasma physiological phospholipids content of about 98.5% (estimated on dipalmitoylphosphatidylcholine, DPPC) and more than 99.9% on added deuterated internal standard (phosphatidylcholine (15:0–18:1) d7, PC d7). By contrast, a shorter time (inferior to 1 h) allows a lower reduction of phospholipids, which can be estimated between 93 and 95% with respect to not-treated samples. The warm incubation overnight (48 °C) has been historically introduced to uniformly level the lipid in the extracting solvent since different sphingolipids can have high phase transition temperatures. However, we believe that this passage could be shortened since its beneficial effect was not observed (see below) [12]. The traditional liquid–liquid extraction protocols firstly proposed by Folch and Bligh-Dyer [15,16] and the monophasic extraction—here and elsewhere described [13,17,18]—are essentially identical in extraction rate for the content of Cer, dihydroceramides (dhCer), SM and glycosphingolipids (Figure 1C). The prominent polarity of acidic glycosphingolipids—such as the simplest gangliosides (GM3)—does not grant a standardized recovery in the bottom chloroform phase of Folch (mean ± SD, 0.15 ± 0.07 vs. 3.8 ± 0.21 single-phase) and also in the more polar Bligh-Dyer (1.12 ± 0.03 vs. 3.8 ± 0.21 single-phase) protocols. The use of Folch and Bligh-Dyer also emphasizes the recovery of the sphingoid bases especially in their phosphate forms S1P and dihydrosphingosine-1-phosphate (dhS1P). This effect was also noticeable on the internal standard used for this purpose (sphinganine d17:0) whose extraction fate is diminished in Folch by 55% and in Bligh-Dyer by 26% with respect to monophasic extraction.

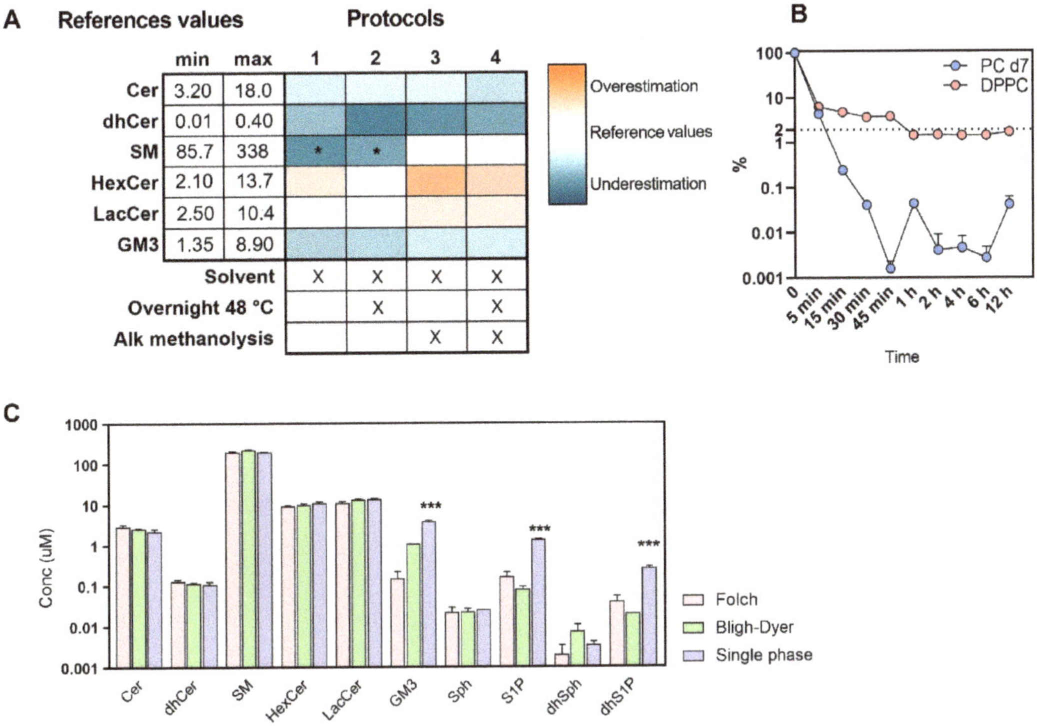

Figure 1. (**A**) Quantification of sphingolipids in a plasma pool from healthy volunteers (n = 20) as a function of different extraction protocols and comparison to the reference values found in the scientific literature. On the left of the heatmap, the range of concentration (μM) of sphingolipids in plasma EDTA from healthy volunteers found in the scientific literature [19–27]. For visualization, data were scaled to reference values and reported as a fold-change logarithm. Those significantly modulated were evaluated by performing repeated measures one-way ANOVA and the Dunnett post hoc test. The different steps in each protocol are schematized under the heatmap and their occurrence is marked with an "X". (**B**) Estimation of plasma phospholipids content after alkaline hydrolysis (KOH 73 mM) over time (0, 1, 2, 4, 6, 12 h) at 38 °C. Data were visualized as a percentage of DPPC and PC d7 with respect to untreated samples. Each point represents the mean of n = 2 technical replicates. (**C**) Comparison of the traditional liquid–liquid extraction for total lipid content (Folch n = 3 and Bligh-Dyer n = 3) and the single-phase extraction (n = 3) for the recovery of sphingolipids in a plasma pool from healthy volunteers (n = 20). Statistical differences were measured by one-way ANOVA and the Dunnett post hoc test against monophasic extraction. p values are schematized as follows: * < 0.05; *** < 0.001.

The effects of times (1/2/4/12 h) and temperatures (room temperature, rt/4/38/48 °C) on the recovery of sphingolipids from plasma were also considered and the results are graphed in Figure 2A. As already postulated above, the overnight extraction (12 h) seems to be futile or even counterproductive, thus we believe that this passage could be shortened between 1 and 2 h. The incubation at 48 °C is overall worthless and detrimental, especially on complex sphingolipids (e.g., Cer, dhCer and hexosylceramides, HexCer). The only species which strongly benefit from this long and hot period of extraction are phosphate forms of sphingoid bases (+48% at 48°; +25% at 38° vs. baseline condition 1 h at rt). We demonstrated that the recovery of either: (a) 1 h at 38 °C; (b) 2 h at rt; or (c) 2 h at 4 °C is essentially superior and interchangeable between them since their mean recovery is +5% (Figure 2B) in respect to baseline (1 h at rt). The extraction with a temperature between 38 and 48 °C and prolonged from 2 to 4 h revealed a slightly decrease in the recovery of plasma sphingolipids. The results presented in this paragraph are summed up in a final protocol proposed and outlined in Figure 2C.

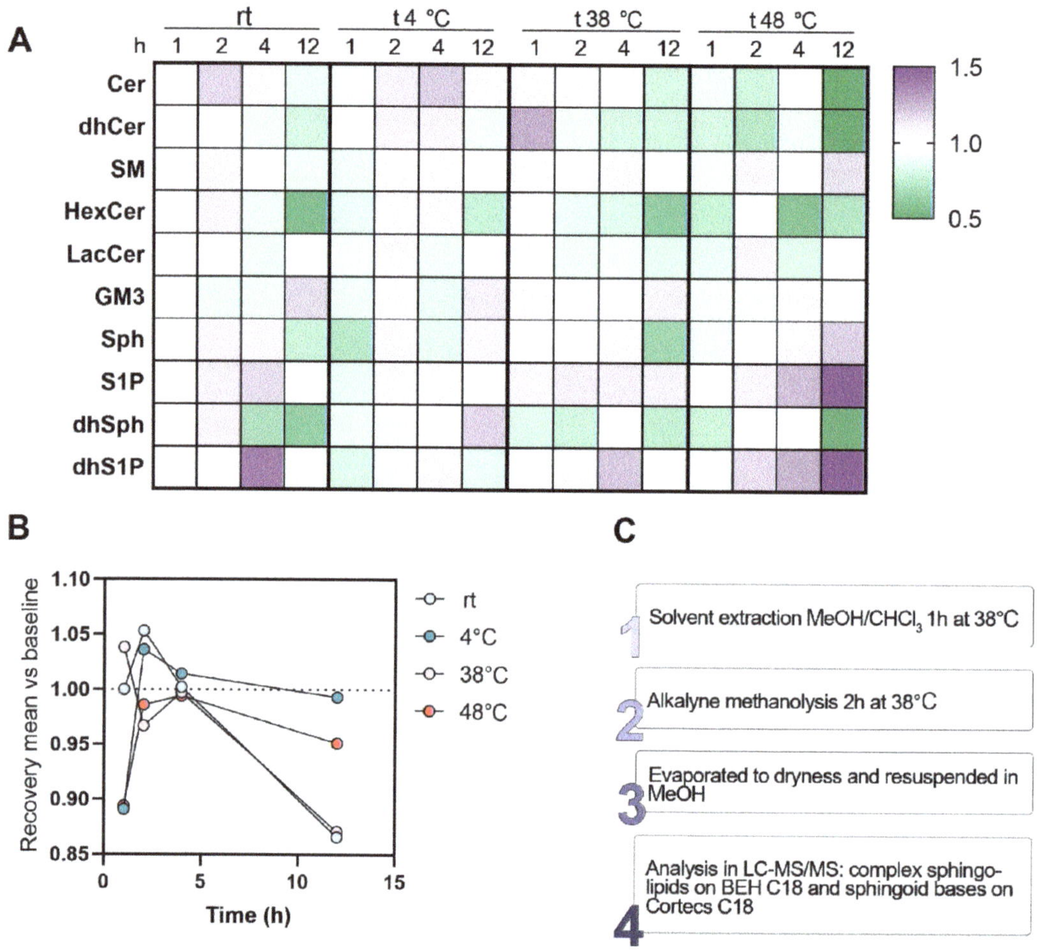

Figure 2. (**A**) Effects of time (1/2/4/12 h) and temperature (rt/4/38/48 °C) on the recovery of sphingolipids from plasma using a single-phase extraction (n = 2 per each condition). Data were scaled for visualization on the recovery obtained for 1 h at rt (22 °C). (**B**) The effects of times (1/2/4/12 h) and temperatures (room temperature, rt/4/38/48 °C) on the recovery of sphingolipids from plasma. Data were scaled for visualization on the recovery obtained for 1 h at rt (22 °C, baseline). (**C**) Scheme of the final steps included in the protocol.

2.2. *Choosing the Best Analytical Condition*

One of the main issues in the analysis of Sph and other sphingoid bases is that their levels in plasma and other biological matrices are not always high enough to allow a precise quantitation. Bearing in mind that the concentrations of sphingoid bases in human plasma range from 0.006 µM to 1.56 µM [19–27] (Supplementary Figure S1), it is critical to be aware of any possible interferents in the analysis. As particularly appraisable in Figure 3, choosing the appropriate column and chromatography conditions can make a huge difference in sphingoid bases analysis. In fact, many interfering signals of Sph are detected along with the chromatogram. While a short chromatography (Figure 3A) may seem an optimal choice for the analysis of sphingoid bases, the interferences over Sph are not even detected, and lengthening the runtime (Figure 3B), on the other hand, does not allow a clear distinction of Sph from its interfering signals. We resolved this issue by switching the column from an Acquity BEH C18 to a Cortecs C18; in fact, while Sph-interfering signals are still detected, they are completely separated from Sph (approximately three minutes apart), allowing an as close to reality as possible quantitation. Moreover, the use of a relative long elution program also enabled a sensible reduction of carry-over of phosphate derivatives, which can be displayed in run times inferior than 10 min.

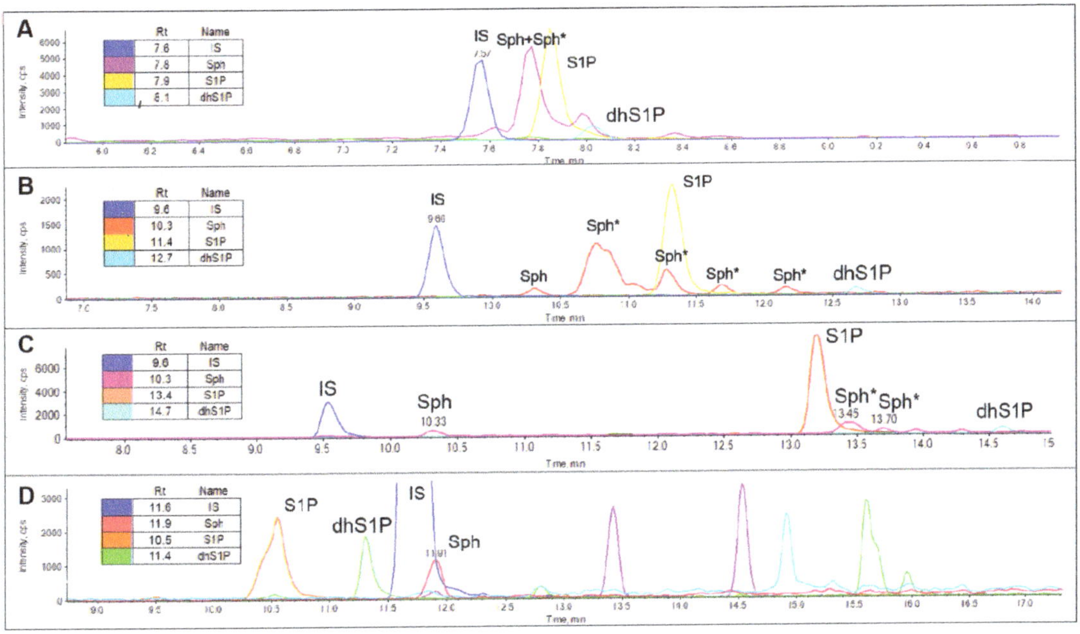

Figure 3. Chromatograms of plasma-free sphingoid bases analyzed by (**A**) Acquity BEH C18 with a short chromatography elution program, (**B**) Acquity BEH C18 with a long chromatography elution program, (**C**) Cortecs C18 with a long chromatography elution program and (**D**) Acquity BEH C18 with a long chromatography elution program after chemical derivatization with phenylisothiocyanate. In each panel * indicates the interferences on sphingosine transition. dhSph is not always appreciable since its concentration is markedly lower than other sphingosine bases. See Materials and Methods for the detailed LC-MS/MS conditions.

2.2.1. Sphingoid Bases Derivatization

In order to fix the issue of Sph-interfering signals, we evaluated whether a derivatization of the extract could be determined. As displayed in Figure 3D, the interfering signals of Sph completely disappeared and the chromatographic separation was excellent for all analytes. The detection of derivatives of Sph and dhSph is increased by the mean of 1.5–2.5-fold (Supplementary Table S1). On the other hand, though, the signal intensity of S1P is reduced by approximately 50%, which interferes with the intent of detecting sphingoid bases even in matrices and systems that may not be particularly enriched in these species (Supplementary Table S1). However, their quantification in plasma—which maintains a relatively high sphingoid bases concentration—can be achieved undeniably by either derivatizing their amine function or not, as displayed in Supplementary Figure S2. In the analytes considered here, the derivatization indeed unveiled slightly higher concentrations vs. the same samples not derivatized.

2.3. Performance in Human Plasma and Red Blood Cells

When we adopted the final extraction protocol (Materials and Methods, Section 3.4, protocol 4), for both complex sphingolipids and sphingoid bases (long chromatography on Cortecs C18), the concentration range of the analytes fell perfectly into those described in the literature [19–27] and the reproducibility of the methods was validated (Tables 1 and 2). In Figure 4, the attained ranges are shown. In Tables 3 and 4, furthermore, the results are expressed in numerical form and the percentage of analyzed species. Red blood cells (RBCs) sphingolipid concentrations [28–32] are introduced in Figure 5. The main sphingolipid in RBCs remains SM (87.5%) and Cer (5.8%) but with respect to the glycosphingolipids, lactosylceramides (LacCer) are prevalent (4% RBCs vs. 2.2% plasma), whereas in plasma the mono HexCer are predominant (3.0% plasma vs. 0.4% RBCs). The low-abundant dhCer are fairly detectable in plasma, accounting for less than 0.2% of total sphingolipids, but contrarily, in RBCs, they are more abundant, estimated at 1.4%. In Supplementary Table S2, the concentrations of sphingolipids in RBCs are reported in pmol/10^6 cells.

Table 1. Intra- (n = 5 independent extraction replicates) and inter-days (n = 10 independent extraction replicates) precision for the analysis of the whole panel of plasma sphingolipids on a plasma pool from healthy volunteers (n = 20). The analyses of the sphingolipids and sphingoid bases were performed as described in Sections 3.3.1 and 3.3.2 (long chromatography), respectively.

Class	CV% Intra-Day (n = 5)	CV% Inter-Days (n = 10)
Cer	1.6	7.9
dhCer	3.8	7.6
SM	3.8	7.9
HexCer	7.2	6.0
LacCer	7.8	12.8
GM3	7.6	14.0
Sph	4.2 (4.8)	4.8 (13.1)
S1P	5.0 (2.1)	3.8 (12.4)
dhSph	10.1 (11.1)	11.2 (14.8)
dhS1P	8.1 (2.4)	6.8 (11.9)

Table 2. Intra- (n = 5 independent extraction replicates) and inter-days (n = 10 independent extraction replicates) precision for the analysis of the whole panel of RBCs sphingolipids on an RBCs pool from healthy volunteers (n = 20). The analyses of the sphingolipids and sphingoid bases were performed as described in Sections 3.3.1 and 3.3.2 (long chromatography), respectively.

Class	CV% Intra-Day (n = 5)	CV% Inter-Days (n = 10)
Cer	7.4	10.5
dhCer	7.2	15.0
SM	8.4	9.3
HexCer	13.4	10.5
LacCer	12.4	9.8
GM3	11.2	11.3
Sph	9.6	13.9
S1P	9.4	8.3
dhSph	9.9	8.0
dhS1P	7.7	10.0

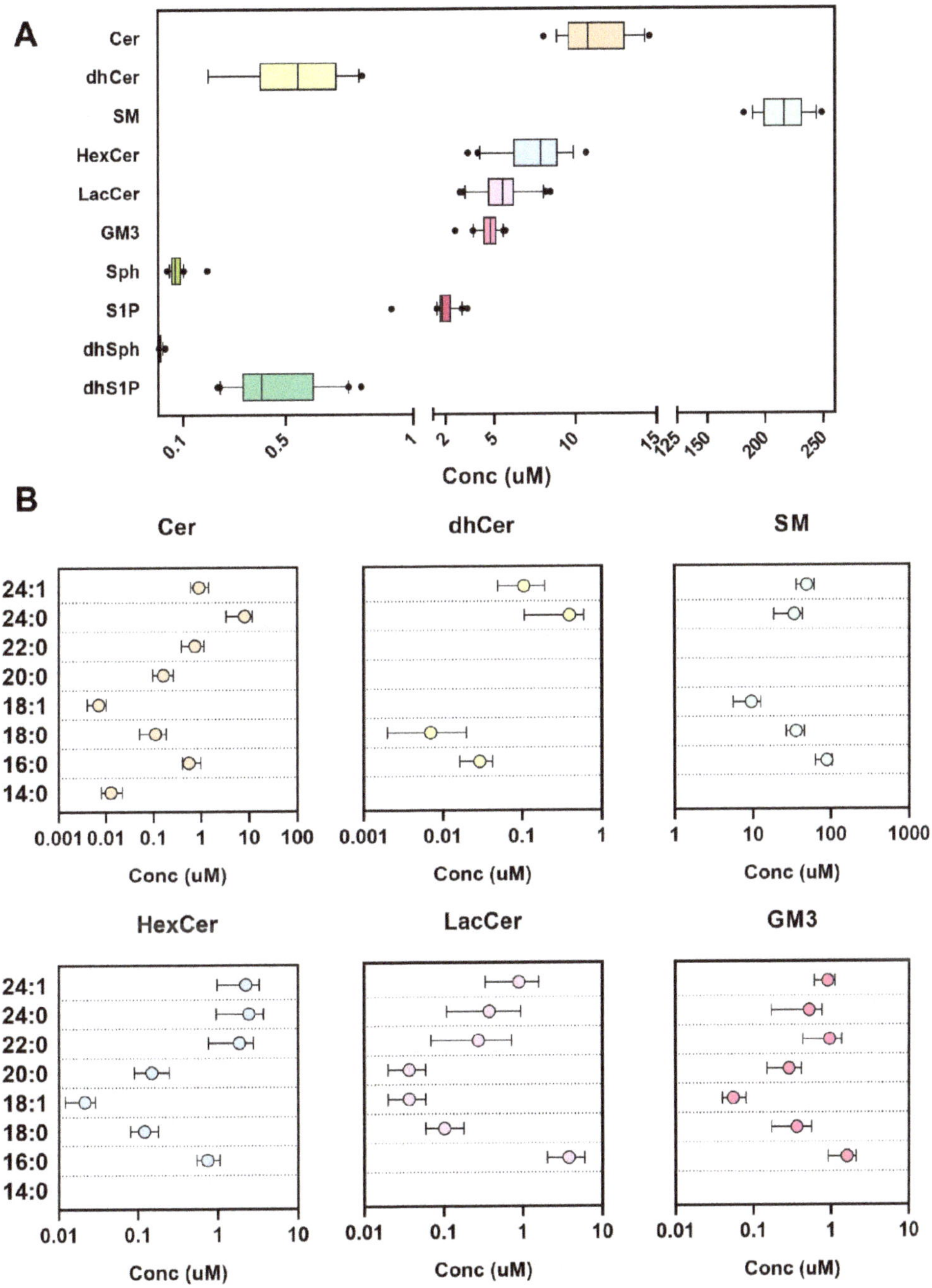

Figure 4. Plasma sphingolipids concentration in healthy volunteers ($n = 20$). (**A**) Concentrations of complex sphingolipids and free sphingoid bases (min–max, line at mean, dots represent the 10–90th percentile) as the sum of the species in each class and (**B**) divided according to their fatty acid composition (mean ± SD).

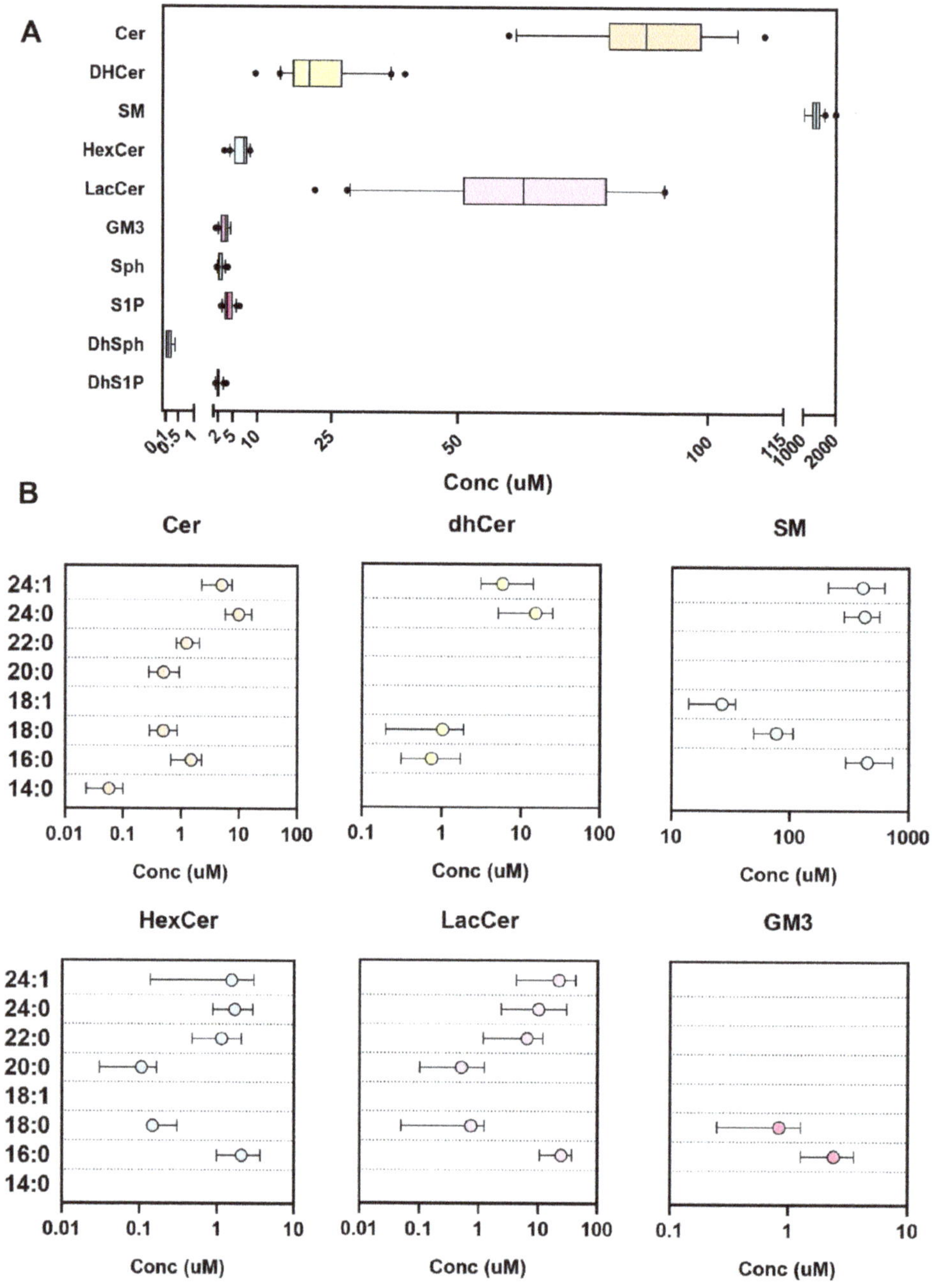

Figure 5. RBCs sphingolipid concentration in healthy volunteers (n = 20). (**A**) Concentrations of complex sphingolipids and free sphingoid bases (min–max, line at mean, dots represent the 10–90th percentile) as the sum of the species in each class and (**B**) divided according to their fatty acid composition (mean ± SD).

Table 3. Plasma EDTA sphingolipids levels (μM) in healthy volunteers (n = 20) expressed as min–max, mean ± SD and percentage over total sphingolipid content.

Conc (μM)	Min	Max	Mean ± SD (n = 20)	%
Cer	8.1	14.6	11.2 ± 2.0	4.5
dhCer	0.2	0.8	0.54 ± 0.20	0.2
SM	181.7	248.9	217.9 ± 19.3	87.2
HexCer	3.4	10.7	7.6 ± 1.9	3.0
LacCer	2.9	8.5	5.5 ± 1.4	2.2
GM3	2.6	5.7	4.7 ± 0.7	1.9
Sph	0.04	0.2	0.07 ± 0.034	0.03
S1P	0.9	3.3	2.0 ± 0.593	0.8
dhSph	0.003	0.03	0.01 ± 0.006	0.004
dhS1P	0.23	0.80	0.45 ± 0.177	0.2

Table 4. RBCs sphingolipid levels (μM) in healthy volunteers (n = 20) expressed as min–max, mean ± SD and percentage over total sphingolipid content.

Conc (μM)	Min	Max	Mean ± SD (n = 20)	%
Cer	59.9	148.5	92.8 ± 21.1	5.9
dhCer	9.6	39.3	22.7 ± 8.0	1.4
SM	854.6	2000	1387 ± 237.3	87.4
HexCer	3.3	8.5	6.7 ± 1.5	0.4
LacCer	21.6	115.9	64.2 ± 23.8	4.0
GM3	1.5	4.6	3.3 ± 0.9	0.2
Sph	1.7	4.0	2.5 ± 0.64	0.1
S1P	2.6	6.4	4.1 ± 1.041	0.3
dhSph	0.1	0.4	0.2 ± 0.112	0.01
dhS1P	1.5	3.7	2.2 ± 0.563	0.2

3. Materials and Methods

3.1. Biological Samples from Healthy Volunteers

All subjects, who voluntarily agreed to participate in the study, were informed and authorization was obtained by signing a letter of consent. These participants were selected from a wider clinical trial that was approved by the institutional local ethical committee (Ospedale San Paolo, Milano, Italy). Blood from twenty volunteers was collected in the fasting state using K_2EDTA as an anticoagulant, and the resulting plasma was obtained by centrifugation for 15 min at 1400× g. The recruited volunteers ranged in age from 18 to 85 and they were not diagnosed for cardiometabolic, liver or kidney diseases. Each volunteer was tested for complete blood count and their results had to fall within the medical laboratory's physiological parameters in order to be included in the research. Prior to the analysis, plasma and RBCs were stored at −80 °C. All the procedures adopted in the present study were respectful of the ethical standards in the Helsinki Declaration. In order to study the method performances (Table 2), the implementations of different extraction protocols (Figure 1) and the effects of time and temperature on the recovery of sphingolipids (Figure 2), a pool of all the plasma and RBCs gathered (n = 20) was made and stored or processed as other samples. Otherwise, the use of individual samples was applied in the study of the sphingolipids' physiological range in the biological matrix (Figures 4 and 5, Tables 3 and 4).

3.2. Chemicals and Reagents

The chemicals methanol, chloroform, formic acid, acetic acid, ammonium acetate, ammonium formate, dibutylhydroxytoluene (BHT), phenylisothiocyanate (PITC) and 4-nitrophenylisothiocyanate (NO_2PITC) were all at analytical grade and were purchased from Sigma-Aldrich (St. Louis, MO, USA). All aqueous solutions were prepared using purified water at a Milli-Q grade (Burlington, MA, USA). Lipid standards were purchased from Avanti Polar (supplied by Sigma-Aldrich, St. Louis, MO, USA).

3.3. *LC-MS/MS*

The LC-MS/MS consisted of an LC Dionex 3000 UltiMate (ThermoFisher Scientific, Waltham, MA, USA) coupled to a tandem mass spectrometer AB Sciex 3200 QTRAP (AB Sciex, Concord, ON, Canada) equipped with electrospray ionization TurboIonSpray™ source operating in positive mode (ESI+).

3.3.1. Sphingolipids and Glycosphingolipids

The instrument parameters were: CUR 25, GS1 45, GS2 50, capillary voltage 5.5 kV and source temperature 300 °C. Spectra were acquired by multiple reaction monitoring, scanning for each analyte, the transitions reported in Supplementary Table S3. To chromatographically isolate the analytes, we used a reverse-phase Acquity BEH C8 column 1.7 µm, 2.1 × 100 mm (Waters, Milford, MA, USA) equipped with pre-column, using as mobile phases (A) water + 0.2% formic acid + 2 mM ammonium formate and (B) methanol + 0.2% formic acid + 1 mM ammonium formate. The flow rate was 0.3 mL/min and the column temperature was set to 30 °C. The elution gradient (%B) was set as follows: 0–3 min (80–90%), 3.0–6.0 min (90%), 6.0–19.0 min (90–99%), 19.0–20.0 min (99–80%), held until 24 min. Five microliters of clear supernatant were directly injected into LC-MS/MS. Due to the lack of authentic standards for every fatty acid chain, those which are not available were quantified as a reference of the closest sphingolipids subspecies.

3.3.2. Free Sphingoid Bases

The instrument parameters were: CUR 25, GS1 45, GS2 55, capillary voltage 5.5 kV and source temperature 500 °C. Spectra were acquired by multiple reaction monitoring, scanning for each analyte, the transitions reported in Supplementary Table S4. Two columns were tested: reverse-phase Acquity BEH C18 column 1.7 µm, 2.1 × 100 mm (Waters, MA, USA) and reverse-phase Cortecs C18 1.6 µm, 2.1 × 100 mm (Waters, MA, USA). Both columns were equipped with pre-column and the mobile phase was (A) water + 0.2% formic acid + 2 mM ammonium formate and (B) methanol + 0.2% formic acid + 1 mM ammonium formate.

Short chromatography (BEH C18). The elution gradient (%B) was set as follows: 0–2 min (20%), 2–4 min (20–99%), 4–7 min (99%), 7–7.5 min (99–20%), held until 10 min [33]. The flow rate was 0.3 mL/min and the column temperature was set to 30 °C.

Long chromatography (BEH C18). The elution gradient (%B) was set as follows: 0–12 min (70–85%), 12.0–12.2 min (85–99%), 12.2–15.0 min (99%), 15.0–15.2 min (99–70%), held until 20 min. Five microliters of clear aqueous supernatant were directly injected into LC-MS/MS. The flow rate was 0.3 mL/min and the column temperature was set to 30 °C.

Long chromatography (Cortecs C18). The elution gradient (%B) was set as follows: 0–12 min (70–85%), 12.0–12.2 min (85–99%), 12.2–15.0 min (99%), 15.0–15.2 min (99–70%), held until 20 min. Three microliters of clear aqueous supernatant were directly injected into LC-MS/MS. The flow rate was 0.2 mL/min and the column temperature was set to 30 °C.

3.3.3. Sphingoid Bases as Phenylthiourea Derivatives

The instrument parameters were: CUR 25, GS1 45, GS2 55, capillary voltage 5.5 kV and source temperature 500 °C. Spectra were acquired by multiple reaction monitoring, scanning for each PITC or NO_2PITC derivative using the transitions reported in Supplementary Tables S5 and S6, respectively. To chromatographically isolate the analytes, we used a reverse-phase Cortecs C18 1.6 µm, 2.1 × 100 mm (Waters, MA, USA) equipped with pre-column using as mobile phase (A) water + 0.2% formic acid + 2 mM ammonium formate and (B) methanol + 0.2% formic acid + 1 mM ammonium formate. The flow rate was 0.3 mL/min and the column temperature was 40 °C. The elution gradient (%B) was set as below: 0–16.0 min (70–99%), 16.0–17.0 min (99%), 17.0–17.2 min (99–70%), held until 20 min. Three microliters of clear supernatant were directly injected into LC-MS/MS.

3.4. Extraction Procedures

Protocol 1. Plasma (25 µL) was diluted with water (75 µL) before being mixed with a methanol/chloroform solution (850 µL, 2:1, *v*/*v*). The lipids were extracted by ice-sonication and thermo-shaking (1 h, 1000 rpm, rt) of the plasma samples. The organic phase was separated via centrifugation (15 min at 20,000× *g*) and evaporated under a stream of nitrogen. The residues were dissolved in 100 µL of methanol + 0.1 mM BHT and withdrawn in a glass vial.

Protocol 2. Plasma (25 µL) was diluted with water (75 µL) before being mixed with a methanol/chloroform solution (850 µL, 2:1, *v*/*v*). The lipids were extracted by ice-sonication and thermo-shaking (overnight, 1000 rpm, 48 °C) of the plasma samples. The organic phase was separated via centrifugation (15 min at 20,000× *g*) and evaporated under a stream of nitrogen. The residues were dissolved in 100 µL of methanol + 0.1 mM BHT and withdrawn in a glass vial.

Protocol 3. Plasma (25 µL) was diluted with water (75 µL) before being mixed with a methanol/chloroform solution (850 µL, 2:1, *v*/*v*). The lipids were extracted by ice-sonication and thermo-shaking (1 h, 1000 rpm, rt) of the plasma samples. They went through alkaline methanolysis (75 µL KOH 1M, 2 h at 38 °C) and were then neutralized by the addition of glacial acetic acid (4 µL). The organic phase was separated via centrifugation (15 min at 20,000× *g*) and evaporated under a stream of nitrogen. The residues were dissolved in 100 µL of methanol + 0.1 mM BHT and were withdrawn in a glass vial.

Protocol 4. Plasma (25 µL) was diluted with water (75 µL) and added with a methanol/chloroform mixture (850 µL, 2:1, *v*/*v*). The lipids were extracted by ice-sonication and thermo-shaking (overnight, 1000 rpm, 48 °C) of the plasma samples. They went through alkaline methanolysis (75 µL KOH 1M, 2 h at 38 °C) and were then neutralized by the addition of glacial acetic acid (4 µL). The organic phase was separated via centrifugation (15 min at 20,000× *g*) and evaporated under a stream of nitrogen. The residues were dissolved in 100 µL of methanol + 0.1 mM BHT and were withdrawn in a glass vial.

3.5. Derivatization of Free Sphingoid Bases

The amine group reacted with PITC to mainly produce the phenylthiourea [34] derivatives of sphingoid bases. An aliquot of the final extract (25 µL) was withdrawn into a new glass vial and PITC derivatization was performed by adding a solution of PITC/pyridine (25 µL, 100 mM PITC in methanol/pyridine 1:1, *v*/*v*). The vial was capped and heated at 80 °C for 1 h. Prior to analysis, pure formic acid (5 µL) was added. The best conditions for derivatization were investigated as reported in Supplementary Table S7. NO_2PITC derivatives were obtained with the same protocol, adding to the final extract (25 µL) a solution of NO_2PITC/pyridine (25 µL, 100 mM NO_2PITC in methanol/pyridine 1:1, *v*/*v*). Supplementary Tables S5 and S6 report the mass spectrometry conditions for PITC and NO_2PITC derivatives.

3.6. Condition for Alkaline Methanolysis

The recovery of low abundant sphingolipids is commonly accomplished through alkaline methanolysis which causes the lysis of the ester linkage while retaining the intact amide bond. The percentage of intact phospholipids was used to monitor the reaction over time (1, 2, 6, 12 h). The instrument parameters were: CUR 25, GS1 40, GS2 45, capillary voltage 5.5 kV and source temperature 400 °C. Spectra were acquired by multiple reaction monitoring, scanning for DPPC (m/z 734.6 > 184.1) and the internal standard PC d7 (m/z 753.6 > 184.1). To chromatographically isolate the analytes, we used a reverse-phase Acquity BEH C8 column 1.7 µm, 2.1 × 100 mm (Waters, MA, USA) equipped with pre-column, using as mobile phases (A) water + 0.2% formic acid + 2 mM ammonium formate and (B) methanol + 0.2% formic acid + 1 mM ammonium formate. The flow rate was 0.3 mL/min and the column temperature was 35 °C. The elution gradient (%B) was set as below: 0–14 min (80–99%), 14–20 min (99%), 20–20.1 min (99–80%), held until 25 min. Five microliters of clear supernatant were directly injected into LC-MS/MS.

3.7. Comparison between Traditional Biphasic and Monophasic Extractions

The performances of the operating protocol described in Section 3.6 were juxtaposed with the micro-scaled versions of the classical liquid–liquid extraction protocols first proposed by Folch and Bligh-Dyer [15,16]. The comparison between the three extraction protocols (Folch, Bligh-Dyer and monophase extraction) was assessed in triplicate using the same plasma pool, obtained by combining a suitable amount of each individual sample (n = 20).

3.8. Time and Temperature for Isolating Sphingolipids from a Biological Matrix

The same plasma pool, already mentioned above (25 μL), was diluted with water (75 μL) before being mixed with a methanol/chloroform solution (850 μL, 2:1, v/v); it was ice-sonicated and extracted by following this scheme: (1) ambient temperature extraction (22 °C) for either 1/2/4 or 12 h; (2) cold extraction (4 °C) for either 1/2/4 or 12 h; (3) warm extraction (38 °C) for either 1/2/4 or 12 h; (4) hot extraction (48 °C) for either 1/2/4 or 12 h. Then, the samples went through alkaline methanolysis (75 μL KOH 1M, 2 h at 38 °C) and were then neutralized by the addition of glacial acetic acid (4 μL). The organic phase was separated via centrifugation (15 min at 20,000× g) and evaporated under a stream of nitrogen. The residues were dissolved in 100 μL of methanol + 0.1 mM BHT and withdrawn in a glass vial.

3.9. Operating Protocol for Plasma Samples

Plasma (25 μL) was diluted with water (75 μL) before being mixed with a methanol/chloroform solution (850 μL, 2:1, v/v). The lipids were extracted by ice-sonication and thermo-shaking (1 h, 1000 rpm, 38 °C) of the plasma samples. They went through alkaline methanolysis (75 μL KOH 1M, 2 h at 38 °C) and were then neutralized by the addition of glacial acetic acid (4 μL). The organic phase was separated via centrifugation (15 min at 20,000× g) and evaporated under a stream of nitrogen. The residues were dissolved in 100 μL of methanol + 0.1 mM BHT and withdrawn in a glass vial.

3.10. Red Blood Cells Protocol

RBCs (10 μL) were lysed by hypotonic shock in double-distilled water (490 μL). An aliquot of the lysed solution (25 μL, which on average corresponds to 2.5×10^6 cells or 0.5 μL of the initial sample) was diluted with water (75 μL) before being mixed with a methanol/chloroform solution (850 μL, 2:1, v/v). The lipids were extracted by ice-sonication and thermo-shaking (1 h, 1000 rpm, 38 °C). They went through alkaline methanolysis (75 μL KOH 1M, 2 h at 38 °C) and were then neutralized by the addition of glacial acetic acid (4 μL). The organic phase was separated via centrifugation (15 min at 20,000× g) and evaporated under a stream of nitrogen. The residues were dissolved in 100 μL of methanol + 0.1 mM BHT and withdrawn in a glass vial.

3.11. Methods Performances

The methods performances were tested using the same plasma pool obtained by combining suitable amounts of each sample (n = 20). The precision of the methods was calculated as the coefficient of variation (CV%) by extracting five times the same pool sample in a day (intra-day) and another five times the day after (inter-day).

3.12. Statistical Analysis

The software used for the visualization of the results and the univariate statistical analysis was GraphPad Prism 9.0 (GraphPad Software, Inc., La Jolla, California, USA). For repeated measures comparison among different groups, repeated measured one-way

ANOVA with Dunnett post hoc test was performed. In all tests, $p < 0.05$ was considered statistically significant.

4. Conclusions

In this work, we assessed whether different extraction and analytical protocols could affect the results attained from a targeted sphingolipidomics analysis. The single-phase extraction followed by an alkaline methanolysis seems to be crucial for acquiring as accurate as possible results, while its duration and temperature might not be as significant. Another pivotal aspect in the analyses of sphingolipids is represented by the choice of appropriate columns for distinctively analyzing complex sphingolipids and sphingoid bases. On the other hand, derivatization of the sphingoid bases, while effective on paper, especially on non-phosphorylated species, does not allow a consistent improvement for the analysis of phosphorylated sphingoid bases. For this purpose, the use of a proper column, in this case a Cortecs C18, coupled with a full-length chromatography, seems to be much more convenient, in order to efficiently separate Sph from its interfering peaks and still appreciate all other sphingoid bases.

Supplementary Materials: The following supporting information can be downloaded at: https://www.mdpi.com/article/10.3390/metabo12050450/s1. Figure S1: Sphingoid bases concentrations in human plasma, according to scientific literature [18–26] (see paper for the references), Figure S2: Comparison between the sphingoid bases concentrations evidenced with (D, n = 20) or without (ND, n = 20) derivatization with phenylisothiocyanate, Table S1: Differences in signal intensities, expressed as fold-change on underivatized analytes, between the same concentration of sphingoid bases (1 μM) after derivatization, Table S2: RBCs sphingolipids levels (pmol/10^6 cells) in healthy volunteers (n = 20) were expressed as min–max and mean ± SD, Table S3: Mass spectrometry parameters for the analysis of complex sphingolipids. In bold are reported the internal standards (IS) used for each package of lipids, Table S4: Mass spectrometry parameters for the analysis of free sphingoid bases, Table S5: Mass spectrometry parameters for the analysis of sphingoid bases as phenylthiourea derivatives after reaction with phenylisothiocyanate, Table S6: Mass spectrometry parameters for the analysis of sphingoid bases as nitrophenylthiourea derivatives after reaction with 4-nitrophenylisothiocyanate, Table S7: The yield of derivatization products after different times and temperatures of reaction. Each experiment was conducted by adding the same amount of reagents, derivatizing agent (PITC) and catalyzer (Pyridine).

Author Contributions: All authors contributed extensively to the work presented in this manuscript and approved the submitted version. Conceptualization: R.P. and M.D.C.; Methodology: C.M. and M.D.C.; Formal analysis and investigation: C.M., A.Z. and M.D.C.; Writing—original draft preparation: C.M. and M.D.C.; Writing—review and editing: G.R., A.C. and R.P.; Supervision: R.P. and M.D.C. All authors have read and agreed to the published version of the manuscript.

Funding: This research received no external funding.

Institutional Review Board Statement: The study was conducted in accordance with the Declaration of Helsinki and was approved on 17 January 2019 by the Ethics Committee of Milano Area 1, Ospedale San Paolo Milano, Italy with approval code 0001249.

Informed Consent Statement: Informed consent was obtained from all subjects involved in the study.

Data Availability Statement: The data presented in this study are available in article and Supplementary Materials.

Acknowledgments: We thank Mariangela Scavone and Elena Bossi and all other members of the Laboratory of Hemostasis and Thrombosis, Department of Health Sciences, Università degli Studi di Milano, Milan, Italy and the Department of Medicina III of ASST-Santi Paolo (Milan) for providing us with the biological samples. The authors acknowledge the support of the APC central fund of the University of Milan.

Conflicts of Interest: The authors declare no conflict of interest.

Abbreviations

BHT, dibutylhydroxytoluene; Cer, ceramides; dhCer, dihydroceramides; dhS1P, dihydrosphingosine-1-phosphate; dhSpH, dihydrosphingosine; DPPC, dipalmitoylphosphatidylcholine; GM3, gangliosides; HexCer, hexosylceramides; LacCer, lactosylceramides; NO2PITC, 4-nitrophenylisothiocyanate; PC, phosphatidylcholine; PC d7, phosphatidylcholine (15:0–18:1) d7; PITC, phenylisothiocyanate; RBC, red blood cell; S1P, sphingosine-1-phosphate; Sph, sphingosine; SM, sphingomyelins.

References

1. Merrill, A.H.; Schmelz, E.M.; Dillehay, D.L.; Spiegel, S.; Shayman, J.A.; Schroeder, J.J.; Riley, R.T.; Voss, K.A.; Wang, E. Sphingolipids—The Enigmatic Lipid Class: Biochemistry, Physiology, and Pathophysiology. *Toxicol. Appl. Pharmacol.* **1997**, *142*, 208–225. [CrossRef]
2. Futerman, A.H. Sphingolipids. In *Biochemistry of Lipids, Lipoproteins and Membranes: Sixth Edition*; Elsevier Inc.: Amsterdam, The Netherlands, 2016; pp. 297–326. ISBN 978-0-44463-438-2.
3. Dowhan, W.; Bogdanov, M.; Mileykovskaya, E. Functional Roles of Lipids in Membranes. In *Biochemistry of Lipids, Lipoproteins and Membranes: Sixth Edition*; Elsevier Inc.: Amsterdam, The Netherlands, 2016; pp. 1–40. ISBN 978-0-44463-438-2.
4. Hannun, Y.A.; Obeid, L.M. Sphingolipids and Their Metabolism in Physiology and Disease. *Nat. Rev. Mol. Cell Biol.* **2018**, *19*, 175–191. [CrossRef] [PubMed]
5. Mullen, T.D.; Jenkins, R.W.; Clarke, C.J.; Bielawski, J.; Hannun, Y.A.; Obeid, L.M. Ceramide Synthase-Dependent Ceramide Generation and Programmed Cell Death: Involvement of Salvage Pathway in Regulating Postmitochondrial Events. *J. Biol. Chem.* **2011**, *286*, 15929–15942. [CrossRef] [PubMed]
6. Wang, G.; Spassieva, S.D.; Bieberich, E. Ceramide and S1P Signaling in Embryonic Stem Cell Differentiation. *Methods Mol. Biol.* **2018**, *1697*, 153–171. [CrossRef] [PubMed]
7. Kohno, M.; Momoi, M.; Oo, M.L.; Paik, J.-H.; Lee, Y.-M.; Venkataraman, K.; Ai, Y.; Ristimaki, A.P.; Fyrst, H.; Sano, H.; et al. Intracellular Role for Sphingosine Kinase 1 in Intestinal Adenoma Cell Proliferation. *Mol. Cell. Biol.* **2006**, *26*, 7211–7223. [CrossRef] [PubMed]
8. Patwardhan, G.A.; Beverly, L.J.; Siskind, L.J. Sphingolipids and Mitochondrial Apoptosis. *J. Bioenerg. Biomembr.* **2016**, *48*, 153–168. [CrossRef] [PubMed]
9. Canela, N.; Herrero, P.; Mariné, S.; Nadal, P.; Ras, M.R.; Rodríguez, M.Á.; Arola, L. Analytical Methods in Sphingolipidomics: Quantitative and Profiling Approaches in Food Analysis. *J. Chromatogr. A* **2016**, *1428*, 16–38. [CrossRef]
10. Haynes, C.A.; Allegood, J.C.; Park, H.; Sullards, M.C. Sphingolipidomics: Methods for the Comprehensive Analysis of Sphingolipids. *J. Chromatogr. B Anal. Technol. Biomed. Life Sci.* **2009**, *877*, 2696–2708. [CrossRef]
11. Morano, C.; Roda, G.; Paroni, R.; Dei Cas, M. Tip-Tip Filtration Ameliorates Single-Phase Extraction Methods for Plasma Large-Scale Lipidomics Analysis. *J. Chromatogr. B Anal. Technol. Biomed. Life Sci.* **2022**, *1189*, 123099. [CrossRef]
12. Merrill, A.H.; Sullards, M.C.; Allegood, J.C.; Kelly, S.; Wang, E. Sphingolipidomics: High-Throughput, Structure-Specific, and Quantitative Analysis of Sphingolipids by Liquid Chromatography Tandem Mass Spectrometry. *Methods* **2005**, *36*, 207–224. [CrossRef]
13. Dalmau, N.; Jaumot, J.; Tauler, R.; Bedia, C. Epithelial-to-Mesenchymal Transition Involves Triacylglycerol Accumulation in DU145 Prostate Cancer Cells. *Mol. Biosyst.* **2015**, *11*, 3397–3406. [CrossRef] [PubMed]
14. Houjou, T.; Yamatani, K.; Nakanishi, H.; Imagawa, M.; Shimizu, T.; Taguchi, R. Rapid and Selective Identification of Molecular Species in Phosphatidylcholine and Sphingomyelin by Conditional Neutral Loss Scanning and MS3. *Rapid Commun. Mass Spectrom.* **2004**, *18*, 3123–3130. [CrossRef] [PubMed]
15. Folch, J.; Lees, M.; Sloane Stanley, G.H. A Simple Method for the Isolation and Purification of Total Lipides from Animal Tissues. *J. Biol. Chem.* **1957**, *226*, 497–509. [CrossRef]
16. Bligh, E.G.; Dyer, W.J. A Rapid Method of Total Lipid Extraction and Purification. *Can. J. Biochem. Physiol.* **1959**, *37*, 911–917. [CrossRef] [PubMed]
17. Wong, M.W.K.; Braidy, N.; Pickford, R.; Sachdev, P.S.; Poljak, A. Comparison of Single Phase and Biphasic Extraction Protocols for Lipidomic Studies Using Human Plasma. *Front. Neurol.* **2019**, *10*, 879. [CrossRef]
18. Dei Cas, M.; Zulueta, A.; Mingione, A.; Caretti, A.; Ghidoni, R.; Signorelli, P.; Paroni, R. An Innovative Lipidomic Workflow to Investigate the Lipid Profile in a Cystic Fibrosis Cell Line. *Cells* **2020**, *9*, 1197. [CrossRef]
19. Hammad, S.M.; Harden, O.C.; Wilson, D.A.; Twal, W.O.; Nietert, P.J.; Oates, J.C. Plasma Sphingolipid Profile Associated with Subclinical Atherosclerosis and Clinical Disease Markers of Systemic Lupus Erythematosus: Potential Predictive Value. *Front. Immunol.* **2021**, *12*, 694318. [CrossRef]
20. Bui, H.H.; Leohr, J.K.; Kuo, M.S. Analysis of Sphingolipids in Extracted Human Plasma Using Liquid Chromatography Electrospray Ionization Tandem Mass Spectrometry. *Anal. Biochem.* **2012**, *423*, 187–194. [CrossRef]
21. Xu, Y.; Li, H.; Han, Y.; Wang, T.; Wang, Y.; Gong, J.; Gao, K.; Chen, W.; Li, W.; Zhang, H.; et al. A Simple and Rapid Method for Extraction and Measurement of Circulating Sphingolipids Using LC-MS/MS: A Targeted Lipidomic Analysis. *Anal. Bioanal. Chem.* **2022**, *414*, 2041–2054. [CrossRef]

22. Scherer, M.; Böttcher, A.; Schmitz, G.; Liebisch, G. Sphingolipid Profiling of Human Plasma and FPLC-Separated Lipoprotein Fractions by Hydrophilic Interaction Chromatography Tandem Mass Spectrometry. *Biochim. Biophys. Acta Mol. Cell Biol. Lipids* **2011**, *1811*, 68–75. [CrossRef]
23. Chan, R.B.; Perotte, A.J.; Zhou, B.; Liong, C.; Shorr, E.J.; Marder, K.S.; Kang, U.J.; Waters, C.H.; Levy, O.A.; Xu, Y.; et al. Elevated GM3 Plasma Concentration in Idiopathic Parkinson's Disease: A Lipidomic Analysis. *PLoS ONE* **2017**, *12*, e0172348. [CrossRef] [PubMed]
24. Hammad, S.M.; Pierce, J.S.; Soodavar, F.; Smith, K.J.; Al Gadban, M.M.; Rembiesa, B.; Klein, R.L.; Hannun, Y.A.; Bielawski, J.; Bielawska, A. Blood Sphingolipidomics in Healthy Humans: Impact of Sample Collection Methodology. *J. Lipid Res.* **2010**, *51*, 3074–3087. [CrossRef]
25. Huang, Q.; Liu, D.; Xin, B.; Cechner, K.; Zhou, X.; Wang, H.; Zhou, A. Quantification of Monosialogangliosides in Human Plasma through Chemical Derivatization for Signal Enhancement in LC-ESI-MS. *Anal. Chim. Acta* **2016**, *929*, 31–38. [CrossRef] [PubMed]
26. Schmidt, H.; Schmidt, R.; Geisslinger, G. LC-MS/MS-Analysis of Sphingosine-1-Phosphate and Related Compounds in Plasma Samples. *Prostagland. Other Lipid Mediat.* **2006**, *81*, 162–170. [CrossRef]
27. Quehenberger, O.; Armando, A.M.; Brown, A.H.; Milne, S.B.; Myers, D.S.; Merrill, A.H.; Bandyopadhyay, S.; Jones, K.N.; Kelly, S.; Shaner, R.L.; et al. Lipidomics Reveals a Remarkable Diversity of Lipids in Human Plasma1. *J. Lipid Res.* **2010**, *51*, 3299–3305. [CrossRef]
28. Knapp, M.; Lisowska, A.; Zabielski, P.; Musiał, W.; Baranowski, M. Sustained Decrease in Plasma Sphingosine-1-Phosphate Concentration and Its Accumulation in Blood Cells in Acute Myocardial Infarction. *Prostagland. Other Lipid Mediat.* **2013**, *106*, 53–61. [CrossRef] [PubMed]
29. Chipeaux, C.; de Person, M.; Burguet, N.; Billette de Villemeur, T.; Rose, C.; Belmatoug, N.; Héron, S.; Le Van Kim, C.; Franco, M.; Moussa, F. Optimization of Ultra-High Pressure Liquid Chromatography—Tandem Mass Spectrometry Determination in Plasma and Red Blood Cells of Four Sphingolipids and Their Evaluation as Biomarker Candidates of Gaucher's Disease. *J. Chromatogr. A* **2017**, *1525*, 116–125. [CrossRef]
30. Dupuis, L.; Chipeaux, C.; Bourdelier, E.; Martino, S.; Reihani, N.; Belmatoug, N.; Billette de Villemeur, T.; Hivert, B.; Moussa, F.; Le Van Kim, C.; et al. Effects of Sphingolipids Overload on Red Blood Cell Properties in Gaucher Disease. *J. Cell. Mol. Med.* **2020**, *24*, 9726–9736. [CrossRef]
31. Awojoodu, A.O.; Keegan, P.M.; Lane, A.R.; Zhang, Y.; Lynch, K.R.; Platt, M.O.; Botchwey, E.A. Acid Sphingomyelinase Is Activated in Sickle Cell Erythrocytes and Contributes to Inflammatory Microparticle Generation in SCD. *Blood* **2014**, *124*, 1941–1950. [CrossRef]
32. Świderska, M.; Knapp, P.; Bodnar, L.; Błachnio-Zabielska, A.; Reszeć, J.; Chabowski, A. Blood Bioactive Sphingolipids in Patients with Advanced Serous Epithelial Ovarian Cancer-Mass Spectrometry Analysis Obstetrics and Gynecology. *Arch. Med. Sci.* **2021**, *17*, 53–61. [CrossRef]
33. Burrello, J.; Biemmi, V.; Dei Cas, M.; Amongero, M.; Bolis, S.; Lazzarini, E.; Bollini, S.; Vassalli, G.; Paroni, R.; Barile, L. Sphingolipid Composition of Circulating Extracellular Vesicles after Myocardial Ischemia. *Sci. Rep.* **2020**, *10*, 16182. [CrossRef] [PubMed]
34. Santa, T. Isothiocyanates as Derivatization Reagents for Amines in Liquid Chromatography/Electrospray Ionization-Tandem Mass Spectrometry. *Biomed. Chromatogr.* **2010**, *24*, 915–918. [CrossRef] [PubMed]

metabolites

Article

Lipid Serum Profiling of Boar-Tainted and Untainted Pigs Using GC×GC–TOFMS: An Exploratory Study

Kinjal Bhatt [1,*], Thibaut Dejong [1], Lena M. Dubois [1], Alice Markey [2], Nicolas Gengler [2], José Wavreille [3], Pierre-Hugues Stefanuto [1] and Jean-François Focant [1]

1 Organic and Biological Analytical Chemistry Group (OBiAChem), MolSys, University of Liège, 4000 Liège, Belgium
2 TERRA Teaching and Research Center, Gembloux Agro-Bio Tech, University of Liège, 5030 Gembloux, Belgium
3 Animal Production Unit, Walloon Agricultural Research Centre, 5030 Gembloux, Belgium
* Correspondence: kinjal.bhatt@uliege.be

Abstract: Mass spectrometry (MS)-based techniques, including liquid chromatography coupling, shotgun lipidomics, MS imaging, and ion mobility, are widely used to analyze lipids. However, with enhanced separation capacity and an optimized chemical derivatization approach, comprehensive two-dimensional gas chromatography (GC×GC) can be a powerful tool to investigate some groups of small lipids in the framework of lipidomics. This study describes the optimization of a dedicated two-stage derivatization and extraction process to analyze different saturated and unsaturated fatty acids in plasma by two-dimensional gas chromatography–time-of-flight mass spectrometry (GC×GC–TOFMS) using a full factorial design. The optimized condition has a composite desirability of 0.9159. This optimized sample preparation and chromatographic condition were implemented to differentiate between positive (BT) and negative (UT) boar-tainted pigs based on fatty acid profiling in pig serum using GC×GC–TOFMS. A chemometric screening, including unsupervised (PCA, HCA) and supervised analysis (PLS–DA), as well as univariate analysis (volcano plot), was performed. The results suggested that the concentration of PUFA ω-6 and cholesterol derivatives were significantly increased in BT pigs, whereas SFA and PUFA ω-3 concentrations were increased in UT pigs. The metabolic pathway and quantitative enrichment analysis suggest the significant involvement of linolenic acid metabolism.

Keywords: lipidomics; fatty acids; boar taint; gas chromatography; GC×GC–TOFMS

Citation: Bhatt, K.; Dejong, T.; Dubois, L.M.; Markey, A.; Gengler, N.; Wavreille, J.; Stefanuto, P.-H.; Focant, J.-F. Lipid Serum Profiling of Boar-Tainted and Untainted Pigs Using GC×GC–TOFMS: An Exploratory Study. *Metabolites* **2022**, *12*, 1111. https://doi.org/10.3390/metabo12111111

Academic Editor: Joana Pinto

Received: 27 October 2022
Accepted: 11 November 2022
Published: 15 November 2022

1. Introduction

Lipidomics, or the comprehensive analysis of lipids, is rapidly expanding and providing critical information to the field of bioscience. Lipids have been studied using mass spectrometry (MS) for decades, but lipidomics is one of the newest members of the "omics" family introduced by Spener [1] and Han and Gross [2]. Although metabolomics mainly focuses on the hydrophilic classes, lipidomics has emerged as an independent omics owing to its structural complexities and hydrophobic and amphiphilic nature, which provides a wide range of biological functions [3,4]. When considering lipidomics, liquid chromatography (LC)–mass spectrometry (MS)-based techniques are widely used. However, enhanced separation capacities and lower limits of detection are challenging for these LC–MS(/MS)-based approaches [4]. Aside from this, chemical derivatization has the potential to make some families of lipids more "gas chromatography (GC)-amenable", allowing more sensitive GC–MS to also be considered.

The gold standard for lipid extraction techniques for biological matrices based on chloroform and methanol was introduced by Folch [5] and Bligh and Dyer [6]. When considering GC, chemical derivatization is essential for the conversion of the extracted fatty acid components of lipids into more volatile and stable derivatives such as methyl esters to analyze saturated and unsaturated fatty acids. There are six frequently used protocols for the derivatization of lipids in plasma using GC [7]. These are potassium hydroxide (KOH) derivatization, trimethylsulfonium hydroxide (TMSH) derivatization, TMSH direct injections, boron trifluoride (BF_3) derivatization [8–12], chlorhydric acid (HCl) derivatization, and sodium hydroxide (NaOH) + BF_3 derivatization. The fatty acid composition of plasma can be divided into four classes: saturated fatty acids (SFA), monounsaturated fatty acids (MUFA), polyunsaturated fatty acids (PUFA (ω-3)), (PUFA (ω-6)), and derivatives. As mentioned, there are two main families of PUFAs: (PUFA (ω-3)) and (PUFA (ω-6)) because of the relevance of PUFAs to human health: polyunsaturated fatty acids (PUFAs) may modulate inflammatory processes and regulate the antioxidant signaling pathway. They impact liver lipid metabolism and physiological responses of other organs, including the heart [13]. Each derivatization technique has pros and cons; however, when selecting a sample preparation protocol, its efficiency over all four classes of fatty acids becomes essential to consider. A systematic comparative study of different derivatizations and extraction efficacies was conducted to determine lipid composition as fatty acid methyl esters (FAMEs) in plasma by Ostermann (Figure 1) [7]. Both the HCl and NaOH+BF_3 derivatization protocols proved to be suited for the analysis of overall fatty acid patterns without discriminating individual classes of FAMEs. However, as demonstrated by Micalizzi et al. [14], NaOH + BF_3 derivatization can be fully automated using a dual head autosampler, providing an upper edge compared to HCl derivatization.

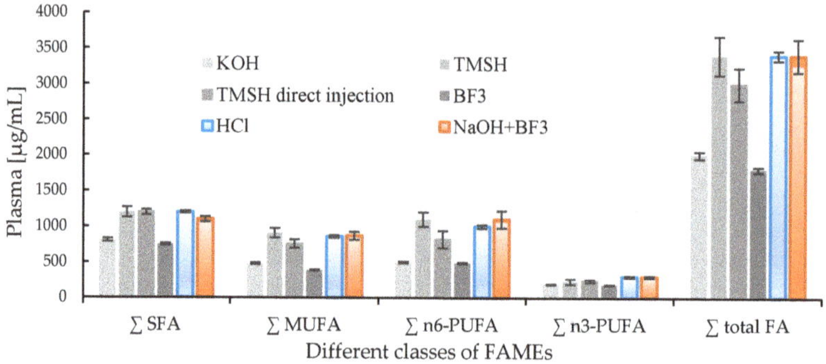

Figure 1. Comparison of derivatization efficacy of different derivatization methods [7].

Untargeted lipid profiling can help to better understand the ongoing biological mechanisms that have an observable effect, such as the production of detectable smell in food products. Boar taint is a pungent, unpleasant smell or taste found in the meat of some uncastrated male pigs. This smell is caused by a complex mixture of molecules released upon heating the meat [15]. The widely known molecules responsible for boar taint are androstenone and skatole [16–18]. The surgical castration of male piglets is a traditional practice to prevent boar taint in meat worldwide. However, it is performed without anesthesia or analgesia, causing great pain to the piglets. Hence, due to increased animal welfare concerns, European pork production stakeholders agreed to prohibit surgical castration of male piglets from 2018. These objectives are yet to be fully achieved successfully [15,19].

This study uses pig serum as a biological sample instead of preferred backfat to analyze boar taint. Currently, boar taint detection techniques used in slaughterhouses are sensory evaluation by the human nose upon heating the fat [20] and spectrophotometric detection at 580 nm. Modern analytical techniques investigated for boar taint identification are UHPLC–HRMS [17], GC–MS [16,21,22], HPLC–FD [23], and Raman spectroscopy [24]. These modern analytical techniques focus on quantification and/or validation of known boar taint compounds, e.g., indole, skatole, and androsterone in porcine adipose tissue [16–24].

In this study, we report the two-stage sample preparation protocol for extracting lipids from 25 μL of plasma/serum for GC×GC–TOFMS. The optimized approach becomes valuable for analytes with low abundance, e.g., PUFAs (ω-3). The protocol has been optimized using human plasma, confirmed with NIST plasma metabolites, and implemented on animal serum, indicating its efficiency and usability. Widely available analytical approaches for identifying boar taint in pork mainly focus on androstenone and skatole molecules using the backfat of the pig. With this protocol, we investigated a new approach, focusing on lipids by studying the fatty acid composition of pig serum responsible for boar taint (boar-tainted) (BT) compared to that without boar taint (untainted) (UT). Ultimately, the lipid profiling of pig serum enables the use of a different type of biological sample instead of backfat; thus, different molecules of fatty acids will provide us with new biological information.

2. Materials and Methods

2.1. Samples and Chemicals

For derivatization, 0.5 M sodium methoxide (CH_3ONa) and boron trifluoride (BF_3) 20% solution in methanolic solution were purchased from ACROS organics and Sigma-Aldrich, respectively. A Supelco 37 FAMEs standard mixture was purchased from Sigma-Aldrich. A 10 ppm FAMEs solution was prepared in dichloromethane. The n-alkanes mixture (C7-30) in hexane was purchased from Millipore Sigma and diluted to 10 ppm in hexane for the calculation of linear retention indices (LRIs).

Pooled human plasma of six humans was purchased from TCS Biosciences (Buckingham, UK). Biological reference standard SRM 1950 "Metabolites in frozen human plasma" was purchased from NIST. To optimize extraction and derivatization conditions, pooled human plasma was used and stored as sub-aliquots at −80 °C to avoid thawing effects. For the identification of analytes, 37 FAMEs standard mixture, NIST SRM 1950, and n-alkane standard were analyzed.

The pig blood samples (n = 40) (sex = male, age = 6 months ± 15 days) were collected in 16 × 125 mm BD Vacutainer® SST™ plastic tubes (cat# 367985) to obtain serum. Serum samples identified as boar-tainted (n = 20) and untainted (n = 20) by the human nose at slaughterhouse were stored at −20 °C. The blood samples were collected after the slaughtering of pigs as a part of a large study.

2.2. Instrumental Method

GC×GC–TOFMS analysis was performed with a Pegasus 4D (LECO Corporation, St. Joseph, MI, USA) equipped with Agilent 7890 GC. The analysis was performed using a normal column set configuration, Rxi-5Sil-MS (30 m × 0.25 mm ID × 1.0 μm df), and VF-17ms (2 m × 0.25 mm ID × 0.5 μm df). A guard column of 2 m was installed.

The temperature program for the primary and secondary oven was the same, starting at 50 °C and holding for 2 min, then increasing temperature to 160 °C at 30 °C/min, followed by a ramp of 2 °C/min until it reached 280 °C. At last, the 300 °C temperature was achieved with a ramp of 30 °C/min and held for 2 min. The total run time for the GC method was 69.33 min. The secondary oven temperature offset was +5 °C, and the modulator temperature offset was +15 °C. A mass range of 45 to 700 m/z was collected at an acquisition rate of 150 spectra/s by positive mode electron ionization (EI) at 70 eV. Ion source and transfer line temperatures were maintained at 230 °C and 250 °C, respectively.

2.3. Sample Preparation

As shown in Figure 2, 500 μL of CH_3ONa was added to 25 μL of pooled human plasma for the first stage of derivatization and heated for a specified time (Table 1). After cooling, 500 μL of methanolic BF_3 solution was added and again heated (Table 1). In the end, 300 μL of heptane was added for liquid–liquid extraction. The upper heptanoic solution was collected and injected into the GC×GC–TOFMS system. Pig serum samples of 25 μL were analyzed using optimized derivatization and extraction protocol.

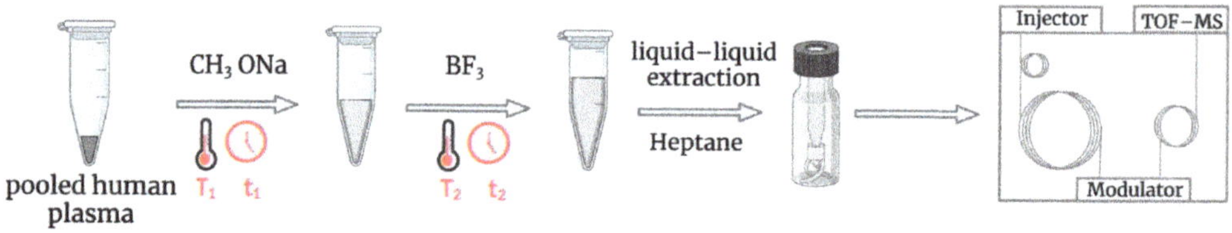

Figure 2. Schematic diagram of two-stage chemical derivatization and extraction approach.

Table 1. Factors and levels tested for derivatization and extraction optimization using DoE. The optimized conditions are in bold.

Factors	−1	0	+1
T_1 (°C)	**85**	95	105
t_1 (min)	5	15	**25**
T_2 (°C)	**85**	95	105
t_2 (min)	5	15	**25**

2.4. Data Processing

The data processing for the optimization of the sample preparation conditions was performed in ChromaTOF® (ver. 4.72, LECO Corp., St. Joseph, MI, USA). The putative identification of analytes was conducted with a spectral similarity library search against the NIST17 mass spectral library. Analytes were quantified at specific m/z: 74, 55, and 67 for FAMEs with zero to two double bonds, respectively, while FAMEs with three to six double bonds were quantified at 79 m/z. The composite desirability and response optimization plot for the Design of Experiments (DoE) were analyzed on Minitab (Ver. 20.2.0). The pig plasma data were processed using GC Image™ (ver. 2021r). Data pre-processing of normalization to sample median, square root transformation, and mean centering were conducted prior to applying chemometric tools. The chemometric tests, unsupervised screening (PCA, HCA), univariate analysis profile (volcano plot), multivariate supervised analysis (PLS-DA), pathway analysis, and enrichment analysis were performed using MetaboAnalyst 5.0 (Xia Lab, McGill University, Montréal, QC, Canada) [25]. The pathway topology analysis is measured with relative betweenness centrality for the node importance and globaltest for enrichment analysis.

3. Results

3.1. Optimization of Derivatization and Extraction Conditions via Experimental Design

To improve measurement efficiency and obtain clean chromatographic separation of the lipids, a two-step sample extraction and derivatization approach was optimized using the DoE. An amount of 25 µL of pooled human plasma was used for the optimization process (Figure 2). In the first step, the addition of CH_3OH helps in the derivatization of fatty acids bound in sources such as cholesterol in plasma (base-catalyzed transesterification). The second step, the addition of BF_3, helps in the esterification of free fatty acids (acid-catalyzed esterification). Liquid–liquid extraction was performed using heptane at the end of the second derivatization. However, the optimization of temperature (T_1, T_2) and time (t_1, t_2) for sample preparation can help us achieve overall maximum extraction efficiency. A two-level full factorial design for 16 different conditions with an additional three center points for a total of 19 runs was evaluated (Table 1).

The formation of a structured chromatographic separation gives an advantage over the conventional GC approach. As shown in Figure 3b, the separation of FAMEs in the first dimension occurs as the number of carbon atoms or volatility decreases, while in the second dimension, as the polarity of FAMEs increases, the number of double bonds increases. The elution pattern of FAMEs also depends on the position of double bonds. The parallel-aligned ω compounds are separating at an obtuse angle, with higher-ω FAMEs eluting before lower ones. The increased film thickness of the ^{1}D and ^{2}D columns (1.0 μm, 0.5 μm) enables structured chromatographic elution without a wrap-around effect by retaining the compound on the ^{2}D column longer. Thus, the structured chromatographic separation of FAMEs can become a valuable tool for identifying unknown analytes.

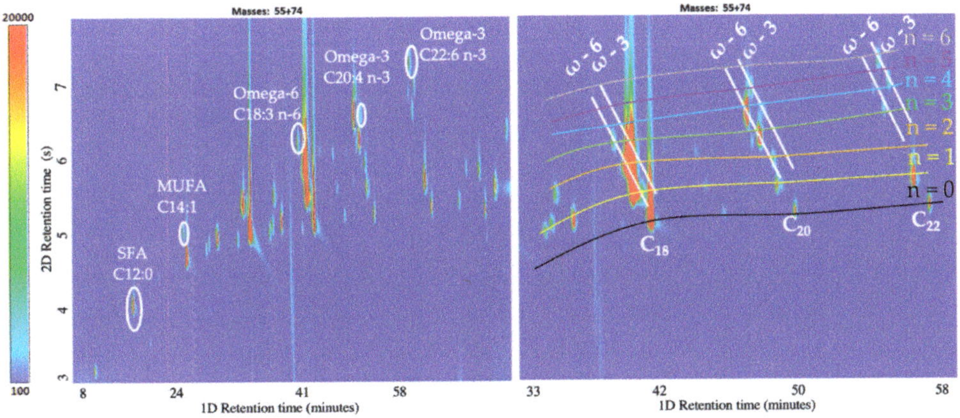

(a) (b)

Figure 3. (**a**) Representative of each class selected for response optimization: zoomed-in Contour plot of pooled Human plasma at *m*/*z*: 55 + 74. (**b**) Zoomed-in contour plot of pooled human plasma for C_{18} to C_{22} region at *m*/*z*: 55 + 74.

As per the certificate of analysis (CoA) of NIST SRM 1950 (Figure S1), due to the vastly varying concentration range of fatty acids in blood plasma, a number of analytes—C16:0, C18:1 n-9, C18:2 n-6—become oversaturated in the chromatogram. However, it remains possible to achieve the separation of all FAMEs. Therefore, to avoid bias while optimizing the DoE parameters, a representative of each class was selected instead of a summation of the entire individual classes.

For the response optimization, five analytes were selected, covering the different classes of FAMEs, including SFA, MUFA, omega-3, and omega-6 (Figure 3a) (Table 2). The center points injected three times at a randomized interval had an overall %RSD of 8.47, indicating the reproducibility of the method. The optimal composite desirability is 0.9159 for the optimized sample preparation condition (Figure S2). Thus, the optimized derivatization method is efficient in extracting all the classes of FAMEs without creating a bias for a specific class.

Table 2. Selected representative of each class for response optimization.

	SFA	MUFA	Omega-6	Omega-3	Omega-3
Analyte	C12:0	C14:1	C18:3 n-6	C20:4 n-3	C22:6 n-3
$^{1}t_R$ (min)	16.39	23.46	39.46	48.26	55.06
$^{2}t_R$ (s)	4.07	5.08	6.36	6.986	7.39

3.2. *Identification of Pigs Responsible for Boar Taint by Lipid Profiling of Serum Using Optimized Derivatization and Separation Conditions*

In total, 40 pig serum samples were analyzed using optimized sample preparation and chromatographic conditions, out of which 20 were identified as boar-tainted and 20 as untainted pigs at the slaughterhouse by the human nose. As previously observed in human plasma (Figure 3), it was possible to observe SFA, MUFA, PUFA, and cholesterol derivatives (Figure 4) in pig serum. For a better illustration, the contour plots were reconstructed in Python.

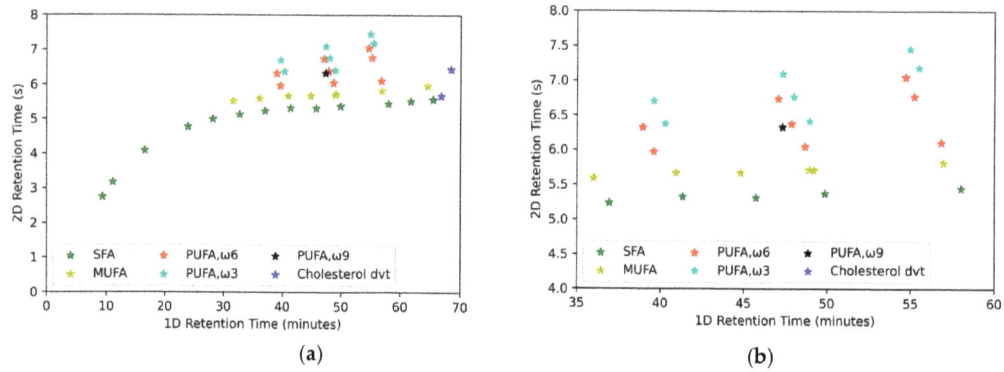

Figure 4. (**a**) Reconstructed contour plot of pig serum. (**b**) Zoomed-in reconstructed contour plot for C_{18} to C_{22} region for pig serum.

In pig serum samples, out of 39 features, 13 SFAs, 8 MUFAs, 7 PUFAs ω-3, 9 PUFAs ω-6, ω-9, and 2 cholesterol derivatives are present (Table S1). Unsupervised principal component analysis (PCA) was performed to visualize a potential clustering trend between the boar-tainted and untainted pig serum samples. As seen from the PCA scores plot (Figure 5a), PC 1 and PC 2 contributed 62.4% variance. There is one outlier outside the 95% confidence interval identified by a Grubbs test, possibly because of the less regulated sample collection conditions at a slaughterhouse. A significant clustering trend was observed between the two groups, which indicated that the fatty acid derivatives were able to differentiate the boar-tainted pigs from untainted pigs.

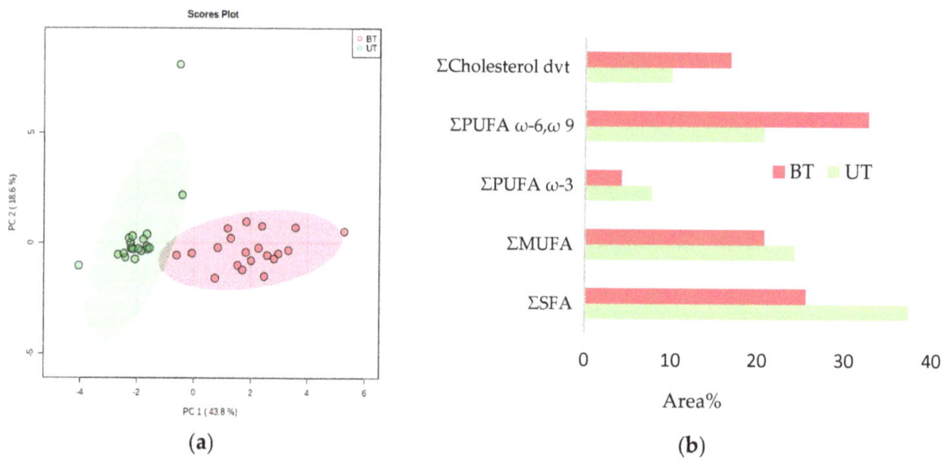

Figure 5. (**a**) PCA score plot. (**b**) % Area contribution of fatty acids per class for boar–tainted (BT) and untainted (UT) pig serum samples.

The hierarchical clustering result is shown as a heat map (Figure 6). Using Euclidean distance measure and clustering algorithm Ward.D for the top 25 features, it is shown that SFA and PUFA (ω-3) are present in higher concentrations in untainted pig serum. In contrast, PUFA (ω-6), (ω-9), and cholesterol derivatives are in higher abundance in boar-tainted pig serum.

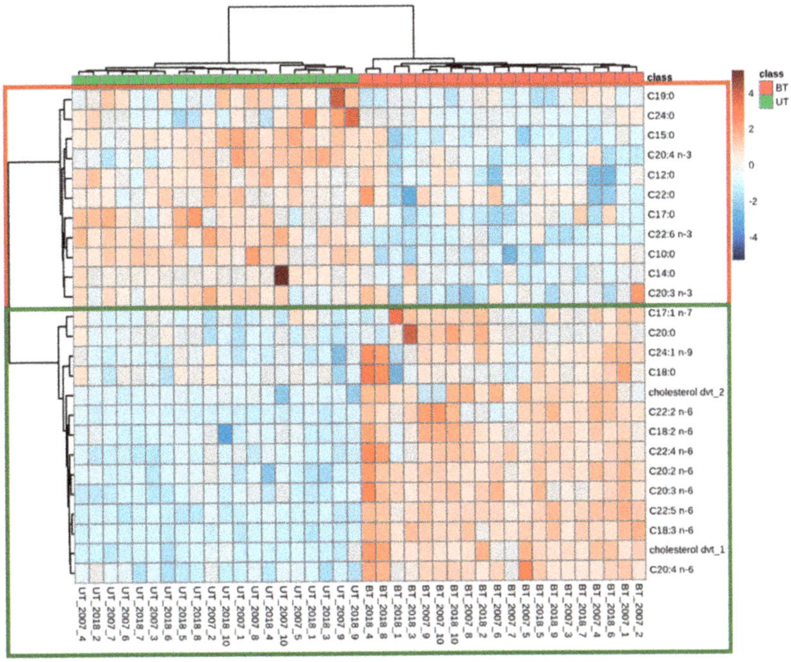

Figure 6. Heat map using top 25 features of boar–tainted (BT) and untainted (UT) pig serum.

The volcano plot combines a fold change (FC) analysis and a *t*-test. For this test, the *t*-test threshold was set at 0.05, and the direction of comparison was boar-tainted pig serum divided by untainted pig serum (BT/UT). The volcano plot (Figure 7a) has 11 important features, of which two SFAs are downregulated in boar-tainted pigs, while six PUFAs ω-6 and two cholesterol derivatives are upregulated in boar-tainted pigs. Moreover, the partial least squares discriminant analysis (PLS–DA) had 8 features out of 39 features with a threshold of variable importance in projection (VIP) score > 0.9.

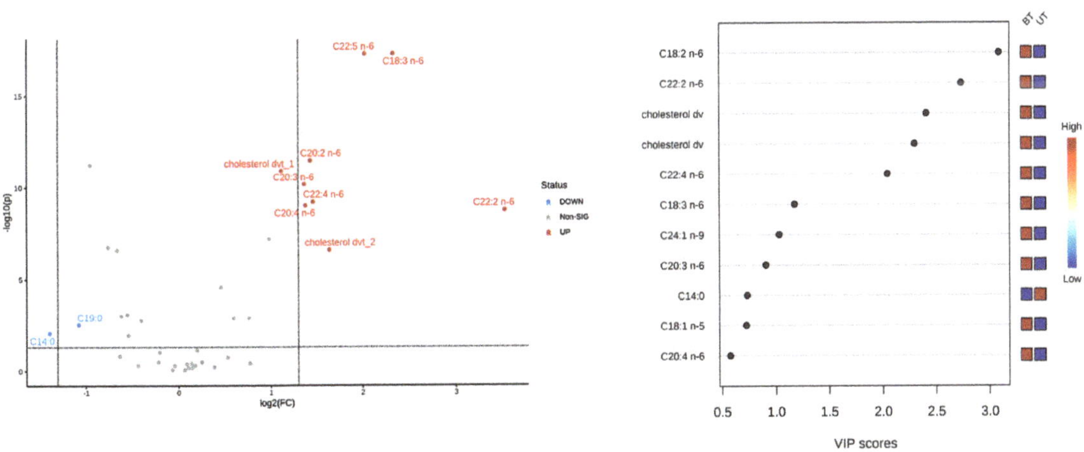

(**a**) (**b**)

Figure 7. (**a**) Volcano plot of differentially expressed features (BT/UT). (**b**) PLS–DA: VIP score graph.

4. Discussion

A comprehensive analytical method workflow for analyzing lipids in plasma/serum has been detailed herein. The method includes liquid–liquid extraction of lipids from 25 μL plasma/serum optimized to maintain a wide selectivity towards multiple classes of FAMEs (SFA, MUFA, and PUFA (ω-3 and ω-6)). A micro-volume extraction optimized using pooled human plasma (Figure 3), tested on NIST plasma metabolites (Figure S3), and utilized on pig serum (Figure 4) illustrates that the optimized sample preparation protocol is efficient for both human plasma and animal serum for lipid extraction, opening the possibility to translate this analytical protocol to other plasma/serum-related studies.

GC×GC–TOFMS is a powerful technique for lipidomics as it provides structured chromatographic separation, which adds value to identifying untargeted lipidomics. The optimized method presented here provides valuable insight on FAMEs identification without requiring in-depth MS/MS investigations.

For the first time, lipid profiling of serum for boar-tainted and untainted pigs is analyzed, identifying distinguished FAMEs class composition. The results pertaining to the observed significant presence of PUFA ω-6 and cholesterol derivatives in boar-tainted pig serum are supported by multivariate analysis, while SFA and PUFA ω-3 are significant in untainted pig serum. These differences in the lipid composition are opening new investigation routes to better understand bore taint deviation. Moreover, 36 features were subjected to pathway and enrichment analysis (Figure 8). The linoleic acid metabolism pathway was the key metabolic pathway, with a pathway impact of 1 supported by quantitative enrichment analysis (Figure 8).

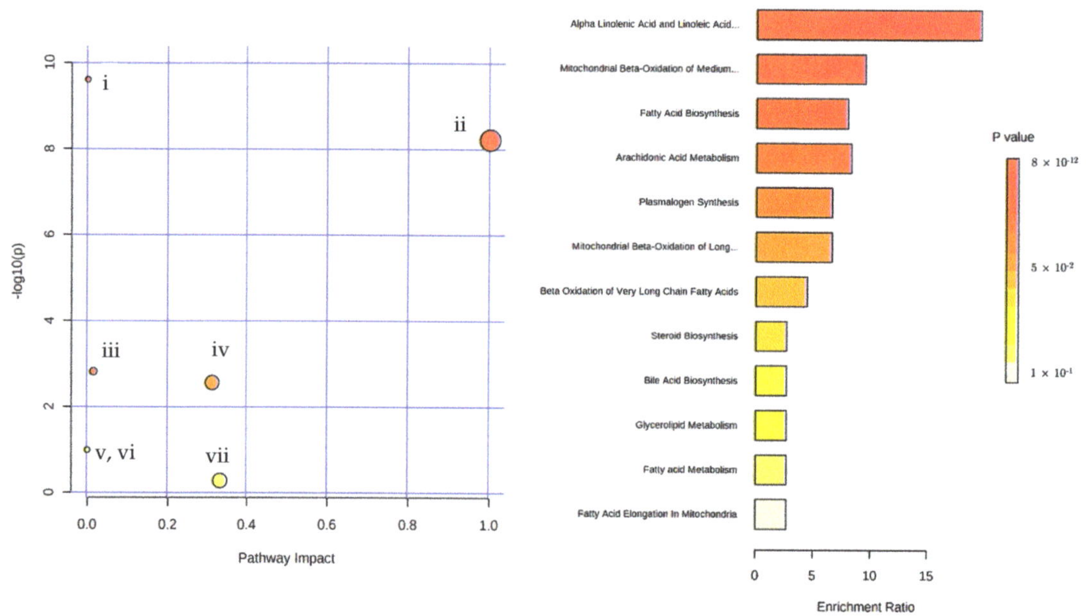

(a) (b)

Figure 8. (**a**) Summary of pathway analysis: (i) Biosynthesis of unsaturated fatty acids; (ii) Linoleic acid metabolism; (iii) Fatty acid biosynthesis; (iv) Arachidonic acid metabolism; (v) Fatty acid elongation; (vi) Fatty acid degradation; (vii) alpha-Linolenic acid metabolism. (**b**) Quantitative enrichment analysis (QEA): Metabolite set enrichment overview. * Features 16, 38, and 39 (13-Octadecenoic acid, methyl ester; Cholesta-3,5-diene; Cholesta-2,4-diene) were excluded due to lack of metabolite ID (HMDB ID) conversion match.

A ω-3-rich feed for pigs is recommended to maintain pork as a good nutritional source of ω-3 fatty acids for humans [26]. Moreover, the outcome of lipid profiling of pig serum gives rise to an intriguing possibility of the importance of PUFA (ω-3) not only as a nutritional essential, but also in the involvement of reducing boar taint in pigs.

Supplementary Materials: The following supporting information can be downloaded at: https://www.mdpi.com/article/10.3390/metabo12111111/s1, Figure S1: Concentration range of lipids in human plasma as per NIST CoA, Figure S2: Optimization plot of derivatization condition, Figure S3: (a) Contour plot of FAMEs standard mixture. (b) Zoomed-in contour plot of FAMEs standard mixture, Figure S4: (a) Contour plot of NIST SRM 1950 standard. (b) Zoomed-in contour plot of NIST SRM 1950 standard, Figure S5: (a) Contour plot of pig serum. (b) Zoomed-in contour plot of pig serum, Table S1: Fatty acid composition (% Area) of boar-tainted (BT, n = 20) and untainted (UT, n = 20) pig serum.

Author Contributions: Conceptualization, K.B. and P.-H.S.; methodology, K.B., T.D. and P.-H.S.; software, K.B., T.D. and P.-H.S.; validation, K.B. and T.D.; formal analysis, K.B. and P.-H.S.; investigation, K.B., L.M.D., P.-H.S. and J.-F.F.; resources, K.B., P.-H.S. and J.-F.F.; data curation, K.B., T.D., A.M., N.G., J.W., P.-H.S. and J.-F.F.; writing—original draft preparation, K.B., P.-H.S. and J.-F.F.; writing—review and editing, K.B., T.D., L.M.D., A.M., N.G., J.W., P.-H.S. and J.-F.F.; visualization, K.B. and P.-H.S.; supervision, P.-H.S. and J.-F.F.; project administration, K.B. and L.M.D.; funding acquisition, J.-F.F. All authors have read and agreed to the published version of the manuscript.

Funding: This research was funded by the FWO/FNRS Belgium EOS Grant 30897864, "Chemical Information Mining in a Complex World". The authors acknowledge the support of the Walloon Government (Service Public de Wallonie, Namur, Belgium) through the NoWallOdor project (Grant agreements D31-1396 and D65-1430). The author Lena M. Dubois is currently working as an application scientist at LECO Instrument GmbH, Deutschland.

Institutional Review Board Statement: The animal study protocol was approved by the Ethics Committee of the University of Liège (Protocol code #2307; approved on 21 December 2020).

Informed Consent Statement: Not applicable.

Data Availability Statement: The data presented in this study are available on request from the corresponding author. The raw data was obtained on LECO ChromaToF software (exported with baseline correction in cdf format) and analysed on GC Image software. In order to reproduce the results the user will require specific software. Therefore, to maintain the data integrity the data will be provided upon request.

Acknowledgments: K.B. would like to thank Chiara Emilia Cordero and Simone Squara for their insightful discussion on data processing. The authors would like to thank LECO Corp. for its continuous support.

Conflicts of Interest: The authors declare no conflict of interest.

References

1. Spener, F.; Lagarde, M.; Géloën, A.; Record, M. What Is Lipidomics? *Eur. J. Lipid Sci. Technol.* **2003**, *105*, 481. [CrossRef]
2. Han, X.; Gross, R.W. Global Analyses of Cellular Lipidomes Directly from Crude Extracts of Biological Samples by ESI Mass Spectrometry: A Bridge to Lipidomics. *J. Lipid Res.* **2003**, *44*, 1071–1079. [CrossRef] [PubMed]
3. Wang, R.; Li, B.; Lam, S.M.; Shui, G. Integration of Lipidomics and Metabolomics for In-Depth Understanding of Cellular Mechanism and Disease Progression. *J. Genet. Genom.* **2020**, *47*, 69–83. [CrossRef] [PubMed]
4. Holčapek, M.; Liebisch, G.; Ekroos, K. Lipidomic Analysis. *Anal. Chem.* **2018**, *90*, 4249–4257. [CrossRef] [PubMed]
5. Folch, J.; Lees, M.; Stanley, G.H.S. A Simple Method for the Isolation and Purification of Total Lipides from Animal Tissues. *J. Biol. Chem.* **1957**, *226*, 497–509. [CrossRef]
6. Bligh, E.G.; Dyer, W.J. A rapid method of total lipid extraction and purification. *Can. J. Biochem. Physiol.* **1959**, *37*, 911–917. [CrossRef] [PubMed]
7. Ostermann, A.I.; Müller, M.; Willenberg, I.; Schebb, N.H. Determining the Fatty Acid Composition in Plasma and Tissues as Fatty Acid Methyl Esters Using Gas Chromatography—A Comparison of Different Derivatization and Extraction Procedures. *Prostaglandins Leukot. Essent Fat. Acids* **2014**, *91*, 235–241. [CrossRef]
8. Morrison, W.R.; Smith, L.M. Preparation of Fatty Acid Methyl Esters and Dimethylacetals from Lipids with Boron. *Fluorid. Methanol.* **1964**, *5*, 600–608. [CrossRef]

9. Metcalfe, L.D.; Schmitz, A.A.; Pelka, J.R. Rapid Preparation of Fatty Acid Esters from Lipids for Gas Chromatographic Analysis. *Anal. Chem.* **1966**, *38*, 514–515. [CrossRef]
10. Lands, W.E.M.; Libelt, B.; Morris, A.; Kramer, N.C.; Prewitt, T.E.; Bowen, P.; Schmeisser, D.; Davidson, M.H.; Burns, J.H. Maintenance of Lower Proportions of (n − 6) Eicosanoid Precursors in Phospholipids of Human Plasma in Response to Added Dietary (n − 3) Fatty Acids. *Biochim. Et Biophys. Acta (BBA) Mol. Basis Dis.* **1992**, *1180*, 147–162. [CrossRef]
11. Ma, J.; Folsom, A.R.; Shahar, E.; Eckfeldt, J.H. Plasma Fatty Acid Composition as an Indicator of Habitual Dietary Fat Intake in Middle-Aged Adults. The Atherosclerosis Risk in Communities (ARIC) Study Investigators. *Am. J. Clin. Nutr.* **1995**, *62*, 564–571. [CrossRef] [PubMed]
12. Schuchardt, J.P.; Schmidt, S.; Kressel, G.; Willenberg, I.; Hammock, B.D.; Hahn, A.; Schebb, N.H. Modulation of Blood Oxylipin Levels by Long-Chain Omega-3 Fatty Acid Supplementation in Hyper- and Normolipidemic Men. *Prostaglandins Leukot. Essent. Fat. Acids* **2014**, *90*, 27–37. [CrossRef] [PubMed]
13. Djuricic, I.; Calder, P.C. Beneficial Outcomes of Omega-6 and Omega-3 Polyunsaturated Fatty Acids on Human Health: An Update for 2021. *Nutrients* **2021**, *13*, 2421. [CrossRef] [PubMed]
14. Micalizzi, G.; Ragosta, E.; Farnetti, S.; Dugo, P.; Tranchida, P.Q.; Mondello, L.; Rigano, F. Rapid and Miniaturized Qualitative and Quantitative Gas Chromatography Profiling of Human Blood Total Fatty Acids. *Anal. Bioanal. Chem.* **2020**, *412*, 2327–2337. [CrossRef] [PubMed]
15. Duarte, D.A.S.; Schroyen, M.; Mota, R.R.; Vanderick, S.; Gengler, N. Recent Genetic Advances on Boar Taint Reduction as an Alternative to Castration: A Review. *J. Appl. Genet.* **2021**, *62*, 137–150. [CrossRef]
16. Fischer, J.; Elsinghorst, P.W.; Bücking, M.; Tholen, E.; Petersen, B.; Wüst, M. Development of a Candidate Reference Method for the Simultaneous Quantitation of the Boar Taint Compounds Androstenone, 3α-Androstenol, 3β-Androstenol, Skatole, and Indole in Pig Fat by Means of Stable Isotope Dilution Analysis-Headspace Solid-Phase Microextraction-Gas Chromatography/Mass Spectrometry. *Anal. Chem.* **2011**, *83*, 6785–6791. [CrossRef]
17. Bekaert, K.M.; vanden Bussche, J.; François, S.; Tuyttens, F.A.M.; de Brabander, H.F.; Vandendriessche, F.; Vanhaecke, L. A Validated Ultra-High Performance Liquid Chromatography Coupled to High Resolution Mass Spectrometry Analysis for the Simultaneous Quantification of the Three Known Boar Taint Compounds. *J. Chromatogr. A* **2012**, *1239*, 49–55. [CrossRef]
18. Buttinger, G.; Wenzl, T. Validation by Collaborative Trial of a Method for the Determination by GC–MS and LC–MS/MS of Boar Taint Marker Compounds in Pork Tissue. *Food Chem X* **2020**, *6*, 100083. [CrossRef]
19. Burgeon, C.; Debliquy, M.; Lahem, D.; Rodriguez, J.; Ly, A.; Fauconnier, M.L. Past, Present, and Future Trends in Boar Taint Detection. *Trends Food Sci. Technol.* **2021**, *112*, 283–297. [CrossRef]
20. Trautmann, J.; Gertheiss, J.; Wicke, M.; Mörlein, D. How Olfactory Acuity Affects the Sensory Assessment of Boar Fat: A Proposal for Quantification. *Meat. Sci.* **2014**, *98*, 255–262. [CrossRef]
21. Sørensen, K.M.; Engelsen, S.B. Measurement of Boar Taint in Porcine Fat Using a High-Throughput Gas Chromatography-Mass Spectrometry Protocol. *J. Agric. Food Chem.* **2014**, *62*, 9420–9427. [CrossRef] [PubMed]
22. Verplanken, K.; Wauters, J.; van Durme, J.; Claus, D.; Vercammen, J.; de Saeger, S.; Vanhaecke, L. Rapid Method for the Simultaneous Detection of Boar Taint Compounds by Means of Solid Phase Microextraction Coupled to Gas Chromatography/Mass Spectrometry. *J. Chromatogr. A* **2016**, *1462*, 124–133. [CrossRef] [PubMed]
23. Hansen-Møller, J. Rapid High-Performance Liquid Chromatographic Method for Simultaneous Determination of Androstenone, Skatole and Indole in Back Fat from Pigs. *J. Chromatogr. B Biomed. Sci. Appl.* **1994**, *661*, 219–230. [CrossRef]
24. Liu, X.; Schmidt, H.; Mörlein, D. Feasibility of Boar Taint Classification Using a Portable Raman Device. *Meat. Sci.* **2016**, *116*, 133–139. [CrossRef] [PubMed]
25. Xia, J.; Psychogios, N.; Young, N.; Wishart, D.S. MetaboAnalyst: A Web Server for Metabolomic Data Analysis and Interpretation. *Nucleic Acids Res.* **2009**, *37*, W652–W660. [CrossRef] [PubMed]
26. Dugan, M.E.R.; Vahmani, P.; Turner, T.D.; Mapiye, C.; Juárez, M.; Prieto, N.; Beaulieu, A.D.; Zijlstra, R.T.; Patience, J.F.; Aalhus, J.L. Pork as a Source of Omega-3 (n-3) Fatty Acids. *J. Clin. Med.* **2015**, *4*, 1999–2011. [CrossRef] [PubMed]

metabolites MDPI

Article

HILIC-MS for Untargeted Profiling of the Free Glycation Product Diversity

Yingfei Yan [1,*], Daniel Hemmler [1,2] and Philippe Schmitt-Kopplin [1,2,*]

1 Research Unit Analytical BioGeoChemistry (BGC), Helmholtz Zentrum München, Ingolstädter Landstrasse 1, 85764 Neuherberg, Germany
2 Comprehensive Foodomics Platform, Chair of Analytical Food Chemistry, TUM School of Life Sciences, Technical University Munich, Maximus-von-Imhof-Forum 2, 85354 Freising, Germany
* Correspondence: yingfei.yan@helmholtz-muenchen.de (Y.Y.); schmitt-kopplin@helmholtz-muenchen.de (P.S.-K.)

Abstract: Glycation products produced by the non-enzymatic reaction between reducing carbohydrates and amino compounds have received increasing attention in both food- and health-related research. Although liquid chromatography mass spectrometry (LC-MS) methods for analyzing glycation products already exist, only a few common advanced glycation end products (AGEs) are usually covered by quantitative methods. Untargeted methods for comprehensively analyzing glycation products are still lacking. The aim of this study was to establish a method for simultaneously characterizing a wide range of free glycation products using the untargeted metabolomics approach. In this study, Maillard model systems consisting of a multitude of heterogeneous free glycation products were chosen for systematic method optimization, rather than using a limited number of standard compounds. Three types of hydrophilic interaction liquid chromatography (HILIC) columns (zwitterionic, bare silica, and amide) were tested due to their good retention for polar compounds. The zwitterionic columns showed better performance than the other two types of columns in terms of the detected feature numbers and detected free glycation products. Two zwitterionic columns were selected for further mobile phase optimization. For both columns, the neutral mobile phase provided better peak separation, whereas the acidic condition provided a higher quality of chromatographic peak shapes. The ZIC-cHILIC column operating under acidic conditions offered the best potential to discover glycation products in terms of providing good peak shapes and maintaining comparable compound coverage. Finally, the optimized HILIC-MS method can detect 70% of free glycation product features despite interference from the complex endogenous metabolites from biological matrices, which showed great application potential for glycation research and can help discover new biologically important glycation products.

Keywords: non-enzymatic glycation; Maillard reaction products; HILIC-MS; untargeted analysis; advanced glycation end products

Citation: Yan, Y.; Hemmler, D.; Schmitt-Kopplin, P. HILIC-MS for Untargeted Profiling of the Free Glycation Product Diversity. *Metabolites* **2022**, *12*, 1179. https://doi.org/10.3390/metabo12121179

Academic Editor: Joana Pinto

Received: 3 November 2022
Accepted: 21 November 2022
Published: 25 November 2022

1. Introduction

Non-enzymatic glycation has received increasing attention over the past few decades in both food chemistry and in vivo studies [1]. This type of reaction was initially discovered by a French chemist, Louis C. Maillard, in 1912, referring to the reaction between reducing carbohydrates and amino compounds. The spontaneous condensation reaction between the amino group and carbonyls first forms unstable Schiff bases, which rearrange to more stable Amadori products (ARPs). Consecutive degradation of ARPs produces highly reactive dicarbonyls, such as deoxyosone, glyoxal, methylglyoxal, etc. Dicarbonyls react with the amino group and yield advanced glycation end products (AGEs). For instance, the reactions between glyoxal, and methylglyoxal with lysine form carboxymethyl-lysine (CML), and carboxyethyl-lysine (CEL), respectively, which are often used as makers for the Maillard reaction (MR). Eventually, a heterogeneous mixture of Maillard reaction products

(MRPs) with diverse structures is produced from minor initial precursors through complex consecutive reaction cascades [2,3].

MR not only contributes to the aroma, taste, and color for foods during thermal processing, but also has close biological associations with aging and diseases, such as chronic hyperglycemia, diabetes, etc. The accumulation of endogenous free glycation products usually indicates metabolic disorders in vivo. Increased levels of free AGEs in plasma were associated with increased levels of diabetes complications [4,5]. Phenylalanine-glucose ARP was found as a biomarker for phenylketonuria, an inherited disorder causing the build-up of phenylalanine in the body [6]. The accumulation of free glycerate-modified amino acids, forming through the non-enzymatic reaction between amino acids and a highly reactive glycolytic intermediate, was detected in the brain of Parkinson's disease protein PARK7 knockout mouse [7]. The above-mentioned points suggested the important biological roles of endogenous AGEs and that a reliable method for analyzing these free glycation products is a basis to reveal their functions.

Common methods for glycation product analysis involve enzyme-linked immunosorbent assay (ELISA) and analytical instruments, including LC-MS [8–10] and GC-MS [11]. Among these methods, LC-MS provides a sensitive, selective, and high-throughput analysis without the need for specific antibodies or derivatization. Based on the structural prototypes, relevant in vivo AGEs can be classified into various categories depending on the amino acids involved and if crosslinks between amino acid residues occur [12]. AGEs with lysine residues, arginine residues, and crosslinks between two residues are the most investigated, such as CML, CEL, glyoxal-hydroimidazolone (G-H), methylglyoxal-hydroimidazolone (MG-H), glyoxal-lysine dimer (GOLD), and pentosidine. Up to twenty selected free glycation products, including ARPs and AGEs, could be quantified in targeted approaches in biological matrices, such as plasma, saliva, and urine [13,14]. However, AGEs have very diverse structures and biologically important glycation products are not always these well-studied AGEs. Thus, there is a need to establish a reliable method for comprehensively qualifying and quantifying free glycation products in biological samples, including the many hitherto unknown AGEs. Commercially available glycation product standards are very limited in number and structural diversity. Moreover, the synthesis of AGE standards is very time-consuming and challenging. Maillard model systems are valuable alternatives to limited standards for method development. They can reproducibly produce a mixture of hundreds of different free glycation products from only a few initial precursors with low-cost [15,16]. Although some reaction pathways for forming these glycation products are different in vivo compared with model systems [1], most glycation compounds can also be formed in model systems. Therefore, in current study, we used the MR model system to optimize the method based on untargeted strategy rather than using a limited number of commercially available reference standards.

Most free glycation products are highly polar and, therefore, have limited retention on reverse phase (RP) columns which are the most commonly used stationary phase in LC. To solve this issue, ion-pair reagents, such as heptafluorobutyric acid, nonafluoropentanoic acid were often used in previous studies to provide enough retention and efficient separation for free glycation products on RP columns [8,9,17]. However, ion-pair reagent has several drawbacks, including ionization suppression for mass spectrometry (MS), contamination of the instrument, and reduced column lifetime. Hydrophilic interaction liquid chromatography (HILIC) is a promising option for analyzing polar compounds. Until now, very limited studies used HILIC for the quantification of AGEs and they all focused on the targeted quantification of specific AGEs in foods [18–20]. To the best of our knowledge, systematic evaluation of HILIC columns and conditions for profiling free glycation products in an untargeted way is still lacking.

This study aimed to establish a method for the simultaneous characterization of diverse free glycation products using an untargeted metabolomics approach. Five different HILIC columns were compared. The mobile phase composition was further investigated to maximize performance and robustness of the analytical method. Methods were evaluated in terms of the number of detected features, distribution of detected features, precision, number of detected known glycation products, and their peak shapes. The ability of the optimized method was finally evaluated using biological matrices such as plasma, feces, and urine.

2. Materials and Methods

2.1. Reagents and Materials

Twenty proteinogenic L-amino acids (>97%), L-cystine (≥98%), D-(+)-glucose (≥99.5%), acetic acid (LC-MS grade), ammonium formate (10 M stock solution), and ammonium acetate (5 M stock solution) were purchased from Sigma–Aldrich (Steinheim, North Rhine-Westphalia, Germany). Acetonitrile (ACN, LC-MS grade) was purchased from Merck (Darmstadt, Hesse, Germany). Formic acid (98%, for mass spectrometry) was ordered from Honeywell Fluka (Charlotte, NC, USA). Purified water (18.2 MΩ) was obtained from a Milli-Q integral water purification system (Billerica, MA, USA). ESI-L low-concentration tuning mix was supplied by Agilent (Santa Clara, CA, USA). Lyophilized human plasma (P9523) was purchased from Sigma-Aldrich (Saint Louis, MO, USA) and reconstituted with double distilled water. One feces and urine sample was collected from a healthy male volunteer. Plasma, feces, and urine samples were kept frozen at −80 °C until use.

2.2. Maillard Model Systems Preparation

Twenty-one amino acids were reacted with glucose, respectively, to obtain MRPs with a wide range of physicochemical properties. Equal molar mixtures of glucose (0.1 M) and each amino acid (0.1 M) were prepared in Milli-Q water and heated in closed glass vials at 100 °C for seven hours to make Maillard model system samples. The same volume of each model system was mixed together, referred to as the model systems mixture sample (MSM1), and then diluted by 1:5 (v/v) with 90% acetonitrile for LC-MS/MS analysis. Model system samples of lysine (Glc-Lys), arginine (Glc-Arg), and histidine (Glc-His) were diluted 1:10 (v/v) with 90% acetonitrile for analysis. All samples were analyzed in triplicate.

2.3. Biological Sample Preparation

Biological sample extracts were mixed with model systems to evaluate the optimized HILIC-MS method. Plasma, urine, and feces samples were thawed on ice and vortexed for 60 s before sample preparation. 25 μL plasma was mixed with 500 μL ACN and vortexed for 5 min. 50 μL urine was diluted using 1 mL ACN and stored on ice for 5 min. The feces sample was first centrifuged at 14,000× g rpm at 4 °C for 10 min. Thereafter, approximately 50 mg of the pellet was homogenized with ACN at a fixed ratio of 1:20 (mg/mL) and sonicated in an ice bath for 30 min. All samples were then centrifuged at 14,000× g rpm at 4 °C for 10 min. The supernatants were transferred to another tube and then mixed with the model system solution later.

Equal volumes of Glc-Lys and Glc-Arg were mixed to get a model system mixture (named as MSM2). Before being spiked with biological extracts, MSM2 was mixed with water to obtain the low (1:25, v/v), medium (1:5, v/v), and high (undiluted) model system solution. In total, 90 μL of each type of biological extract supernatant was mixed with 10 μL of MSM2 solution in three concentration levels separately for matrix effect evaluation. The biological extracts without MSM2 (90 μL biological extracts mixed with 10 μL water) and MSM2 without biological extracts (10 μL MSM2 mixed with 90 μL ACN) were used as controls.

2.4. LC-MS/MS

Samples were analyzed using a Waters Acquity (Milford, MA, USA) UPLC system coupled with a Bruker maXis Quadrupole time-of-flight (QTOF) MS (Bremen, Germany). The injection volume was 5 μL for each sample. The MS analyses were performed in positive electrospray mode with a mass range of 50–1500 m/z. The ion source settings were: nebulizer gas pressure 2 bar, capillary voltage 4500 V, dry gas flow 10 L/min, and dry gas temperature 200 °C. Mass spectra were acquired with a scan rate of 5 Hz in data-dependent mode, where the three highest MS1 ions of each precursor scan were chosen for MS/MS with a collision energy of 30 eV. The TOF analyzer was calibrated using ESI-L Low Concentration Tuning Mix (Agilent, Santa Clara, CA, USA). Additionally, the same diluted Tuning Mix (1:4 (v/v) with 75% ACN) was injected from 0.1 to 0.3 min of every measurement using a switching valve for internal recalibration.

2.5. Chromatographic Conditions Optimization

This study aimed to establish a method for analyzing free glycation products instead of in-depth investigating the mechanisms of HILIC. Therefore, to find appropriate parameters in a straightforward and time-consuming manner, we chose a univariate method optimization approach, rather than testing all factors at all levels.

Firstly, five different HILIC columns were tested: iHILIC-Fusion (100 × 2.1 mm, 1.8 µm, 100 Å, zwitterionic, HILICON AB, Umea, Sweden), ZIC-cHILIC (100 × 2.1 mm, 3 µm, 100 Å, zwitterionic, Merck, Darmstadt, Germany), ZIC-HILIC (100 × 2.1 mm, 3.5 µm, 100 Å, zwitterionic, Merck, Darmstadt, Germany), BEH-HILIC (150 × 2.1 mm, 1.7 µm, 130 Å, unbonded ethylene bridged hybrid (BEH) particle substrates, Waters, Eschborn, Germany), BEH-Amide (100 × 2.1 mm, 1.7 µm, 130 Å, BEH amide, Waters, Eschborn, Germany). All columns were compared using the same eluents and gradient based on the recommendation of our previous study [21]. Eluent A consisted of 25 mM ammonium acetate (AA) in 30% ACN (pH 4.6), and Eluent B consisted of 5 mM AA in 95% ACN (pH 4.6). The binary gradient was: 0 min, 99.9% B; 2 min, 99.9% B; 9.5 min, 0.1% B; 12 min, 0.1% B; 12.1 min, 99.9% B. Columns were equilibrated at 99.9% B for 3 min after each run to reach the initial status. The flow rate was 0.5 mL/min with the column temperature set at 40 °C.

After the selection of the optimum stationary phase for analyzing MRPs, the effect of pH and mobile phase additives were further evaluated to choose the optimum mobile phase composition. To avoid ion suppression caused by high salt concentration, 5 mM ammonium formate (AF) or AA was used as the mobile phase modifier. Mobile phases A consisted of 5 mM salt in 30% ACN (acidic: with 0.1% corresponding acid; neutral: without corresponding acid) and mobile phases B consisted of 5 mM salt in 95% ACN (acidic: with 0.1% corresponding acid; neutral: without the corresponding acid) was tested. The detailed information for mobile phase optimization is shown in Table S1. Finally, the gradient was improved using the optimal mobile phase. The ultimate chromatographic condition was: eluent A 5% ACN and eluent B 95% ACN, both with 5 mM AF and 0.1% formic acid (FAcid). The gradient was: 0 min, 99.9% B; 2 min, 99.9% B; 13 min, 56% B; 14 min, 30% B; 14.1 min, 10% B; 16 min, 10% B; 16.1 min, 99.9% B.

2.6. Data Processing

Raw data were calibrated and converted to mzXML files by Bruker DataAnalysis 5.0 software (Bremen, Germany). The data preprocessing of the converted files, including peak picking, peak alignment, peak correspondence, and MS2 spectra finding were done based on the XCMS package (4.1.2) in R (version 4.1.0) [22]. The feature grouping, isotopes finding, and adducts annotation were processed by the CAMERA package (version 4.1.1) [23]. The detailed settings for XCMS and CAMERA are shown in Tables S2 and S3, respectively. The in-source fragment (ISF)-finding algorithm was adapted from the ISFrag package (version 0.1.0) [24]. Feature cleaning was carried out by in-house script in R. Principal component analysis (PCA) was completed using the FactoMineR package (version 2.4) [25].

3. Results

In this study, several chromatographic conditions for the analysis of free glycation products were evaluated through untargeted and targeted comparisons using typical MR model systems. The capability of the optimized method was further assessed by analyzing model systems spiked with biological matrices, including plasma, urine, and feces.

3.1. Data Cleaning for Reliable Feature Lists

The main aim of this study is to establish a method for simultaneously characterizing a wide range of free glycation products using the untargeted metabolomics approach. In contrast to the common method development workflow, where the occurrence of fixed reference standard compounds was compared, complex MR systems were chosen for method optimization. As reported by previous studies, the model system consisting of single amino acids and sugar can produce a multitude of glycation products [15,16], which has better coverage of potential glycation products compared to using the very limited commercially available glycation standards. Using the untargeted approach to analyze the model system is capable of generating an overall description of compounds in the MR mixture. Besides MRPs, there are also amino acid, glucose, and minor degradation products of reactants in the model system. The feature list created from the untargeted data processing algorithm contains redundant peaks derived from one compound, including isotope peaks, ISFs, various adducts, multimers, containments and artifacts, which makes the number of features directly derived from peak detection not accurate for method comparison.

Hence, we further attributed feature relationships by an in-house R script following the CAMERA annotation [23]. A typical output of MS feature annotations associated with one compound was exemplified with lysine. As shown in Figure 1A,B, there were 30 co-eluted features with the same chromatographic peak shape (Pearson correlation coefficient > 0.8, $p < 0.001$) in the correlation group containing the lysine $[M + H]^+$ signal. Detailed information and interpretations of each feature are shown in Table S4. By using the current data processing workflow, the 30 features can be divided into 6 categories, including 4 low intensity noise, 5 isotope peaks, 7 artifacts caused by the saturation of the detector [26], 6 adducts, 6 ISFs, and 2 unidentified features. Among them, 22 features were produced by lysine. The approximately 95% feature inflation during untargeted LC-MS analysis was also reported in previous studies, 869 features were detected after the injection of 51 standards [27], and 10,000–30,000 features were observed by analyzing 900 unique metabolites [28].

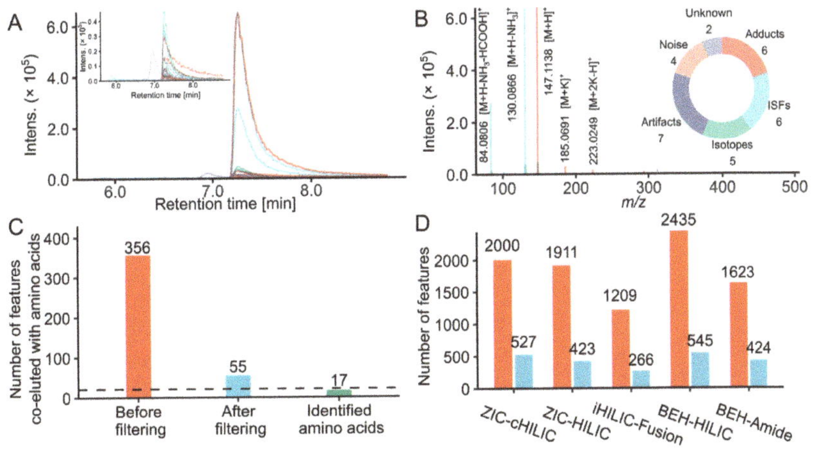

Figure 1. Feature filtering results of model systems. (**A**) Extracted ion chromatograms of lysine and its co-eluted MS features grouped by CAMERA in a glucose and lysine model system analyzed by iHILIC-Fusion. (**B**) Interpretation of the MS features grouped with lysine. Top right: Pie chart of the proportions of peaks categorized by the annotations. (**C**) The number of features co-eluted and grouped with amino acids in the model systems mixture of twenty-one amino acids (MSM1) before and after filtering analyzed by iHILIC-Fusion. (**D**) The total feature number detected in MSM1 before (red) and after filtering (blue) analyzed by different HILIC columns.

To remove such unreliable and redundant features, we filtered feature lists by following steps: remove background ions, noise with low intensity, artifacts caused by saturation, isotope peaks, ISFs, redundant adducts. To evaluate the approach, we checked the features correlated with 21 amino acids in the MSM1 before and after filtering. As shown in Figure 1C, there were 356 features co-eluted and grouped with amino acids with high peak shape similarity before data cleaning. The filtering caused a reduction of 84.5% features, only 55 features were kept. Among these, 17 features were confirmed as amino acids. L-cysteine and L-aspartic acid were not detected because their intensities were below the limit of detection. L-leucine and L-isoleucine were not separated and identified as one compound. Only L-proline was not identified because it was incorrectly annotated as a potassium adduct. This indicates the current workflow can remove redundant features originated from same compounds and keep the real signal at the same time. The effect of the filtering process on whole datasets measured by different LC conditions was also compared and demonstrated using the MSM1 sample (Figure 1D). Around 22% to 26% of the total features were kept. The percentages of the overall removed feature detected in MSM1 were lower compared to the features co-eluted with amino acids. Because amino acids in the MSM1 produced more artifacts (~77% features caused by detector saturation related to amino acids) and redundant features (e.g., ISFs, multiple adducts) due to their high concentration. The feature inflation is less pronounced for most MRPs with relatively low concentration. Moreover, the data filtering did not cause significant alterations to the overall trends, suggesting the comparison results from feature number were acceptable.

3.2. Selection of HILIC Columns

3.2.1. Non-Targeted Evaluation of the Column Selection

Three types of HILIC columns, including one bare silica (BEH HILIC), one amide (BEH Amide), and three zwitterionic columns (iHILIC-Fusion, ZIC-cHILIC and ZIC-HILIC), were selected for testing the selectivity of different stationary phases. According to our previous study on the thorough evaluation of metabolites coverage under different mobile phases, amino acids and their analogs preferred acidic conditions for both zwitterionic, and amide HILIC columns [21]. So, we used 30% ACN with 25 mM AA and 95% ACN with 5 mM AA at pH 4.6 as the starting mobile phase A and B to screen the columns.

We compared the performance of different columns by analyzing model systems. The aim was to maximize the number of detected MRP features, which were characterized by unique *m/z* and retention time with good reproducibility. The MSM1 was chosen for evaluating the selectivity and coverage of columns for MRPs that derivate from all proteinogenic amino acids, like the ARPs. In addition, lysine, arginine and histidine glucose model systems were analyzed separately. Because of the higher reactivity for these amino acids and N-containing side chains, most free AGEs reported in in vivo studies are derived from them [14,29].

Results for features in each model system detected by the tested columns are summarized in Figure 2. The features were classified into two types: features with and features without MS2 spectra. Features with MS2 spectra promise both downstream statistical analysis and the possibility of compound structural identification. The higher number of MS2 spectra also suggests a better separation of LC when total feature numbers are comparable. Among all tested columns, ZIC-cHILIC detected the highest overall number of features using the same mobile phase (Figure 2A). Particularly for Glc-Lys, the feature number detected by ZIC-cHILIC was more than 1.5-fold higher than for other columns (except for ZIC-HILIC). Independent of the tested columns, more features were detected in Glc-Lys and Glc-Arg compared with the Glc-His model system, suggesting the higher reactivity towards Glc for Lys and Arg. However, features observed in MSM1 were lower than expected, and less than those in Glc-Lys and Glc-Arg. This could be ascribed to the low concentration of most MRPs in each model system and the high concentration of reactant amino acids.

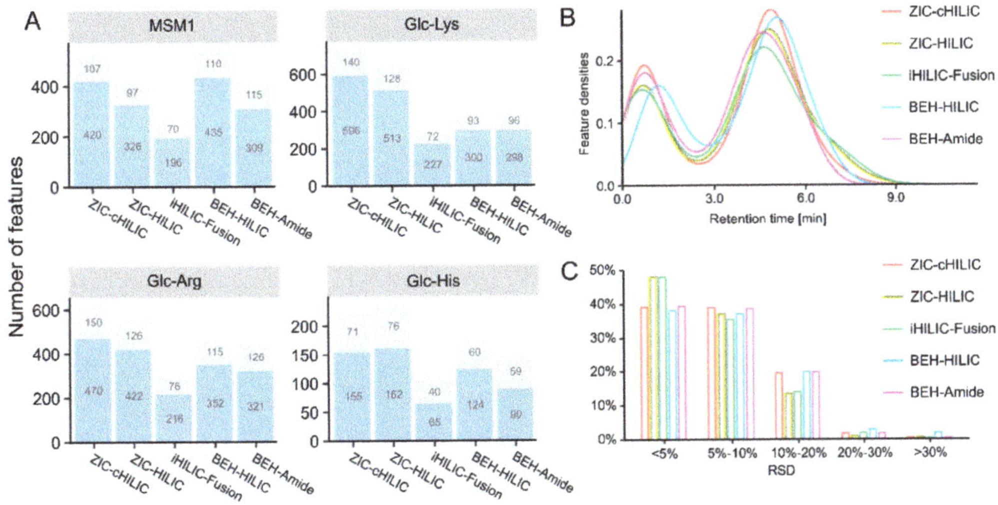

Figure 2. Results of column selection. (**A**) Bar plots representing the number of features detected by different columns in each model system. Features were categorized into two types: features with MS2 (light blue) and features without MS2 but recognizable extracted ion chromatograms (dark blue). MSM1: the model systems mixture of twenty-one amino acids. (**B**) Retention time distribution of all detected features in model systems. (**C**) Relative standard derivation (RSD) distribution of all feature intensities.

The retention time (RT) density plot (Figure 2B) shows the distribution of all features detected in four model systems by different HILIC columns along the RT. For accurate quantitation and simpler spectra complexity, it is preferable for fewer features to elute during the void volume. All HILIC columns showed good retention for detected features, resulting in a higher density between 3 min and 9.5 min compared with a RT of less than 2.5 min. Features were eluted ~0.35 min later for BEH-HILIC compared with other columns because BEH-HILIC has a greater column length (150 mm) and, consequently, a larger void volume. In general, ZIC-HILIC and iHILIC-Fusion columns showed more dispersed separation for features across the RT and fewer features eluting between 0 and 2 min. The precision was evaluated by calculating the relative standard derivation (RSD) of the intensity of the three analytical replicates for all detected features. As shown in Figure 2C, the RSD for more than 95% of features was less than 20%, showing good reproducibility for all tested columns.

3.2.2. Selectivity of Columns for Analyzing Amino Acids and Glycation Products

Amino acids and known glycation products were then subjected to a detailed comparison (Figure 3). The ability of HILIC columns for analyzing amino acids in MSM1 was evaluated based on the retention time, peak shape, and MS intensity. All amino acids can be detected with good retention using ZIC-HILIC, ZIC-cHILIC, and BEH-Amide. Cystine, which is the oxidized dimer of cysteine, can only be detected with good peak shape and sensitivity by BEH-Amide. The peak width of amino acids analyzed by ZIC-HILIC and iHILIC-Fusion was broader compared with the other three columns and tended to tail especially for basic amino acids (Table S5). This is likely due to the stronger electrostatic attraction between the net positive charge of basic amino acids and negatively charge at the distal end of sulfobetaine in ZIC-HILIC [30], and the slightly negative net surface charge in iHILIC-Fusion [31], respectively.

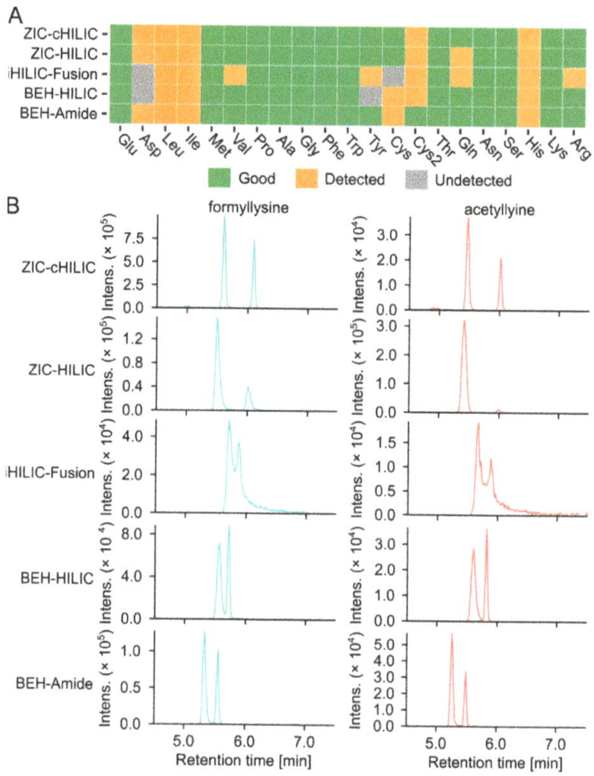

Figure 3. Targeted evaluation of column performance. (**A**) The individual score of amino acids analyzed by different HILIC columns. (**B**) Representative chromatograms of the putative advanced glycation end products isomers, formyllysine ($C_7H_{14}N_2O_3$, $[M + H]^+ = 175.1077$, blue) and acetyllysine ($C_8H_{16}N_2O_3$, $[M + H]^+ = 189.1234$, red), separated by different columns. EICs were extracted with ± 0.005 Da.

The free glycation products, including AGEs and ARPs, that can be produced by the model systems were collected from the literature to build a library (Table S6). The analytical capability of the five tested HILIC columns was also benchmarked by the detectability of these glycation products. We screened the feature table of each column for glycation products using the theoretical m/z with an error < 10 ppm. Considering that the MR can produce multiple isomers of glycated amino acids, including stereoisomers, regioisomers, and anomers [32], all the isomers were summed to compare the number of matching features per unique m/z (Table S7). The ARPs were barely detected in the MSM1 and were not evaluated in the column selection section. This could be due to the ionic suppression caused by the high concentration of salts in the aqueous mobile phase (25 mM AA) [31,33]. Three types of zwitterionic columns can detect higher glycation candidates than bare HILIC and amide columns. ZIC-HILIC detected the highest number of potential AGEs features (24 matched features of 15 unique m/z) in Glc-Lys and GLc-Arg, followed by ZIC-cHILIC, 23 matched glycation product candidates and 12 unique m/z. In comparison, only 12 matched candidates (9 unique m/z) were detected with BEH HILIC. Glyoxal-lysine dimer (GOLD) and methylglyoxal-lysine dimer (MOLD) can be only detected by ZIC-HILIC. Representative extracted ion chromatograms (EICs) of AGEs are shown in (Figure 3B) to present the selectivity and performance of the column. The best separation of isomers of formyllysine and acetyllysine was achieved by the ZIC-cHILIC column.

Altogether, ZIC-HILIC and ZIC-cHILIC showed better fits for analyzing free glycation products through the column selection. These two columns were compared in detail for later optimization with a lower salt concentration in mobile phases.

3.3. Mobile Phase Optimization

3.3.1. Non-Targeted Evaluation of the Mobile Phase pH and Modifiers

We next compared the performance of ZIC-cHILIC and ZIC-HILIC columns under neutral and acidic mobile phase conditions for two different modifiers (AF and AA), respectively. Among the eight investigated conditions, the ZIC-cHILIC column provided the highest feature number for all four samples using 5 mM AF as modifiers under the neutral condition (Figure 4A). Interestingly, ZIC-cHILIC always detected more features than ZIC-HILIC independent of the mobile phase condition and sample type. This could be due to the smaller particle size of the ZIC-cHILIC column (3 μm) compared with the ZIC-HILIC (3.5 μm) resulting in better separation performance [30]. Moreover, ZIC-cHILIC was reported to have better selectivity for polar amino-containing compounds than ZIC-HILIC, like aminoglycosides [34]. The feature count increased in all four model systems using neutral rather than acidic mobile phases. Based on the retention time distribution (Figure 4B,C), both ZIC-cHILIC and ZIC-HILIC showed improved separation under neutral conditions compared to acidic conditions. The better separation also explained the higher feature number at neutral mobile phases. Same results were also reported in previous literature [30,35,36], which may be attributed to the decreased ion-exchange interaction for both ZIC-HILIC and ZIC-cHILIC caused by the protonation of silanols with acidic mobile phases [37]. For the effect of additives, independent of the existence of the corresponding acid, a higher feature count was observed for AF compared with AA. The feature number detected in Glc-Lys and Glc-Arg model systems was higher than Glc-His, which was consistent with previous column selection results. Overall, the mobile phase composition had a similar impact on the feature number detected in each of the four tested model systems.

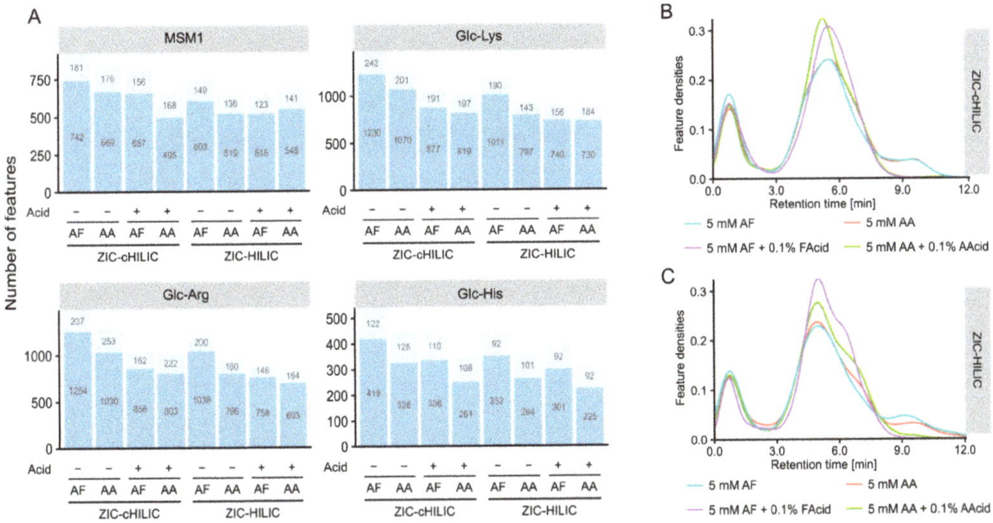

Figure 4. Effects of mobile phase composition on feature coverage and distribution. (**A**) Bar plots representing the number of detected features analyzed by mobile phases consisted of 5 mM ammonium formate (AF) or 5 mM ammonium acetate (AA) with and without its corresponding acid (0.1%, v/v). (**B**) Density distribution of retention times across the chromatographic run for ZIC-cHILIC column and (**C**) ZIC-HILIC column.

The precision of the tested conditions is shown in Figure S1. All conditions showed good precision with more than 95% features observed with an intensity RSD below 20%. The highest percentage of features with RSD less than 20% was observed for the ZIC-HILIC with 5 mM AF and 0.1% FAcid. For the features with less precision (RSD > 20%), a slightly higher percentage was observed under the neutral condition compared to the acidic condition.

3.3.2. Effect of Mobile Phase on Detections of Amino Acids and Glycation Products

The performance of all tested chromatographic conditions for analyzing amino acids is summarized in Figure 5A and Table S8. ZIC-cHILIC and ZIC-HILIC operated with mobile phase containing 5 mM AF and 0.1% FAcid can detect the highest number of amino acids with good peak shape. Mobile-phase pH has a higher impact on the peak shape of amino acids compared to column chemistry. Basic amino acids, including Lys, Arg, and His, have broad peaks under neutral conditions. One reason is the electrostatic attraction between the positively charged side chain of basic amino acid and the negatively charged silica under the neutral condition [31]. For His, the pKa of its side-chain group is 5.97. Under the neutral condition, the side chain of His is half deprotonated causing the broad peak shape. Cys and Cys2 were detected with good peak shape only under acidic conditions as well. The baseline separation of Ile and Leu was not achieved for all tested methods; however, a better separation was obtained by ZIC-cHILIC compared with ZIC-HILIC, which merged into one peak analyzed by ZIC-HILIC under most tested mobile phases.

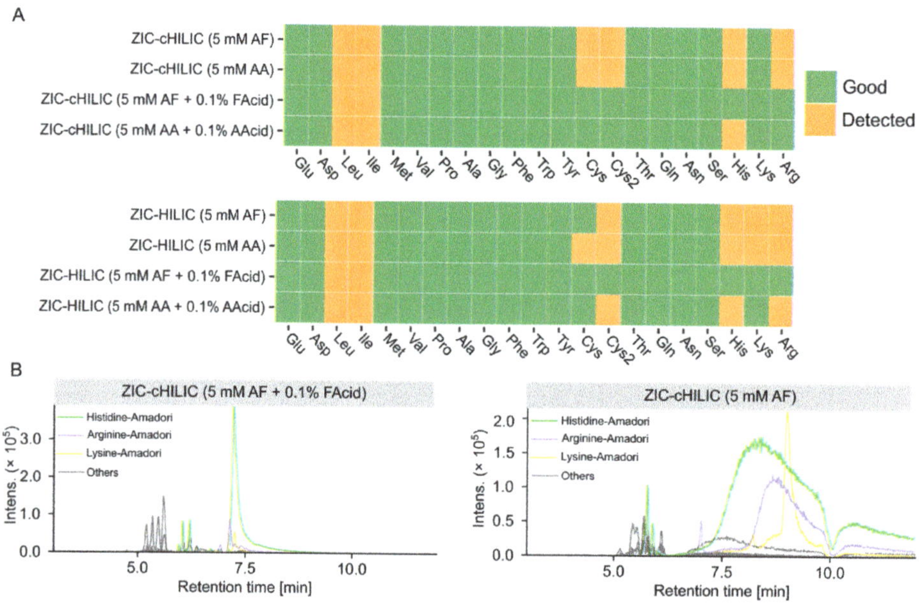

Figure 5. Effects of mobile phase on peak shapes of the amino acids and ARPs. (**A**) Individual score of amino acids analyzed by different mobile phases. (**B**) Extracted ion chromatograms of all 21 ARPs under acidic and neutral conditions (extracted with theoretical $[M + H]^+ \pm 0.005$ Da, detailed information of theoretical *m/z* is shown in Table S6).

ZIC-cHILIC operated under mobile phase containing 5 mM AF showed the best coverage for glycation products because 71 glycation product candidates could be matched (Table S9). Generally, more matched glycation candidates were observed under neutral conditions compared to acidic conditions, showing similar trends as revealed by the above performed comparison of untargeted feature numbers. This could be explained by the better separation of the column operated under neutral conditions and more isomers could be resolved. However, when we checked the peak shape of detected glycation products, we found most ARPs showed broad and split peaks under the neutral mobile phase (Figure 5B), which may be due to the equilibrium between Amadori product anomers [38]. Particularly for ARPs of basic amino acids, the EICs were extremely board analyzed by neutral mobile phase. Poor peak shapes tend to interfere with the peak picking algorithm and lead to unreliable peak detection results, but also more easily cause column carryover and interfere with later analysis.

Considering both quantity and quality of detected features, the ZIC-cHILIC operated under 5 mM AF and 0.1% FAcid provided the best results for amino acid glycation product discovery in terms of peak shape and compound coverage. Finally, the gradient was optimized based on the feature distribution along the chromatographic run. As most of the compounds eluted from 4 min to 7.5 min (Figure 4B), in order to achieve a better separation, a longer gradient was used to change mobile phase composition from 25% B to 75% B. The final gradient program is listed in Section 2.

3.4. Evaluation of the Optimized HILIC-MS Method Using Biological Samples

Compared with model systems, biological samples contain complex endogenous metabolites, large amounts of salts (urine), high diversity of lipids (plasma), and gut microbiota metabolites (feces). These metabolites and salts can affect both chromatographic separation and electrospray ionization [39,40]. Therefore, we further evaluated the performance of the optimized method on analyzing glycation products with complex biological extracts. Free endogenous glycations are aggravated by hyperglycemia, oxidative stress, and other metabolic diseases, whereas glycation products in the biological samples from healthy individuals are subtle without considering the dietary intake of AGEs [1,9,14]. Thus, we used the mixtures of model systems and biological extracts to investigate the effect of biological matrices on HILIC-MS analysis as proof of concept. To avoid detector saturation caused by reactants (amino acids and glucose) and keep the glycation products with low intensity detectable, only the mixture of lysine and arginine model systems (MSM2) was used as an additive to plasma, urine, and feces at three concentration levels: low (25-fold diluted), medium (5-fold diluted) and high (not diluted). For method evaluation, three types of samples were analyzed, including biological sample extractions, MSM2, and MSM2-spiked biological sample extractions. The interference of endogenous compounds on the detectability of glycation products was evaluated by comparing the number of detected MRP features in MSM2 with and without biological extracts. The reproducibility of the method was visualized by PCA score plots.

Based on PCA score plots (Figure 6A), three replicate injections of all sample types clustered together indicating good reproducibility. The first component discriminated the samples according to the concentration of spiked MRPs. For feces and plasma datasets, the samples spread from left to right along with the *x*-axis as their concentration increased, and the urine was the other way around. For all three PCA analyses, the first component can explain more than 50% of the variance, showing the optimized HILIC-MS method is capable of showing concentration differences of MRPs among samples. The second component discriminated the MSM2 samples versus those without biological extracts. The corresponding loading plots (Figure S2) indicate that most features were important in explaining the variability among different samples as they were positioned close to the correlation circle. In the loading plots, unique features detected in biological samples are located in the same region as the biological samples in the score plots (Figure 6A). Likewise, for MSM2 unique features in the loading plots were found in the same region as the MSM2 samples in the score plots (Figure 6A). The position of MSM2-spiked biological samples in score plots was in the middle position of biological samples-specific features and MSM2-specific features in loading plots. This supports the separation of groups of biological samples, MSM2 samples, and MSM2-spiked biological samples displayed in the score plots (Figure 6A).

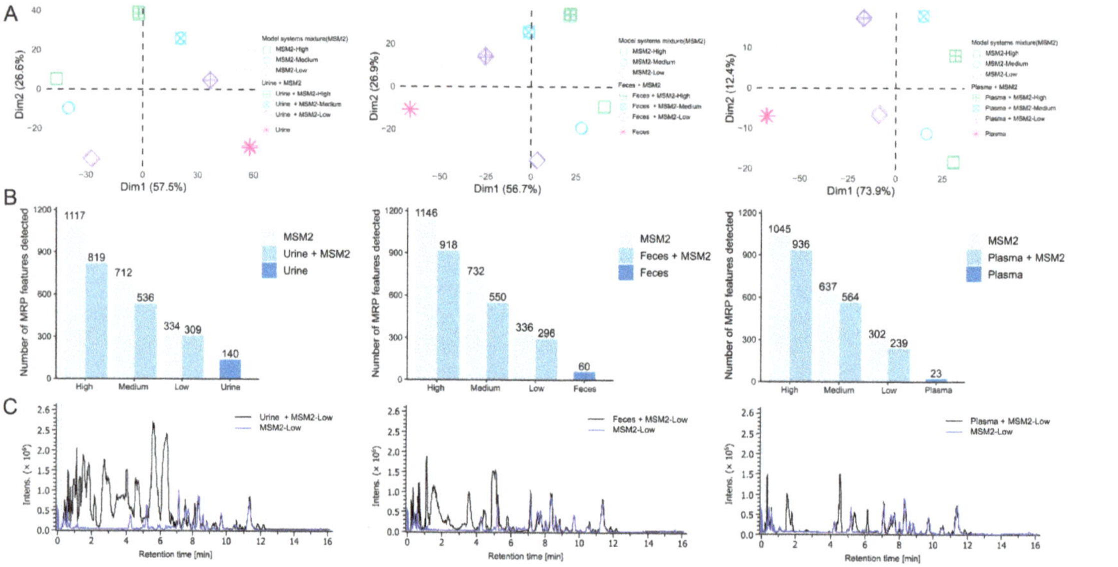

Figure 6. Detectability of glycation products in the biological matrices. (**A**) Principal component analysis (PCA) score plots of biological samples, a model system mixture (lysine and arginine; MSM2), and model system-spiked biological samples, in an order of urine, feces, and plasma. Prior to spiking, the model system mixture was either not diluted (**high**) or diluted 1:5 (**medium**) or 1:25 (**low**). Measurement was performed using the optimized HILIC method described in Section 2. (**B**) Bar plots representing the number of detected MRP features depending on the dilution level and biological extracts. (**C**) Representative base peak chromatograms of MSM2-Low and MSM2-spiked biological samples.

Using human plasma, feces, and urine, we demonstrated that more than 70% of MRP features can be detected even in complex biological matrices (Figure 6B). Among these sample types, plasma has the least matrix effect compared with urine and feces: ~80% of MRP features can be detected for all three levels of mixture. The most suppression was caused by urine samples, only 73% MRPs could be recovered for the mixture with high concentration MSM2. This could be attributed to the more and higher concentration of polar compounds in urine and feces samples compared to plasma, resulting in ion suppression [41,42]. The chromatogram of plasma was relatively empty in contrast to feces and urine in our dataset (as shown in Figure 6C). Most MRPs eluted between 6 and 12 min. Representative EICs of AGEs and amino acids showed that the peak shape and isomer separation were not influenced by the in vivo metabolites (Figure S3). Interestingly, there are several features could be detected in both biological extracts and MSM2. We manually checked that some features were glucose, amino acids as well as their degradation products. Importantly, feces and urine had a higher number of matching features compared to plasma. This also explains the higher detection rate of MRP features in mixed samples at low levels compared to high levels. This suggested that urine and feces are better matrices for discovering glycation product candidates compared to plasma. Additionally, levels of free AGEs in urine and feces are more prone to be affected by dietary AGEs compared to plasma. More than 80% of dietary ARPs were not absorbed and degraded by gut microflora [43]. A significant increase of 40% free urinary CML excretion versus 7% higher plasma CML was detected in healthy people urine and plasma after the 2.5 times higher AGEs diet [44]. That should be considered during the experiment design and data interpretation of finding in vivo free glycation markers.

4. Conclusions

In the present study, we systematically optimized HILIC-MS methods for untargeted profiling of free glycation products using model systems. The performance of the methods was evaluated from both targeted and untargeted aspects. For untargeted comparison, the number of detected features, the distribution of features along the chromatographic window, and precision were assessed. The number of detected amino acids and matched known glycation products in model systems as well as their peak shapes were checked to further confirm the analytical ability of the method. With regard to the number of detected MRP features and matched AGEs, ZIC-HILIC and ZIC-cHILIC columns had better performance than the other three HILIC columns. Further mobile phase optimization of the selected two columns both showed neutral conditions can provide better peak separation and acidic conditions supported higher quality of chromatographic peak shapes. Considering the coverage and reproducibility, ZIC-cHILIC operated under an acidic condition with 5 mM AF and 0.1% FAcid was chosen as the final method. The performance of the optimized method with complex biological extracts proved that it still has good reproducibility and coverage for glycation products in the presence of endogenous metabolites from plasma, urine, and feces. Overall, the proposed method can be used to discover potential glycation markers associated with health and disease.

Supplementary Materials: The following supporting information can be downloaded at: https://www.mdpi.com/article/10.3390/metabo12121179/s1, additional experimental details, data processing parameters, peak shape scores, and detected free glycation products (Tables S1–S9); the precision of feature intensities analyzed by different mobile phases, loading plots of principal component analysis, and representative EICs of amino acids and putative AGEs (Figures S1–S3).

Author Contributions: Conceptualization, Y.Y., P.S.-K. and D.H.; methodology, Y.Y. and D.H.; software, Y.Y.; validation, Y.Y. and D.H.; formal analysis, Y.Y.; investigation, Y.Y.; resources, P.S.-K.; data curation, Y.Y.; writing—original draft preparation, Y.Y.; writing—review and editing, Y.Y., P.S.-K. and D.H.; visualization, Y.Y.; supervision, P.S.-K. and D.H.; project administration, P.S.-K. and D.H.; funding acquisition, Y.Y., P.S.-K. and D.H. All authors have read and agreed to the published version of the manuscript.

Funding: This research received no external funding.

Institutional Review Board Statement: The study was conducted in accordance with the Declaration of Helsinki. Ethical review was waived for this study, due to the fact that urine and feces collection is non-invasive.

Informed Consent Statement: Informed consent was obtained from the subject involved in the study.

Data Availability Statement: The data presented in this study are available in the article and supplementary materials.

Acknowledgments: The authors are thankful to China Scholarship Council (CSC) for the financial support of Yingfei Yan.

Conflicts of Interest: The authors declare no conflict of interest.

References

1. Zeng, C.; Li, Y.; Ma, J.; Niu, L.; Tay, F.R. Clinical/Translational Aspects of Advanced Glycation End-Products. *Trends Endocrinol. Metab.* **2019**, *30*, 959–973. [CrossRef] [PubMed]
2. Hellwig, M.; Henle, T. Baking, ageing, diabetes: A short history of the Maillard reaction. *Angew. Chem. Int. Ed.* **2014**, *53*, 10316–10329. [CrossRef] [PubMed]
3. Hodge, J.E. Dehydrated Foods, Chemistry of Browning Reactions in Model Systems. *J. Agric. Food Chem.* **1953**, *1*, 928–943. [CrossRef]
4. Beisswenger, P.J.; Howell, S.K.; Russell, G.B.; Miller, M.E.; Rich, S.S.; Mauer, M. Early progression of diabetic nephropathy correlates with methylglyoxal-derived advanced glycation end products. *Diabetes Care* **2013**, *36*, 3234–3239. [CrossRef] [PubMed]
5. Koska, J.; Saremi, A.; Howell, S.; Bahn, G.; De Courten, B.; Ginsberg, H.; Beisswenger, P.J.; Reaven, P.D.; Investigators, V. Advanced Glycation End Products, Oxidation Products, and Incident Cardiovascular Events in Patients with Type 2 Diabetes. *Diabetes Care* **2018**, *41*, 570–576. [CrossRef] [PubMed]

6. van Outersterp, R.E.; Moons, S.J.; Engelke, U.F.H.; Bentlage, H.; Peters, T.M.A.; van Rooij, A.; Huigen, M.; de Boer, S.; van der Heeft, E.; Kluijtmans, L.A.J.; et al. Amadori rearrangement products as potential biomarkers for inborn errors of amino-acid metabolism. *Commun. Biol.* **2021**, *4*, 367. [CrossRef] [PubMed]
7. Heremans, I.P.; Caligiore, F.; Gerin, I.; Bury, M.; Lutz, M.; Graff, J.; Stroobant, V.; Vertommen, D.; Teleman, A.A.; Van Schaftingen, E.; et al. Parkinson's disease protein PARK7 prevents metabolite and protein damage caused by a glycolytic metabolite. *Proc. Natl. Acad. Sci. USA.* **2022**, *119*, e2111338119. [CrossRef]
8. Poojary, M.M.; Zhang, W.; Greco, I.; De Gobba, C.; Olsen, K.; Lund, M.N. Liquid chromatography quadrupole-Orbitrap mass spectrometry for the simultaneous analysis of advanced glycation end products and protein-derived cross-links in food and biological matrices. *J. Chromatogr. A* **2020**, *1615*, 460767. [CrossRef] [PubMed]
9. Hohmann, C.; Liehr, K.; Henning, C.; Fiedler, R.; Girndt, M.; Gebert, M.; Hulko, M.; Storr, M.; Glomb, M.A. Detection of Free Advanced Glycation End Products In Vivo during Hemodialysis. *J. Agric. Food Chem.* **2017**, *65*, 930–937. [CrossRef]
10. Sillner, N.; Walker, A.; Hemmler, D.; Bazanella, M.; Heinzmann, S.S.; Haller, D.; Schmitt-Kopplin, P. Milk-Derived Amadori Products in Feces of Formula-Fed Infants. *J. Agric. Food Chem.* **2019**, *67*, 8061–8069. [CrossRef]
11. Lin, J.A.; Wu, C.H.; Yen, G.C. Perspective of Advanced Glycation End Products on Human Health. *J. Agric. Food Chem.* **2018**, *66*, 2065–2070. [CrossRef]
12. Nursten, H.E. *The Maillard Reaction: Chemistry, Biochemistry, and Implications*; Royal Society of Chemistry: London, UK, 2005.
13. Henning, C.; Glomb, M.A. Pathways of the Maillard reaction under physiological conditions. *Glycoconj. J.* **2016**, *33*, 499–512. [CrossRef] [PubMed]
14. Liang, Z.; Chen, X.; Li, L.; Li, B.; Yang, Z. The fate of dietary advanced glycation end products in the body: From oral intake to excretion. *Crit. Rev. Food Sci. Nutr.* **2020**, *60*, 3475–3491. [CrossRef] [PubMed]
15. Hemmler, D.; Roullier-Gall, C.; Marshall, J.W.; Rychlik, M.; Taylor, A.J.; Schmitt-Kopplin, P. Evolution of Complex Maillard Chemical Reactions, Resolved in Time. *Sci. Rep.* **2017**, *7*, 3227. [CrossRef]
16. Hemmler, D.; Roullier-Gall, C.; Marshall, J.W.; Rychlik, M.; Taylor, A.J.; Schmitt-Kopplin, P. Insights into the Chemistry of Non-Enzymatic Browning Reactions in Different Ribose-Amino Acid Model Systems. *Sci. Rep.* **2018**, *8*, 16879. [CrossRef] [PubMed]
17. Guilbaud, A.; Howsam, M.; Niquet-Leridon, C.; Delguste, F.; Boulanger, E.; Tessier, F.J. The LepR(db/db) mice model for studying glycation in the context of diabetes. *Diabetes/Metab. Res. Rev.* **2019**, *35*, e3103. [CrossRef] [PubMed]
18. Zhang, H.; Troise, A.D.; Zhang, H.; Fogliano, V. Cocoa melanoidins reduce the formation of dietary advanced glycation end-products in dairy mimicking system. *Food Chem.* **2021**, *345*, 128827. [CrossRef]
19. Sun, X.; Tang, J.; Wang, J.; Rasco, B.A.; Lai, K.; Huang, Y. Formation of advanced glycation endproducts in ground beef under pasteurisation conditions. *Food Chem.* **2015**, *172*, 802–807. [CrossRef]
20. Akillioglu, H.G.; Lund, M.N. Quantification of advanced glycation end products and amino acid cross-links in foods by high-resolution mass spectrometry: Applicability of acid hydrolysis. *Food Chem.* **2022**, *366*, 130601. [CrossRef] [PubMed]
21. Sillner, N.; Walker, A.; Harrieder, E.M.; Schmitt-Kopplin, P.; Witting, M. Development and application of a HILIC UHPLC-MS method for polar fecal metabolome profiling. *J. Chromatogr. B Anal. Technol. Biomed. Life Sci.* **2019**, *1109*, 142–148. [CrossRef] [PubMed]
22. Smith, C.A.; Want, E.J.; O'Maille, G.; Abagyan, R.; Siuzdak, G. XCMS: Processing mass spectrometry data for metabolite profiling using Nonlinear peak alignment, matching, and identification. *Anal. Chem.* **2006**, *78*, 779–787. [CrossRef] [PubMed]
23. Kuhl, C.; Tautenhahn, R.; Bottcher, C.; Larson, T.R.; Neumann, S. CAMERA: An integrated strategy for compound spectra extraction and annotation of liquid chromatography/mass spectrometry data sets. *Anal. Chem.* **2012**, *84*, 283–289. [CrossRef] [PubMed]
24. Guo, J.; Shen, S.; Xing, S.; Yu, H.; Huan, T. ISFrag: De Novo Recognition of In-Source Fragments for Liquid Chromatography-Mass Spectrometry Data. *Anal. Chem.* **2021**, *93*, 10243–10250. [CrossRef]
25. Le, S.; Josse, J.; Husson, F. FactoMineR: An R package for multivariate analysis. *J. Stat. Softw.* **2008**, *25*, 1–18. [CrossRef]
26. Lommen, A. MetAlign: Interface-Driven, Versatile Metabolomics Tool for Hyphenated Full-Scan Mass Spectrometry Data Preprocessing. *Anal. Chem.* **2009**, *81*, 3079–3086. [CrossRef] [PubMed]
27. Fraisier-Vannier, O.; Chervin, J.; Cabanac, G.; Puech, V.; Fournier, S.; Durand, V.; Amiel, A.; Andre, O.; Benamar, O.A.; Dumas, B.; et al. MS-CleanR: A Feature-Filtering Workflow for Untargeted LC-MS Based Metabolomics. *Anal. Chem.* **2020**, *92*, 9971–9981. [CrossRef] [PubMed]
28. Li, Z.; Lu, Y.; Guo, Y.; Cao, H.; Wang, Q.; Shui, W. Comprehensive evaluation of untargeted metabolomics data processing software in feature detection, quantification and discriminating marker selection. *Anal. Chim. Acta* **2018**, *1029*, 50–57. [CrossRef] [PubMed]
29. Thornalley, P.J. Glycation free adduct accumulation in renal disease: The new AGE. *Pediatr. Nephrol.* **2005**, *20*, 1515–1522. [CrossRef] [PubMed]
30. Zhang, R.; Watson, D.G.; Wang, L.; Westrop, G.D.; Coombs, G.H.; Zhang, T. Evaluation of mobile phase characteristics on three zwitterionic columns in hydrophilic interaction liquid chromatography mode for liquid chromatography-high resolution mass spectrometry based untargeted metabolite profiling of Leishmania parasites. *J. Chromatogr. A* **2014**, *1362*, 168–179. [CrossRef] [PubMed]

31. Iturrospe, E.; Da Silva, K.M.; Talavera Andujar, B.; Cuykx, M.; Boeckmans, J.; Vanhaecke, T.; Covaci, A.; van Nuijs, A.L.N. An exploratory approach for an oriented development of an untargeted hydrophilic interaction liquid chromatography-mass spectrometry platform for polar metabolites in biological matrices. *J. Chromatogr. A* **2021**, *1637*, 461807. [CrossRef] [PubMed]
32. Xing, H.; Yaylayan, V. Insight into Isomeric Diversity of Glycated Amino Acids in Maillard Reaction Mixtures. *Int. J. Mol. Sci.* **2022**, *23*, 3430. [CrossRef] [PubMed]
33. Sonnenberg, R.A.; Naz, S.; Cougnaud, L.; Vuckovic, D. Comparison of underivatized silica and zwitterionic sulfobetaine hydrophilic interaction liquid chromatography stationary phases for global metabolomics of human plasma. *J. Chromatogr. A* **2019**, *1608*, 460419. [CrossRef]
34. Wei, J.; Shen, A.; Wan, H.; Yan, J.; Yang, B.; Guo, Z.; Zhang, F.; Liang, X. Highly selective separation of aminoglycoside antibiotics on a zwitterionic Click TE-Cys column. *J. Sep. Sci.* **2014**, *37*, 1781–1787. [CrossRef] [PubMed]
35. Contrepois, K.; Jiang, L.; Snyder, M. Optimized Analytical Procedures for the Untargeted Metabolomic Profiling of Human Urine and Plasma by Combining Hydrophilic Interaction (HILIC) and Reverse-Phase Liquid Chromatography (RPLC)-Mass Spectrometry. *Mol. Cell. Proteom.* **2015**, *14*, 1684–1695. [CrossRef]
36. Hosseinkhani, F.; Huang, L.; Dubbelman, A.C.; Guled, F.; Harms, A.C.; Hankemeier, T. Systematic Evaluation of HILIC Stationary Phases for Global Metabolomics of Human Plasma. *Metabolites* **2022**, *12*, 165. [CrossRef] [PubMed]
37. Iverson, C.D.; Gu, X.; Lucy, C.A. The hydrophilicity vs. ion interaction selectivity plot revisited: The effect of mobile phase pH and buffer concentration on hydrophilic interaction liquid chromatography selectivity behavior. *J. Chromatogr. A* **2016**, *1458*, 82–89. [CrossRef] [PubMed]
38. Fu, X.; Cebo, M.; Ikegami, T.; Lammerhofer, M. Separation of carbohydrate isomers and anomers on poly-N-(1H-tetrazole-5-yl)-methacrylamide-bonded stationary phase by hydrophilic interaction chromatography as well as determination of anomer interconversion energy barriers. *J. Chromatogr. A* **2020**, *1620*, 460981. [CrossRef] [PubMed]
39. Erngren, I.; Nestor, M.; Pettersson, C.; Hedeland, M. Improved Sensitivity in Hydrophilic Interaction Liquid Chromatography-Electrospray-Mass Spectrometry after Removal of Sodium and Potassium Ions from Biological Samples. *Metabolites* **2021**, *11*, 170. [CrossRef] [PubMed]
40. Ekdahl, A.; Johansson, M.C.; Ahnoff, M. Tracing and separating plasma components causing matrix effects in hydrophilic interaction chromatography-electrospray ionization mass spectrometry. *J. Chromatogr. B Anal. Technol. Biomed. Life Sci.* **2013**, *923–924*, 83–91. [CrossRef] [PubMed]
41. De Paepe, E.; Van Meulebroek, L.; Rombouts, C.; Huysman, S.; Verplanken, K.; Lapauw, B.; Wauters, J.; Hemeryck, L.Y.; Vanhaecke, L. A validated multi-matrix platform for metabolomic fingerprinting of human urine, feces and plasma using ultra-high performance liquid-chromatography coupled to hybrid orbitrap high-resolution mass spectrometry. *Anal. Chim. Acta* **2018**, *1033*, 108–118. [CrossRef] [PubMed]
42. Roca, M.; Alcoriza, M.I.; Garcia-Canaveras, J.C.; Lahoz, A. Reviewing the metabolome coverage provided by LC-MS: Focus on sample preparation and chromatography-A tutorial. *Anal. Chim. Acta* **2021**, *1147*, 38–55. [CrossRef] [PubMed]
43. Henle, T. Dietary advanced glycation end products–a risk to human health? A call for an interdisciplinary debate. *Mol. Nutr. Food Res.* **2007**, *51*, 1075–1078. [CrossRef]
44. Birlouez-Aragon, I.; Saavedra, G.; Tessier, F.J.; Galinier, A.; Ait-Ameur, L.; Lacoste, F.; Niamba, C.N.; Alt, N.; Somoza, V.; Lecerf, J.M. A diet based on high-heat-treated foods promotes risk factors for diabetes mellitus and cardiovascular diseases. *Am. J. Clin. Nutr.* **2010**, *91*, 1220–1226. [CrossRef]

metabolites

MDPI

Article

Combining Feature-Based Molecular Networking and Contextual Mass Spectral Libraries to Decipher Nutrimetabolomics Profiles

Lapo Renai [1,2,*], Marynka Ulaszewska [3,*], Fulvio Mattivi [3,4], Riccardo Bartoletti [5], Massimo Del Bubba [1] and Justin J. J. van der Hooft [2,6,*]

1 Department of Chemistry, University of Florence, Via della Lastruccia 3, Sesto Fiorentino, 50019 Florence, Italy
2 Bioinformatics Group, Wageningen University, 6708 PB Wageningen, The Netherlands
3 Metabolomics Unit, Department of Food Quality and Nutrition, Research and Innovation Centre, Fondazione Edmund Mach (FEM), Via Mach 1, San Michele all'Adige, 38098 Trento, Italy
4 Department of Cellular, Computational, and Integrative Biology (CIBIO), University of Trento, Via Mach 1, San Michele all'Adige, 38098 Trento, Italy
5 Department of Translational Research and New Technologies, University of Pisa, Via Risorgimento 36, 56126 Pisa, Italy
6 Department of Biochemistry, University of Johannesburg, Auckland Park, Johannesburg 2006, South Africa
* Correspondence: lapo.renai@unifi.it (L.R.); maria.ulaszewska@gmail.com (M.U.); justin.vanderhooft@wur.nl (J.J.J.v.d.H.)

Abstract: Untargeted metabolomics approaches deal with complex data hindering structural information for the comprehensive analysis of unknown metabolite features. We investigated the metabolite discovery capacity and the possible extension of the annotation coverage of the Feature-Based Molecular Networking (FBMN) approach by adding two novel nutritionally-relevant (contextual) mass spectral libraries to the existing public ones, as compared to widely-used open-source annotation protocols. Two contextual mass spectral libraries in positive and negative ionization mode of ~300 reference molecules relevant for plant-based nutrikinetic studies were created and made publicly available through the GNPS platform. The postprandial urinary metabolome analysis within the intervention of *Vaccinium* supplements was selected as a case study. Following the FBMN approach in combination with the added contextual mass spectral libraries, 67 berry-related and human endogenous metabolites were annotated, achieving a structural annotation coverage comparable to or higher than existing non-commercial annotation workflows. To further exploit the quantitative data obtained within the FBMN environment, the postprandial behavior of the annotated metabolites was analyzed with Pearson product-moment correlation. This simple chemometric tool linked several molecular families with phase II and phase I metabolism. The proposed approach is a powerful strategy to employ in longitudinal studies since it reduces the unknown chemical space by boosting the annotation power to characterize biochemically relevant metabolites in human biofluids.

Keywords: human urine; liquid chromatography; untargeted mass spectrometry; computational metabolomics; chemometrics; bioinformatics

Citation: Renai, L.; Ulaszewska, M.; Mattivi, F.; Bartoletti, R.; Del Bubba, M.; van der Hooft, J.J.J. Combining Feature-Based Molecular Networking and Contextual Mass Spectral Libraries to Decipher Nutrimetabolomics Profiles. *Metabolites* **2022**, *12*, 1005. https://doi.org/10.3390/metabo12101005

Academic Editor: Joana Pinto

Received: 3 October 2022
Accepted: 18 October 2022
Published: 21 October 2022

1. Introduction

Untargeted tandem mass spectrometry (MS/MS) is one of the most widely used analytical techniques in metabolomics, allowing for the generation of information-rich mass spectral datasets and the identification of metabolic biomarkers in biological complex mixtures [1,2], also thanks to the coupling with separation techniques such as liquid chromatography (LC). Despite the wide application of hyphenated LC-MS/MS platforms, the annotation of biologically relevant metabolites (i.e., biomarkers) is strongly hampered by the complexity of the metabolome and metabolomics data processing and annotation [3]. The annotation process is a pivotal step in untargeted metabolomics that often represents a bottleneck in the process of obtaining biological information and discovering biomarkers. To streamline the metabolite annotation process, metabolomics guidelines have been proposed for the accurate identification and assignment of a metabolite feature [4,5], i.e., through peak picking, mass spectral deconvolution, determination of molecular ions by adduct detection, and fragmentation pattern (MS/MS) analysis [6]. Despite these efforts, the risk of missing relevant information and drawing incorrect conclusions remains relatively high, due to incorrect MS and MS/MS interpretations when matching experimental spectra with available spectral libraries. To aid in structural interpretation, the identification of MS/MS spectral similarities within a given dataset can support the discovery of structurally related metabolites, which plausibly share the same metabolic pathway and/or substructure [7], thus strengthening the biological meaning of the annotations.

In this context, molecular networking (MN) has gained large attention, thanks to the efficient and rapid identification of several molecular families within complex mixtures, providing a visual overview of all the precursor ions, grouped according to their structural relationships, as deduced by their mass fragmentation spectra during an MS/MS experiment [8]. MN uses an unsupervised vector-based computational algorithm to organize molecular ions (i.e., clusters or nodes) into a network of molecular families that share spectral similarities among their MS/MS spectra. At the same time, structural annotation is performed through the Global Natural Products Social Molecular Networking (GNPS) bioinformatics platform [8], which is linked to many mass spectral libraries available as public repository of mass spectra and metadata (i.e., GNPS-MassIVE). Considering the recent growth of public mass spectral libraries, it is expected an increase of the annotation capability (level II or level III) of biologically relevant molecules in comparison with traditional biomarker discovery workflows [9]. MN has been applied in several untargeted LC-MS/MS studies, mainly focusing on phytochemical composition analysis [8], and less frequently on drug metabolism [10], and nutrimetabolomics [11] in human biofluids.

Recently, MN has been extended by its combination with standard feature detection tools into the Feature-Based Molecular Networking (FBMN) workflow that is capable to resolve isomers and incorporate quantitative information (e.g., spectral counts, chromatographic peak areas, etc.), increasing the link between peak picking algorithms and in silico annotation tools [12]. Until now, FBMN has been successfully applied in various fields of metabolomics, allowing level II/level III identification of transformation products of organic micropollutants in water samples [13], native plant constituents [14–16], and endogenous urinary metabolites [17]. However, mass spectral library matching is generally performed by the comparison with mass spectral libraries containing MS/MS spectra acquired under a wide range of instrumental conditions (e.g., time-of-flight, orbitrap, hybrid ion traps, etc.) and collision energies used, with different curation protocols providing different mass accuracy levels [13,16], thus suffering from limited reliability of the annotation due to differences in observed mass fragments and their intensity ratios. This issue can be managed by implementing better contextualized libraries containing reference spectra of study-related compounds and acquired under experimental conditions equal to or comparable to the experimental data being analyzed. Finally, FBMN has the hitherto unexploited potential in biomarker research to provide quantitative data of the structurally annotated (and unannotated) features, thus complementing the traditional biomarker discovery procedure with a chemometric protocol that allows establishing their biological significance.

This research investigates the discovery capacity and the extension of the annotation coverage of the FBMN approach, in comparison with a commonly adopted manual annotation of selected significant m/z features [18]. To this end, the FBMN workflow was applied to deconvoluted and aligned high-resolution LC-MS/MS files of postprandial urine samples from a two-arms intervention study on the intake of *Vaccinium myrtillus* (VM) and *Vaccinium corymbosum* (VC) berry supplements. As far as we are aware, this represents the first nutrimetabolomics application of FBMN to the identification of postprandial endogenous and exogenous metabolites. The MS/MS spectra that were acquired in both negative ionization (NI) and positive ionization (PI) data dependent acquisition modes were compared with the available GNPS libraries. An extensive comparative analysis was done to compare various FBMN parameter settings to arrive at optimal settings for structural annotation purposes in the nutrimetabolomics setting. Furthermore, to support automated nutrimetabolomics annotation workflows, two novel NI and PI contextualized "Nutri-Metabolomics" mass spectral libraries were constructed and made available uniquely on GNPS, each containing MS/MS spectra of about three-hundred food-related human metabolites, acquired under the same mass spectrometric conditions as the study samples. These mass spectral libraries are a fruit of several years of investigations on human responses to dietary interventions at the Edmund Mach Foundation (Italy), and include phase I and phase II human metabolites, as well as food constituents. Special attention was given to microbial metabolites resulting from mixed human and microbiome interaction such as small phenolic acids, phenylacetic acids, phenylpropionic acids, indoles, and carbolines, as well as bile acids. Other classes include sulfate and glucuronides conjugates of common food constituents such as caffeic acid glucuronide, dihydroferulic acid sulfate, isoferulic glucuronide, etc. Several aroma compounds were included to facilitate substructure matching, as those were observed in biological fluids in conjugated form (monoterpenoids, safranal, furfuran, fenchyl alcohol etc.). Finally, the spectral library offers specific advanced glycation end-products including pyrraline, furosine and more.

In the current study, the mass spectral library creation aimed at increasing (i) the accuracy in the annotation thanks to a better match of instrumental metadata such as detector and collision energy, (ii) nutrimetabolomics knowledge on postprandial analysis of biological samples and plant-based food intake. Additionally, the quantitative data within NI and PI FBMN networks were exploited to gain insights into (i) metabolites characterized by different postprandial kinetics and (ii) the relative dietary contribution of VM and VC interventions of the identified metabolites.

2. Materials and Methods

2.1. Chemicals and Reagents

Full purchase details of solvents and standards used are reported in Section S1 of the Supplementary Materials. The complete list of the reference standard adopted to build the "Nutri-Metabolomics" libraries, in both NI and PI modes, is shown in the List of Reference Standards used to build the libraries.xlsx file in the Supplementary Materials.

2.2. Study Design, Sample Extraction, and LC-MS/MS Analysis

The datasets analyzed in this research are part of a more comprehensive clinical intervention trial, based on the hematic and urine biomarker discovery on the intake of VM and VC [18,19]. Urine samples of each volunteer (n = 10 for each intervention) were collected at baseline and 30, 60, 120, 240, and 360 min after VM or VC supplement intake. Pooled urine samples were also collected 24 h and 48 h after supplement intake. Details of supplements characterization (Table S1), as well as study design are reported in Section S2 of the Supplementary Materials. Urine samples were extracted and analyzed as reported elsewhere [19]. The entire procedures of extraction and LC-MS/MS analysis of urine samples are reported in Sections S3 and S4 of the Supplementary Materials, respectively. The entire sample set was acquired in full scan mode, collecting high quality data for an appropriate statistical analysis, as well as in data dependent acquisition (DDA) mode, to leverage large quantities of MS/MS data for structural investigation preserving the kinetic heritage of the study design.

2.3. Data Pre-Processing

The full scan files were processed and analyzed as previously reported [18]. Additionally, the data-dependent spectra files (including blanks) were converted from the *.raw* to *.mzML* MS convert by ProteoWizard (https://proteowizard.sourceforge.io) (accessed on 15 April 2021). Further data processing was performed with MZmine 2 software [20], separately for NI and PI datasets. Data pre-processing included the following steps: mass detection, chromatogram reconstruction and deconvolution, isotope grouping, alignment and gap filling. Subsequently, the aligned feature lists were exported as MS/MS files (.mgf format) and quantification tables (.csv format of aligned features and related chromatographic peak areas), according to GNPS documentation on FBMN (https://ccms-ucsd.github.io/GNPSDocumentation/) (accessed on 21 April 2021).

2.4. Data Availability: MassIVE Repository, Metadata and GNPS Jobs

Data in *.mzML* format are available on-line on GNPS infrastructure (MSV000088336). The metadata describing file/sample properties were entered manually for all samples and organized in two different files according to the acquisition polarity of the uploaded MassIVE datasets, following the GNPS guidelines (https://ccms-ucsd.github.io/GNPS Documentation/metadata/) (accessed on 21 April 2021). In detail, metadata consisted of three descriptive categories, (i) spectrum file name (the same of acquired raw data), (ii) type of supplement (VM or VC), and (iii) related time point after intake. These elements are required to get a correct grouping within FBMN for quantitative analysis (see the Metadata and Library Information.xlsx file in the Supplementary Materials). For the upload on GNPS, metadata files were converted to *.tsv* files. The FBMN analysis are available at the following links:

- PI: https://gnps.ucsd.edu/ProteoSAFe/status.jsp?task=a981ebd40809453ebe1524ff1fc8e265 (accessed on 27 June 2021).
- NI: https://gnps.ucsd.edu/ProteoSAFe/status.jsp?task=0a239e71bb2045c292c4c96f4501249c (accessed on 27 June 2021).

2.5. "Nutri-Metabolomics" Library Building and Implementation

The analytical standards used to build the in-house libraries were acquired in the same MS/MS conditions as study samples (replicated three times), which are reported in Section S4 of the Supplementary Materials. GNPS provides a platform to build MS/MS spectral libraries, requiring good quality MS/MS spectra and annotation spread sheets containing key and machine-readable descriptors such as file name, compound name, SMILES, InChiKey, PubMed. To build the library, only pure analytical standards were used, thus no putative or un-known compounds are present in the files. Two annotation spread sheets were built in NI and PI, containing 319 and 339 injected compounds, respectively (see the Metadata and Library Information.xlsx file in the Supplementary Materials). Analysis of standards included their separation on the chromatographic column; however, a retention time match is not supported in GNPS and therefore this information was used manually when needed. The "Batch Validator Workflow" [21] step was run to evaluate the correct match between spreadsheets (dropped as .csv files), and original spectra. The completed libraries can be found in the public spectral library collection of GNPS named as "Nutri-Metabolomics".

2.6. Molecular Networking Analyses

Molecular networks were obtained following the online workflow on the GNPS web-platform (https://gnps.ucsd.edu/) (accessed on 21 April 2021). FBMN was performed adopting the most suitable basic and advanced networking options, selected through the recommended network qualitative optimization by classical MN (see Section 3.1), for NI and PI dataset exported from MZmine 2 software. The detailed investigation of MN options is reported in Section S5 of the Supplementary Materials. The most appropriate input parameters were set as follows: NI were analyzed using precursor ion mass tolerance

(PIMT) and fragment ion mass tolerance (FIMT) equal to 0.1 Da and 0.01 Da, respectively. The other parameters were set as follows: minimum matched fragment ions = 3, networking cosine score > 0.6, library cosine score > 0.5, and minimum library shared peaks = 3. PI dataset was processed adopting PIMT = 0.05, FIMT = 0.05, minimum matched fragment ions = 3, networking cosine score > 0.5, library cosine score > 0.3, and minimum library shared peaks = 3. Network analysis and quantitative results were investigated and exported adopting Cytoscape environment [22]. Moreover, unknown nodes were annotated with putative molecular structures by manual annotation based on: (i) mass difference between identified and unknown node, (ii) precursor ion mass accuracy, and (iii) fragmentation patterns in MS/MS spectra (see Section S6 of Supplementary Materials).

2.7. Analysis of Postprandial Kinetics

Reinjection of the entire dataset in DDA fashion enabled the exploitation of postprandial kinetics data. To extract the postprandial information from FBMN, the Pearson product-moment correlation (PPMC) analysis was performed, using the "corrplot: A visualization of a correlation matrix" package implemented in R (https://cran.r-project.org/) (accessed on 28 July 2021), thus estimating the linear correlation between the mean chromatographic peak area of identified nodes and time points. The quantitative FBMN data used for the correlation analysis were extracted from the "node table" of the Cytoscape environment, built using the loaded metadata for both NI and PI datasets. Statistically significant (p-value ≤ 0.05) PPMC coefficients (r) were used to discriminate early (1–2 h postprandial) from late (approximately 4 h and more postprandial) occurring postprandial metabolites, which are commonly considered as the result of phase II or phase I metabolism, respectively [23]. Accordingly, positive and negative r-values indicated nodes associated to phase I (late postprandial) and phase II (early postprandial) metabolism, respectively. A limitation of using PPMC within FBMN was the absence of sample normalization as this functionality is currently not available. Findings from this step were compared to those obtained through the PPMC analysis of longitudinal variations of the chromatographic area of aligned features (i.e., outside FBMN), as a control strategy. It should be highlighted that, although the PPMC coefficients can be associated with the metabolism phase, its relation to the specific food intake remains elusive without further biochemical interpretations. Simultaneously, full scan data underwent the conventional data processing, as previously described [17]. Briefly, biomarkers of food intake in postprandial responses were selected by applying selected R packages to full scan data [24], according to the following two-step procedure: (i) verification of increasing trend along time points and (ii) calculation of AUC curves and intra-intervention discrimination. Statistically significant features were annotated manually with use of on-line spectra databases such as mzCloud and HMDB. Details of this procedure are reported in Section S5 of Supplementary Materials.

3. Results and Discussion

The NI and PI datasets were treated following the workflow illustrated in Figure 1, which integrates the PPMC analysis of postprandial kinetics within the FBMN environment. However, since FBMN extracts only the mean values of the chromatographic area as quantitative data for PPMC analysis, thereby losing knowledge of inter-individual variability, the variance of metabolite feature abundance among volunteers was investigated at each time point as a control, before applying the FBMN workflow. Accordingly, the coefficient of variation (CV%) of chromatographic areas of each aligned feature within a same time point was calculated, highlighting a strong variability (CV% approximately in the range of 30–300% and median higher than 100% in most cases). These findings highlighted the importance of evaluating the results of PPMC analysis at the population level.

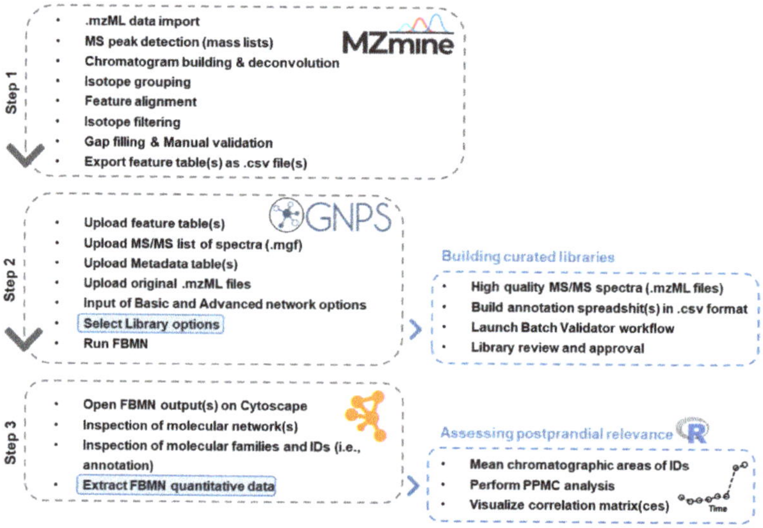

Figure 1. Schematic representation of the available (grey dashed lines) and proposed (blue dashed lines) workflows of data management, applied by the combination of Feature-Based Molecular Networking and the "Nutri-Metabolomics" mass spectral libraries that were manually curated.

3.1. Optimization of the Input Parameters for Network Analysis

Before running FBMN, various networking basic and advanced options must be investigated to find out the most suitable parameters to perform the MN analysis. To properly evaluate the effect of input parameters, the total number of nodes (precursor ions with identical fragmentation pattern, i.e., consensus spectrum), edges (i.e., node connections related to structural similarities), identified compounds (IDs, i.e., annotated through spectral library matching), and spectral families (i.e., the groups or clusters, also referred as molecular families), were analyzed in both NI and PI datasets and the results are reported in Figure S1 of the Supplementary Materials. In this regard, increasing PIMT value, the number of nodes, edges, and spectral families decreased, whereas the number of IDs showed a predominantly increasing trend, mainly due to the less strict conditions as consensus spectra got merged (i.e., considering different isobaric compounds as one) at increasing PIMT. Hence, to keep a reliable number of nodes and spectral families without significantly affecting the number of IDs, PIMT was set at 0.1 Da and 0.05 Da for NI and PI, respectively. FIMT exerted an effect on the output variables like that of PIMT, except for the total number of edges, which increased by increasing values of FIMT. Due to the loss of accuracy in node networking for high FIMT, values of 0.01 and 0.05 were chosen for NI and PI datasets, respectively. The number of minimum matched fragment ions was set at 3 for both NI and PI for two reasons: (i) Its increase exerts a significant reduction of the number of nodes with an ID and their reliability, (ii) many food-derived metabolites have only a few characteristic mass fragments. Cosine scores for networking and library matching affected mainly the number of spectral families and of IDs, respectively. A good compromise between these two outputs was obtained by setting the networking and library matching cosine score thresholds at 0.6 for NI and 0.5 for PI. Finally, the number of minimum library shared peaks was set at 3, because higher values of this parameter were responsible for a drastic reduction of IDs, similarly to what was observed for the number of matched fragments.

3.2. FBMN Annotation of NI and PI Datasets

FBMN workflow applied to NI and PI datasets combined with aligned feature lists and quantitative tables exported from data pre-processing, was able to remove the 57% and 27% of NI and PI redundant IDs (i.e., artefacts like duplicated features) found by classical MN, respectively.

As first result, the effect of including context specific "Nutri-Metabolomics" mass spectral libraries in the annotation workflow was evaluated by applying the FBMN protocol in their presence and absence (i.e., GNPS libraries "only"). Indeed, substantial advantages were observed upon using the dedicated mass spectral libraries, i.e., the increase of (i) 20%, 48%, in the number of IDs (Figure S2A) and (ii) 62.5%, 34%, in the number of IDs with a mass error < 5 ppm (Figure S2B), for NI and PI datasets, respectively. Additionally, the use of the "Nutri-Metabolomics" libraries solved two mis-annotations (i.e., incorrect annotation of nodes) in the NI datasets. These results highlight the importance of applying the FBMN annotation strategy in combination with contextual libraries, i.e., containing true reference standards that are relevant for the application of interest and analyzed under the same instrumental conditions adopted for the analysis of real samples.

The FBMN network of the NI dataset consisted of 545 nodes and 799 edges, with a total number of connected components equal to 307, corresponding to 65 spectral families, whereas molecular networking of the PI dataset resulted in 5079 nodes and 6904 edges, with a total number of connected components equal to 3543 (i.e., 663 spectral families). The ID lists obtained from the library matching in both NI and PI datasets contained 39 and 384 unique annotated compounds, respectively, which were checked for mass accuracy to be around or lower than 5 ppm. Table S2 (see Section S5 of the Supplementary Materials) reports the metabolites identified by library matching (based on cosine score similarity) of nodes within and outside molecular families (the latter are typically called singletons) from both NI (24 IDs) and PI (43 IDs) datasets, characterized by the lowest mass error (Δ ppm).

Even though the FBMN approach has specific methodological inputs and results that differentiate it from commonly used workflows in untargeted nutrimetabolomics, it is interesting to compare the discovery capacity and annotation coverage obtained with other approaches. For this purpose, the FBMN was compared against two widely used annotation protocols: (i) MZmine Library Search and (ii) statistical-based feature selection followed by manual annotation (see Section S5 of the Supplementary Materials) [18]. It should be emphasized that the compared workflows differ substantially as per their rationale. MZmine Library Search workflow matches each row of the NI and PI feature lists (used also for FBMN) against the imported spectral library. To make a consistent comparison with the annotation performed with FBMN, the "ALL-GNPS" library was used. The conventional protocol aims at selecting only statistically significant m/z features from full scan data, followed by manual annotation using the MS/MS spectra often obtained in targeted mode. In contrast with the presented approaches, FBMN explores all available MS/MS data from the DDA metabolomics profiles (taking advantage of all structural annotations that can be made), annotating them against mass spectral libraries. Only then, further statistical analysis is performed to discover their potential postprandial relevance. Thus, the direct comparison of these annotation and prioritization workflows is not and will never be straightforward; yet, here we highlight some relevant aspects.

Table 1 shows the final number of IDs found adopting the three approaches. MZmine Library Search workflow provided the metabolite annotation with 26 and 49 unique IDs in NI and PI datasets, respectively, with cosine similarity scores (isotopic pattern at full scan level) higher than 0.7. The number of IDs identified by this approach was comparable with the results of the applied FBMN workflow, and several metabolite categories were commonly annotated by the two procedures (data not shown), such as hippuric acids, catechols, and derivatives of phenylacetic acid, coumaric acid, indoles, and hydroxybenzoic acid. However, due to the format of our data unsuitable for MS/MS-based mass spectral matching within MzMine (i.e., incomplete mass lists for the MS/MS scans), the MZmine-based approach relied on precursor m/z and isotope pattern matching, thus possibly resulting in a less reliable annotations due to the limited structural information. The statistical-based/manual annotation method resulted in 50 and 106 statistically significant *m/z* features in PI and NI datasets, respectively, corresponding to 24 metabolite features after manual checking. Manual structure elucidation putatively identified 18 metabolites (12 in NI and 6 in PI datasets), while 6 metabolites remained unknown (see Table S3). Using FBMN, a higher number of metabolites was putatively annotated, i.e., 24 IDs in NI, and 43 IDs in PI, when compared to the statistical-based/manual annotation approach. These differences were due to both (i) the automatic query (intrinsic of FBMN) of all publicly available mass spectral libraries, including "Nutri-Metabolomics" ones, and (ii) the different strategies to select the metabolite features to be annotated. In fact, the conventional approach processes the NI and PI datasets to highlight physiologically relevant features, before their annotation is performed by unqueried matching with analytical standards available in on-line spectral libraries. On the contrary, FBMN automatically generates a list of IDs, which is then refined by applying, for example, a mass accuracy threshold, in combination with the use of mass spectral similarity scoring (i.e., modified cosine score), as presented in this study. Despite these methodological differences, hydroxyhippuric acid and dihydrocaffeic acid glucuronide were identified with both approaches. Moreover, the conventional postprandial analysis confirmed the FBMN identification of structurally-related metabolites significantly altered upon berry intake, belonging to furoic and abscisic acid derivatives, hydroxy and/or methoxy benzoic acids. By contrast with FBMN, the conventional protocol for postprandial analysis identified the metabolite categories of valerolactone and valeric acid derivatives (see Section S5 and Figure S3 of the Supplementary Materials), which are well-known colon-derived catabolites of flavanols [25]. These metabolite features were found also inside the FBMN molecular networks; however, they were not structurally characterized as such, due to their absence in the "Nutri-Metabolomics" and other mass spectral libraries. These findings highlighted the importance of expanding the coverage of online spectral repositories to boost metabolite annotations.

Table 1. Number of IDs annotated by Feature-Based Molecular Networking (FBMN) of NI and PI datasets, including the developed "Nutri-Metabolomics" mass spectral libraries, in comparison with the annotation performed with (i) MZmine Library Search using GNPS compatible mass spectral libraries (ALL_GNPS, https://gnps-external.ucsd.edu/gnpslibrary) (accessed on 5 September 2022) and with (ii) the statistical-based approach followed by manual annotation, reported in Section S5 of the Supplementary materials.

Number of IDs	NI	PI
MZmine [1]	26	49
Statistical-based approach & manual annotation	12	6
FBMN	24	43

[1] Library search performed at full scan MS level using *m/z* and isotope pattern matching.

3.3. VM and VC Relative Contributions to the Postprandial Metabolome

Categorization of NI and PI metadata based on VM and VC interventions (see Section 2.4 for details) allowed for the separate storage of spectral counts (i.e., the number of mass spectra recorded for a node) of each ID precursor ion. This information was used here for assessing the VM and VC relative contributions of each ID to the postprandial metabolome, by the representation of a pie chart (see Figures 3 and 4 in Section 3.5). Moreover, a preliminary and descriptive contribution to the annotated urinary metabolome of VM and VC interventions can be estimated. Interestingly, VM and VC interventions exhibited an opposite feature occurrence in the two ionization datasets, highlighting the importance of investigating both polarity modes. In detail, NI IDs resulted in a higher postprandial occurrence after the intake of VC supplement ($62 \pm 6\%$ vs. $38 \pm 4\%$ for VC and VM), whereas for PI dataset, a slight predominance was found for VM ($54 \pm 2\%$ vs. $46 \pm 2\%$ for VM and VC).

3.4. PPMC Analysis of Postprandial Kinetics

Longitudinal data analyzed by FBMN approach allows for additional data exploration to highlight the specificity of food intake as well as the "background" metabolism, since no feature selection is performed. Accordingly, the PPMC analysis was performed on the mean values of chromatographic area of each ID as a function of time. Following this analysis, 65.7% of the annotated metabolites (i.e., 44 IDs on a total of 67) showed a statistically significant trend approximating an increasing or decreasing postprandial response, thus highlighting the reliability of this approach. Among the significant correlated metabolites, 35 IDs showed a positive coefficient (r-values) and were therefore associated to phase I metabolism, whilst 9 IDs were characterized by negative r-values, suggesting a phase II metabolism. Figure 2A,B illustrates two representative postprandial trends of IDs corresponding to significant negative and positive r-values, respectively.

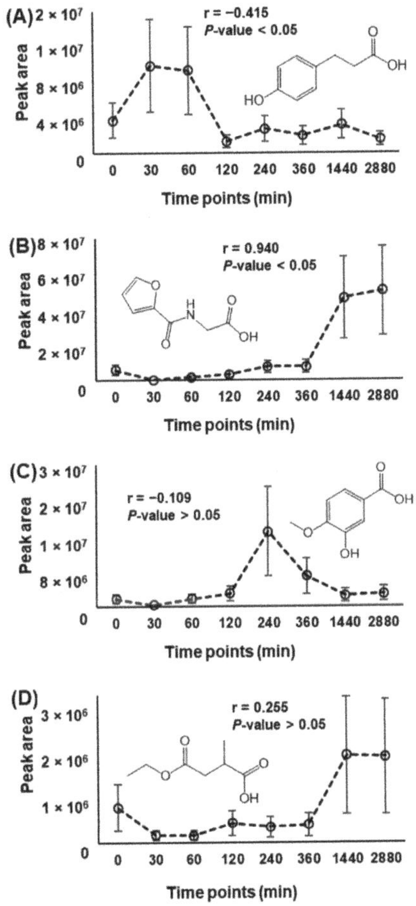

Figure 2. Representative examples of the use of PPMC coefficients (r) to evaluate phase I and II metabolisms of annotated metabolites (Table S2), as described in Section 2.7 of the main text. (**A**) Significant and negative r-value indicating a phase II metabolism. (**B**) Significant and positive r-value indicating a phase I metabolism. (**C**) Non-significant and negative r-value. (**D**) Non-significant and positive r-value.

The remaining 23 metabolites exhibited a non-linear and not significant trend, as shown in the two representative examples of Figure 2C,D. The postprandial behavior of these IDs cannot therefore be assigned through this approach and requires a qualitative investigation (plots of chromatographic areas vs timepoints) and/or a dedicated treatment outside the FBMN environment.

As stated above, these results were based on the correlation analysis of mean values of chromatographic areas, i.e., without considering the dispersion of individual data around the mean. To evaluate the impact of the extent of this dispersion on the statistical significance of linear correlations, the data obtained for each volunteer and for each annotated feature were submitted to PPMC analysis outside the FBMN environment (i.e., using the data from MZmine feature lists). For 25 IDs out of the 44 IDs found to be significant based on mean chromatographic areas, the statistical significance of the r-values was confirmed, notwithstanding the high variability observed in the peak area datasets. These results encourage the applicability of the postprandial analysis proposed here at least as a first immediate screening of the postprandial behaviour of annotated metabolites, capturing their metabolic trends over time. It is of note that this approach would produce more accurate results when a lower dispersion of individual data around the average value is observed; and to achieve this, increasing the sample size may be of help.

3.5. Nutrimetabolomics Outcomes from FBMN Molecular Networks

Figure S4 of the Supplementary Materials shows representative examples of the structural modification involved in phase I and II metabolism of well-known VM and VC native constituents [26], in association with the annotated metabolites and their significant PPMC r-values.

Accordingly, potential metabolic modifications such as conjugations (e.g., glucuronidation) and additions (e.g., methylation) should be expected to undergo in-source hydrolysis and dissociation, leading to accurate annotations, but losing a relevant structural information. To limit these drawbacks, a robust network inspection was performed to ensure a reliable annotation.

Within NI dataset (24 IDs, see Table S2), four singletons were identified through spectral matching with a good mass accuracy: azelaic acid, galacturonic acid, glutamine, and ethoxy-oxobutenoic acid. The occurrence of galacturonic acid and glutamine can be addressed to in-source dissociation of glycosidic and peptidic bonds of metabolite conjugations. Figure 3 illustrates the molecular families in which at least one of the remaining 20 metabolites was annotated. These metabolites were grouped according to their postprandial kinetics, as assessed by statistically significant r-values. In Figure 3, the structure of unknown nodes labelled with a gear was proposed as level III identification by the analysis of their MS/MS spectra the hypothesized scheme of fragmentation (Figure S5 of the Supplementary Materials).

About the 50% of the identified structures was characterized by molecular scaffolds related to cinnamic and dihydrocinnamic acids. Interestingly, among unknown nodes, a relevant number of putative glucuronide derivatives was easily recognized by the occurrence in the MS/MS spectra of peaks at m/z 175.02 and 113.02, typical of glucuronic acid (Figure S5).

Two nodes highlighted in one box in Figure 3 were recognized as a molecular family related to abscisic acid glucuronide derivatives. The ID occurring in this family, was at first addressed as dihydroxy-diphenylphenoxy-trihydroxyoxane-carboxylic acid, with a mass error of about 128 ppm. However, the inspection of its MS/MS spectra (see Figure S6 of the Supplementary Materials) led to a more accurate putative annotation of this node as methoxyabscisic acid glucuronide ($\Delta = 2.6$ ppm). In addition, the hypothesized structure of the linked node was consistent with abscisic acid glucuronide (Figure S5), which was already putatively identified in a previous study by a conventional annotation workflow [18].

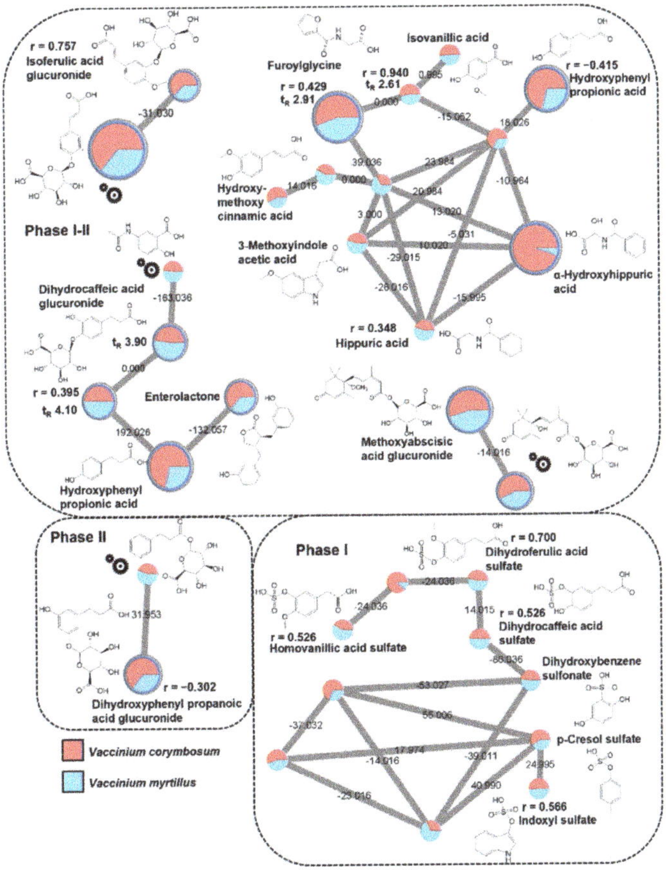

Figure 3. Extracted molecular families of identified metabolites in negative ionization mode listed in Table S2, belonging to the category of (poly)phenolic compounds, abscisic acid, and their glucuronide or sulfate derivatives. Dashed boxes group the identified metabolites according to phase I and II metabolism following PPMC analysis. The "gear" symbols refer to the putative structure identified by manual investigation as reported in Section 2.6 of the main text. Statistically significant Pearson correlation coefficients (r) are reported. Edge labels refer to the mass difference between two nodes.

Among the identified molecular families in the NI dataset, some of them exhibited a mixed metabolic contribution (i.e., phase I–II). In detail, isoferulic acid glucuronide showed a positive and significant PPMC correlation (r = 0.757), but was linked with a node exhibiting an opposite postprandial behavior (peak area vs. time points, data not shown), thus suggesting a phase I-II mixed contribution.

An analogous mixed metabolic contribution can be also proposed for the abscisic acid spectral family since the methoxyabscisic acid glucuronide is most likely associated to phase I due to the methylation of the hydroxyl group, whereas the node putatively associated to the glucuronide derivative of abscisic acid, is related to phase II.

The molecular family containing hydroxyphenyl propionic and hydroxy-methoxy cinnamic acids was characterized by peculiar structural relationships and depicted a heterogeneous metabolic contribution. In fact, the postprandial analysis of the identified nodes evidenced a phase I expression (0.348 < r < 0.940) for most metabolites [19,27], with the only exception of hydroxyphenyl propionic acid, for which a phase II metabolism can be suggested, based on its r-value (−0.415). Even though this metabolic association to phase II metabolism is apparently questionable due the lack of a conjugated group, the analysis of the full scan spectra evidenced an in-source fragmentation of the sulfate derivative of the hydroxyphenyl propionic acid (m/z = 263.02 Da), thus confirming the phase II metabolic attribution. A detailed analysis of this molecular family evidenced also that the compound annotated as hydroxy-methoxycinnamic acid probably underwent in-source dissociation, since in the same t_R range, an ion at m/z 273 fragmented in m/z 229.02 and m/z 193.05, corresponding to losses of 44 Da (neutral loss of CO_2) and 80 Da (loss of SO_3). These findings suggested that the annotated compound was conjugated with sulfate. The other spectral family associated with the phenylpropionic scaffold also included metabolites associated with both phase I (i.e., enterolactone and hydroxyphenylpropionic acid) and phase II (i.e., dihydrocaffeic acid glucuronide) [28]. A clearer postprandial kinetics was highlighted for the molecular family of dihydroxyphenyl propanoic acid glucuronide, being addressed as phase II ($r = -0.302$) metabolites [26]. Several sulfate metabolites occurred in the same molecular family, belonging to the categories of dihydrocinnamic and vanillic acids, phenolic derivatives, and indoles. Thanks to the analysis of postprandial profiles and in agreement with literature findings [29,30], the molecular scaffolds of the identified molecules are probably related to the activity of the gut microbiota. The metabolites occurring in this spectral family can be addressed to the phase I metabolism ($0.526 < r < 0.700$). Ultimately, it should be emphasized that some IDs belonging to the abovementioned molecular families showed structural similarities with previously annotated compounds. For example, compounds 4 and 5 of the NI dataset (Table S2), are characterized by retention time and MS2 fragments like those reported for the related glucuronidated conjugates found in urine by Ancillotti and co-workers [19].

In the PI dataset, twenty-four singletons were identified. In detail, several metabolites, annotated as (poly)phenolics and phenolics derivatives, were linked to phase I metabolism (e.g., dihydroxy-trimethyl-isochromenone, trihydroxybutyrophenone, and dihydroresveratrol) and with mixed contribution of phase I-II (e.g., cinnamic acid and hesperetin) by PPMC analysis. Other plant endogenous compounds, annotated with high accuracy, did not show any significant PPMC. Among them, β-glucopyranosyl-tryptophan and furaneol, as well as abscisic acid and nerol, which are well-known food-intake biomarker [31,32], and plant constituents [33,34], respectively. Some human endogenous compounds were also annotated (i.e., alpha-CEHC, ethylindole carboxylicacid, folinic acid, formylkynurenine, indole acetic acid, sebacic acid, ketodeoxycholic acid, keto-octadecadienoic acid, and hydroxy-methoxybenzophenone), exhibiting different trends against time points ($-0.645 < r < 0.958$), thus resulting in a complex metabolic output potentially associated with the investigated interventions, or resulting from background diet. Finally, PI mode exhibited three singletons that matched the NI annotations (i.e., azelaic acid, furoylglycine and enterolactone) and postprandial behavior interpretation based on PPMC analysis, being their longitudinal trend characterized by high and positive r-values ($0.640 < r < 0.967$).

The other annotated compounds occurred inside molecular families (Table S2), allowing for identifying interesting metabolites. Figure 4 displays the molecular families occurring in the PI dataset with the unknown nodes labelled by "gear" symbols for which were provided hypothesized structures (Figure S7 of the Supplementary Materials). The match with the PI "Nutri-Metabolomics" library identified two nodes as isomers of vanillic acid at different retention times, whereas the remaining nodes were putatively addressed as protocatechuic acid derivatives with high mass accuracy (from −3.26 to −0.06 ppm), by structural elucidation (Figure S7). This molecular family resulted the only one with mixed phase II-phase I contributions. In fact, vanillic acid was characterized by a statistically significant negative r-value, suggesting its direct origin from the supplements intake [35], whereas the two hypothesized protocatechuic acid derivatives exhibited an increasing signal around 6–24 h when their signals were plotted manually, probably originating from microbiota activity [30]. Most identified molecular families were related to phase I metabolism ($0.441 < r < 0.930$) and, interestingly, several identified and hypothesized node structures can be addressed as metabolite of the native polyphenols occurring in the bilberry and blueberry supplements [30]. Furthermore, derivatives of phloroglucinol carboxylic acid (i.e., hydroxy-dimethoxyphenyl-ethanone), cinnamic acid (i.e., coumaric acid, methyl-cinnamate, ferulic and isoferulic acid), and mandelic acid (i.e., methoxy-hydroxymandelate) were recognized. A deeper network inspection revealed the occurrence of in-source fragmentations of the conjugation of cinnamic acid derivatives. In detail, at the same t_R value of the compound annotated as ferulic acid (t_R = 5.14, m/z = 177.05), the feature at m/z 252.09 fragmented originating ions at m/z 177 (methoxycinnamic moiety) and at m/z 85 (H_4SO_3+H^+ sulfate moiety), suggesting that the annotated compound is a sulfate conjugate. Similarly, vanillic acid (t_R = 3.65, m/z = 169.05) could be addressed as sulfate conjugated, since a feature at t_R = 3.7 and m/z = 261 was characterized by fragments at m/z = 99 ($H_3SO_4^+$) and at m/z=122 (probably benzoic acid). Finally, the compound annotated as isoferulic acid (t_R = 4.91, m/z = 177.05), coeluted with a feature at m/z = 263, which is probably a derivative of dihydrocaffeic acid sulfate (annotated in NI dataset), thus supporting the sulfated conjugation of isoferulic acid. Three additional interesting spectral families were identified as β-carboline derivatives (i.e., tetrahydroharmane carboxylic acid and tetrahydro-β-carboline carboxylic acid), previously identified in serum samples from this study [18], xanthine pathway metabolites (i.e., dimethyl-uric acid, caffeine), and terpene derivatives (i.e., curcumenol). Regarding xanthine derivatives, even though the identification of uric acid derivatives is in accordance with literature [36], the occurrence of caffeine has never been reported in association with berries consumption and could be attributed to the consumption of caffeine-rich foods before the fasting period foreseen in the study design and/or within the period of pool samples collection [37]. Additionally, caffeine was annotated with Δ = 6.1 ppm by matching with the Massbank mass spectral library, which includes 64 spectra for caffeine acquired in heterogenous instrumental conditions. Thus, caution should be paid on this annotation. Curcumenol and its hypothesized sesquiterpene derivative were reported in this study as well as dihydroxy-trimethyl-isochromenone and ligustilide isomers (however; they were hardly related to the intake of bilberry and blueberry), as well as dihydroxy-trimethyl-isochromenone and ligustilide isomers. Finally, Gamma-CEHC, an endogenous metabolite of vitamin E [38], occurred inside molecular families, exhibiting a significant and positive r-value (0.441), representing a first report in relation to berry consumption.

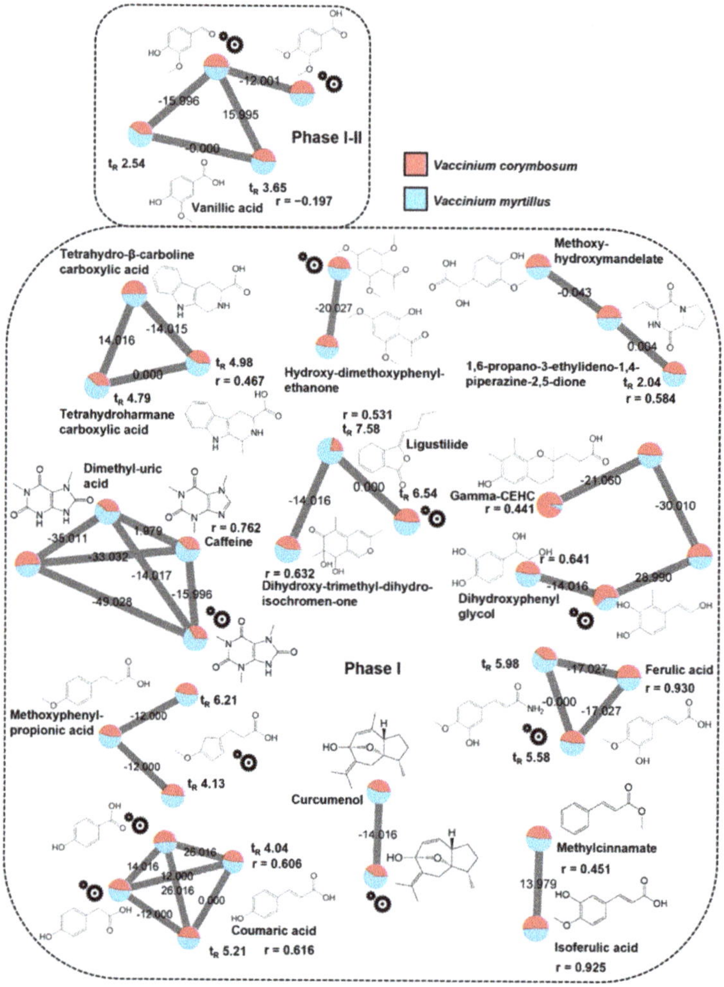

Figure 4. Extracted molecular families of identified metabolites in positive ionization mode listed in Table S2, belonging to the category of (poly)phenolic derivatives and plant endogenous constituents. Dashed boxes group the identified metabolites according to the phase I and II metabolism following PPMC analysis. The "gear" symbols refer to the putative structure identified by manual investigation as reported in Section 2.6 of the main text. Statistically significant Pearson correlation coefficients (r) are reported. Edge labels refer to the mass difference between two nodes.

4. Conclusions

This research investigated for the first time the applicability of the FBMN approach in combination with mass spectral libraries relevant to nutrikinetic studies as well as PPMC analysis to boost the structural annotation of postprandial urinary metabolites and to explore their nutrikinetic behavior within a two-arm intervention study on the intake of VM and VC supplements, as a relevant nutrimetabolomics application.

By using the FBMN approach, 24 and 43 metabolites were annotated with high mass accuracy in NI and PI mode, respectively. The comparison with widely used annotation protocols underlined the great potential of the FBMN workflow in providing the basis for an automated exploratory data analysis workflow resulting in a comprehensive and accurate annotation coverage. The proposed workflow offers a wider exploration of the urinary metabolome and allows for a prioritization strategy based on qualitative information. Additionally, the reliability of the presented approach was confirmed by the annotation of biochemically relevant metabolite categories across the three different annotation protocols followed.

The quantitative information introduced by FBMN approach provided an estimation of the impact of the two bilberry intakes on NI and PI datasets. Furthermore, the PPMC analysis of the chromatographic areas of each identified mass feature in relation to the postprandial timepoint proved to be a successful strategy to assess the kinetic shape recognition related to phase I/phase II metabolism of IDs.

It can therefore be concluded that future integration of contextual mass spectral libraries and PPMC analysis within the FBMN environment would be useful for nutrimetabolomics studies, as well as for other omics applications, where boosting annotation rates and streamlining the metabolite selection procedure are key for the data interpretation. Furthermore, it was demonstrated that the automated FBMN approach offers a versatile and scalable alternative to existing approaches that handle untargeted metabolomics profiles of biofluids for biomarker discovery. Finally, our work clearly evidenced the need for curated and contextualized mass spectral libraries that are fundamental for successful metabolite identification and thus biochemical interpretation of metabolomics profiles.

Supplementary Materials: The following supporting information can be downloaded at: https://www.mdpi.com/article/10.3390/metabo12101005/s1, Supplementary Materials, List of Referece Standards used to build the library, Metadata and Library Information.

Author Contributions: L.R.: conceptualization, investigation, software, formal analysis, data curation, writing—original draft; M.U.: conceptualization, software, data curation, sample analysis and annotation, formal analysis, supervision, writing—review & editing; F.M.: coordinating metabolomics experiments, project administration, writing—review & editing; R.B.: clinical trial execution, project administration; M.D.B.: project administration, funding acquisition, supervision, writing—original draft, review & editing; J.J.J.v.d.H.: conceptualization, formal analysis, writing—review & editing, supervision. All authors have read and agreed to the published version of the manuscript.

Funding: This research received no external funding.

Institutional Review Board Statement: The study was conducted according to the guidelines of the Declaration of Helsinki and approved by the Institutional Review Board (or Ethics Committee) of AREA VASTA CENTRO (protocol code SPE14.178_AOUC, 23 March 2015).

Informed Consent Statement: Informed consent was obtained from all subjects involved in the study.

Data Availability Statement: Data available in MassIVE accessible repository (MSV000088336). The mass spectral libraries presented in this study are openly available in at https://gnps.ucsd.edu/ProteoSAFe/gnpslibrary.jsp?library=GNPS-NUTRI-METAB-FEM-POS (accessed date 9 November 2021) and https://gnps.ucsd.edu/ProteoSAFe/gnpslibrary.jsp?library=GNPS-NUTRI-METAB-FEM-NEG (accessed date 9 November 2021).

Acknowledgments: The authors wish to acknowledge the support of all the parties involved in the PRAF Misura 1.2. e) grant. The authors also want to acknowledge the GNPS community for providing computational metabolomics solutions to the scientific community. Additionally, the authors want to acknowledge the support of the Italian Ministry for Education, University, and Research by the "Progetto Dipartimenti di Eccellenza 2018–2022" to the Department of Chemistry "Ugo Schiff" of the University of Florence.

Conflicts of Interest: J.J.J.v.d.H. is a member of the Scientific Advisory Board of Naicons Srl., Milano, Italy. All other authors declare that they have no conflict of interest.

References

1. Schrimpe-Rutledge, A.C.; Codreanu, S.G.; Sherrod, S.D.; McLean, J.A. Untargeted metabolomics strategies—Challenges and emerging directions. *J. Am. Soc. Mass Spectrom.* **2016**, *27*, 1897–1905. [CrossRef] [PubMed]
2. Pomastowski, P.; Buszewski, B. Complementarity of matrix-and nanostructure-assisted laser desorption/ionization approaches. *Nanomaterials* **2019**, *9*, 260. [CrossRef] [PubMed]
3. Chaleckis, R.; Meister, I.; Zhang, P.; Wheelock, C.E. Challenges, progress and promises of metabolite annotation for LC–MS-based metabolomics. *Curr. Opin. Biotechnol.* **2019**, *55*, 44–50. [CrossRef]
4. Kind, T.; Fiehn, O. Seven Golden Rules for heuristic filtering of molecular formulas obtained by accurate mass spectrometry. *BMC Bioinform.* **2007**, *8*, 105. [CrossRef] [PubMed]
5. Sumner, L.W.; Amberg, A.; Barrett, D.; Beale, M.H.; Beger, R.; Daykin, C.A.; Fan, T.W.-M.; Fiehn, O.; Goodacre, R.; Griffin, J.L. Proposed minimum reporting standards for chemical analysis. *Metabolomics* **2007**, *3*, 211–221. [CrossRef]
6. Beniddir, M.A.; Kang, K.B.; Genta-Jouve, G.; Huber, F.; Rogers, S.; van der Hooft, J.J. Advances in decomposing complex metabolite mixtures using substructure-and network-based computational metabolomics approaches. *Nat. Prod. Rep.* **2021**, *38*, 1967–1993. [CrossRef] [PubMed]
7. van der Hooft, J.J.; de Vos, R.C.; Mihaleva, V.; Bino, R.J.; Ridder, L.; de Roo, N.; Jacobs, D.M.; van Duynhoven, J.P.; Vervoort, J. Structural elucidation and quantification of phenolic conjugates present in human urine after tea intake. *Anal. Chem.* **2012**, *84*, 7263–7271. [CrossRef]
8. Ramos, A.E.F.; Evanno, L.; Poupon, E.; Champy, P.; Beniddir, M.A. Natural products targeting strategies involving molecular networking: Different manners, one goal. *Nat. Prod. Rep.* **2019**, *36*, 960–980. [CrossRef]
9. Aron, A.T.; Gentry, E.C.; McPhail, K.L.; Nothias, L.-F.; Nothias-Esposito, M.; Bouslimani, A.; Petras, D.; Gauglitz, J.M.; Sikora, N.; Vargas, F. Reproducible molecular networking of untargeted mass spectrometry data using GNPS. *Nat. Protoc.* **2020**, *15*, 1954–1991. [CrossRef]
10. van der Hooft, J.J.; Padmanabhan, S.; Burgess, K.E.; Barrett, M.P. Urinary antihypertensive drug metabolite screening using molecular networking coupled to high-resolution mass spectrometry fragmentation. *Metabolomics* **2016**, *12*, 125. [CrossRef]
11. Said, I.H.; Truex, J.D.; Heidorn, C.; Retta, M.B.; Petrov, D.D.; Haka, S.; Kuhnert, N. LC-MS/MS based molecular networking approach for the identification of cocoa phenolic metabolites in human urine. *Food Res. Int.* **2020**, *132*, 109119. [CrossRef] [PubMed]
12. Nothias, L.-F.; Petras, D.; Schmid, R.; Dührkop, K.; Rainer, J.; Sarvepalli, A.; Protsyuk, I.; Ernst, M.; Tsugawa, H.; Fleischauer, M. Feature-based molecular networking in the GNPS analysis environment. *Nat. Methods* **2020**, *17*, 905–908. [CrossRef] [PubMed]
13. Oberleitner, D.; Schmid, R.; Schulz, W.; Bergmann, A.; Achten, C. Feature-based molecular networking for identification of organic micropollutants including metabolites by non-target analysis applied to riverbank filtration. *Anal. Bioanal. Chem.* **2021**, *413*, 5291–5300. [CrossRef]
14. Padilla-González, G.F.; Sadgrove, N.J.; Ccana-Ccapatinta, G.V.; Leuner, O.; Fernandez-Cusimamani, E. Feature-based molecular networking to target the isolation of new caffeic acid esters from yacon (*Smallanthus sonchifolius*, Asteraceae). *Metabolites* **2020**, *10*, 407. [CrossRef] [PubMed]
15. Rivera-Mondragón, A.; Tuenter, E.; Ortiz, O.; Sakavitsi, M.E.; Nikou, T.; Halabalaki, M.; Caballero-George, C.; Apers, S.; Pieters, L.; Foubert, K. UPLC-MS/MS-based molecular networking and NMR structural determination for the untargeted phytochemical characterization of the fruit of *Crescentia cujete* (Bignoniaceae). *Phytochemistry* **2020**, *177*, 112438. [CrossRef]
16. Xie, H.-F.; Kong, Y.-S.; Li, R.-Z.; Nothias, L.-F.; Melnik, A.V.; Zhang, H.; Liu, L.-L.; An, T.-T.; Liu, R.; Yang, Z. Feature-Based Molecular Networking Analysis of the Metabolites Produced by in vitro Solid-State Fermentation Reveals Pathways for the Bioconversion of Epigallocatechin Gallate. *J. Agric. Food Chem.* **2020**, *68*, 7995–8007. [CrossRef]
17. Neto, F.C.; Raftery, D. Expanding Urinary Metabolite Annotation through Integrated Mass Spectral Similarity Networking. *Anal. Chem.* **2021**, *93*, 12001–12010. [CrossRef]
18. Renai, L.; Ancillotti, C.; Ulaszewska, M.; Garcia-Aloy, M.; Mattivi, F.; Bartoletti, R.; Del Bubba, M. Comparison of chemometrics strategies for potential exposure markers discovery and false positive reduction in untargeted metabolomics: Application to the serum analysis by LC-HRMS after intake of *Vaccinium* fruits supplements. *Anal. Bioanal. Chem.* **2022**, *414*, 1841–1855. [CrossRef]
19. Ancillotti, C.; Ulaszewska, M.; Mattivi, F.; Del Bubba, M. Untargeted metabolomics analytical strategy based on liquid chromatography/electrospray ionization linear ion trap quadrupole/orbitrap mass spectrometry for discovering new polyphenol metabolites in human biofluids after acute ingestion of vaccinium myrtillus berry supplement. *J. Am. Soc. Mass Spectrom.* **2019**, *30*, 381–402. [CrossRef]
20. Pluskal, T.; Castillo, S.; Villar-Briones, A.; Orešič, M. MZmine 2: Modular framework for processing, visualizing, and analyzing mass spectrometry-based molecular profile data. *BMC Bioinform.* **2010**, *11*, 395. [CrossRef]
21. Vargas, F.; Weldon, K.C.; Sikora, N.; Wang, M.; Zhang, Z.; Gentry, E.C.; Panitchpakdi, M.W.; Caraballo-Rodríguez, A.M.; Dorrestein, P.C.; Jarmusch, A.K. Protocol for community-created public MS/MS reference spectra within the Global Natural Products Social Molecular Networking infrastructure. *Rapid Commun. Mass Spectrom.* **2020**, *34*, e8725. [CrossRef] [PubMed]
22. Shannon, P.; Markiel, A.; Ozier, O.; Baliga, N.S.; Wang, J.T.; Ramage, D.; Amin, N.; Schwikowski, B.; Ideker, T. Cytoscape: A software environment for integrated models of biomolecular interaction networks. *Genome Res.* **2003**, *13*, 2498–2504. [CrossRef] [PubMed]

23. Pellis, L.; van Erk, M.J.; van Ommen, B.; Bakker, G.C.; Hendriks, H.F.; Cnubben, N.H.; Kleemann, R.; van Someren, E.P.; Bobeldijk, I.; Rubingh, C.M. Plasma metabolomics and proteomics profiling after a postprandial challenge reveal subtle diet effects on human metabolic status. *Metabolomics* **2012**, *8*, 347–359. [CrossRef] [PubMed]
24. Garcia-Aloy, M.; Ulaszewska, M.; Franceschi, P.; Estruel-Amades, S.; Weinert, C.H.; Tor-Roca, A.; Urpi-Sarda, M.; Mattivi, F.; Andres-Lacueva, C. Discovery of intake biomarkers of lentils, chickpeas, and white beans by untargeted LC–MS metabolomics in serum and urine. *Mol. Nutr. Food Res.* **2020**, *64*, 1901137. [CrossRef]
25. Iglesias-Carres, L.; Mas-Capdevila, A.; Bravo, F.I.; Aragonès, G.; Arola-Arnal, A.; Muguerza, B. A comparative study on the bioavailability of phenolic compounds from organic and nonorganic red grapes. *Food Chem.* **2019**, *299*, 125092. [CrossRef] [PubMed]
26. Ancillotti, C.; Ciofi, L.; Rossini, D.; Chiuminatto, U.; Stahl-Zeng, J.; Orlandini, S.; Furlanetto, S.; Del Bubba, M. Liquid chromatographic/electrospray ionization quadrupole/time of flight tandem mass spectrometric study of polyphenolic composition of different Vaccinium berry species and their comparative evaluation. *Anal. Bioanal. Chem.* **2017**, *409*, 1347–1368. [CrossRef]
27. de Mello, V.D.; Lankinen, M.A.; Lindström, J.; Puupponen-Pimiä, R.; Laaksonen, D.E.; Pihlajamäki, J.; Lehtonen, M.; Uusitupa, M.; Tuomilehto, J.; Kolehmainen, M. Fasting serum hippuric acid is elevated after bilberry (*Vaccinium myrtillus*) consumption and associates with improvement of fasting glucose levels and insulin secretion in persons at high risk of developing type 2 diabetes. *Mol. Nutr. Food Res.* **2017**, *61*, 1700019. [CrossRef]
28. Vetrani, C.; Rivellese, A.A.; Annuzzi, G.; Adiels, M.; Borén, J.; Mattila, I.; Orešič, M.; Aura, A.-M. Metabolic transformations of dietary polyphenols: Comparison between in vitro colonic and hepatic models and in vivo urinary metabolites. *J. Nutr. Biochem.* **2016**, *33*, 111–118. [CrossRef]
29. Aravind, S.M.; Wichienchot, S.; Tsao, R.; Ramakrishnan, S.; Chakkaravarthi, S. Role of dietary polyphenols on gut microbiota, their metabolites and health benefits. *Food Res. Int.* **2021**, *142*, 110189. [CrossRef]
30. Stevens, J.F.; Maier, C.S. The chemistry of gut microbial metabolism of polyphenols. *Phytochem. Rev.* **2016**, *15*, 425–444. [CrossRef]
31. Du, X.; Finn, C.E.; Qian, M.C. Volatile composition and odour-activity value of thornless 'Black Diamond' and 'Marion' blackberries. *Food Chem.* **2010**, *119*, 1127–1134. [CrossRef]
32. Gutsche, B.; Grun, C.; Scheutzow, D.; Herderich, M. Tryptophan glycoconjugates in food and human urine. *Biochem. J.* **1999**, *343*, 11–19. [CrossRef] [PubMed]
33. Degu, A.; Ayenew, B.; Cramer, G.R.; Fait, A. Polyphenolic responses of grapevine berries to light, temperature, oxidative stress, abscisic acid and jasmonic acid show specific developmental-dependent degrees of metabolic resilience to perturbation. *Food Chem.* **2016**, *212*, 828–836. [CrossRef] [PubMed]
34. Elsharif, S.A.; Buettner, A. Structure–odor relationship study on geraniol, nerol, and their synthesized oxygenated derivatives. *J. Agric. Food Chem.* **2016**, *66*, 2324–2333. [CrossRef] [PubMed]
35. Colak, N.; Primetta, A.K.; Riihinen, K.R.; Jaakola, L.; Grúz, J.; Strnad, M.; Torun, H.; Ayaz, F.A. Phenolic compounds and antioxidant capacity in different-colored and non-pigmented berries of bilberry (*Vaccinium myrtillus* L.). *Food Biosci.* **2017**, *20*, 67–78. [CrossRef]
36. Lotito, S.B.; Frei, B. Consumption of flavonoid-rich foods and increased plasma antioxidant capacity in humans: Cause, consequence, or epiphenomenon? *Free Radic. Biol. Med.* **2006**, *41*, 1727–1746. [CrossRef]
37. Martínez-López, S.; Sarriá, B.; Gómez-Juaristi, M.; Goya, L.; Mateos, R.; Bravo-Clemente, L. Theobromine, caffeine, and theophylline metabolites in human plasma and urine after consumption of soluble cocoa products with different methylxanthine contents. *Food Res. Int.* **2014**, *63*, 446–455. [CrossRef]
38. Wu, J.; Croft, K. Vitamin E metabolism. *Mol. Asp. Med.* **2007**, *28*, 437–452. [CrossRef]

metabolites

Article

Integration of Liver Glycogen and Triglyceride NMR Isotopomer Analyses Provides a Comprehensive Coverage of Hepatic Glucose and Fructose Metabolism

Ivan Viegas [1], Giada Di Nunzio [2], Getachew D. Belew [2,3], Alejandra N. Torres [2], João G. Silva [2], Luis Perpétuo [2,4], Cristina Barosa [2], Ludgero C. Tavares [5] and John G. Jones [2,*]

1 Centre for Functional Ecology, Department of Life Sciences, University of Coimbra, Coimbra, Portugal
2 Center for Neurosciences and Cell Biology, University of Coimbra, UC-Biotech, Biocant Park, Nucleo 8, Lote 4, 3060-197 Cantanhede, Portugal
3 Biotechnology Department, Addis Ababa Science and Technology University, Addis Ababa, Ethiopia
4 iBiMED, Department of Medical Sciences, University of Aveiro, Aveiro, Portugal
5 CIVG—Vasco da Gama Research Center, University School Vasco da Gama—EUVG, 3020-210 Coimbra, Portugal
* Correspondence: john.griffith.jones@gmail.com

Abstract: Dietary glucose and fructose are both efficiently assimilated by the liver but a comprehensive measurement of this process starting from their conversion to sugar phosphates, involvement of the pentose phosphate pathway (PPP), and conversion to glycogen and lipid storage products, remains incomplete. Mice were fed a chow diet supplemented with 35 g/100 mL drinking water of a 55/45 fructose/glucose mixture for 18 weeks. On the final night, the sugar mixture was enriched with either [U-^{13}C]glucose or [U-^{13}C]fructose, and deuterated water (^2H_2O) was also administered. ^{13}C-isotopomers representing newly synthesized hepatic glucose-6-phosphate (glucose-6-P), glycerol-3-phosphate, and lipogenic acetyl-CoA were quantified by ^{2}H and ^{13}C NMR analysis of post-mortem liver glycogen and triglyceride. These data were applied to a metabolic model covering glucose-6-P, PPP, triose-P, and de novo lipogenesis (DNL) fluxes. The glucose supplement was converted to glucose-6-P via the direct pathway, while the fructose supplement was metabolized by the liver to gluconeogenic triose-P via fructokinase–aldolase–triokinase. Glucose-6-P from all carbohydrate sources accounted for 40–60% of lipogenic acetyl-CoA and 10–12% was oxidized by the pentose phosphate pathway (PPP). The yield of NADPH from PPP flux accounted for a minority (~30%) of the total DNL requirement. In conclusion, this approach integrates measurements of glucose-6-P, PPP, and DNL fluxes to provide a holistic and informative assessment of hepatic glucose and fructose metabolism.

Keywords: pentose phosphate pathway; triose phosphates; acetyl-CoA; lipogenesis; ^{13}C NMR

Citation: Viegas, I.; Di Nunzio, G.; Belew, G.D.; Torres, A.N.; Silva, J.G.; Perpétuo, L.; Barosa, C.; Tavares, L.C.; Jones, J.G. Integration of Liver Glycogen and Triglyceride NMR Isotopomer Analyses Provides a Comprehensive Coverage of Hepatic Glucose and Fructose Metabolism. *Metabolites* **2022**, *12*, 1142. https://doi.org/10.3390/metabo12111142

Academic Editor: Joana Pinto

Received: 14 October 2022
Accepted: 16 November 2022
Published: 19 November 2022

1. Introduction

1.1. Background

The liver is a key site for the metabolism of dietary sugar, with glucose and fructose being the principal species absorbed into the portal vein blood outside of milk products. In mammals and many other organisms, the fate of dietary sugar is heavily influenced in real time by systemic glucose homeostasis, with the main priorities being maintenance of a threshold level of blood glucose for the central nervous system and erythrocyte function, while also minimizing large excursions of blood glucose levels. At the same time, sugar is sensed as a precious and desirable nutrient to be sequestered as rapidly and efficiently as possible [1]. This balance is achieved via a highly flexible and well-regulated hepatic metabolic network. Not only can it rapidly switch between net hepatic glucose production and uptake, but it can also direct temporary sugar surplus into short-term storage as glycogen or into longer-term storage as lipids. Since sugar in nature is typically composed of approximately equimolar amounts of glucose and fructose, for omnivorous mammals, including humans, the hepatic metabolic network has evolved to efficiently utilize both hexoses. As can be seen in Figure 1, glucose-6-phosphate (glucose-6-P) is a key nexus in hepatic sugar metabolism since it is a common product of glucose and fructose metabolized via direct and indirect pathways, respectively. Glucose-6-P is also at the intersection of glycogen synthesis and the pentose phosphate pathway (PPP).

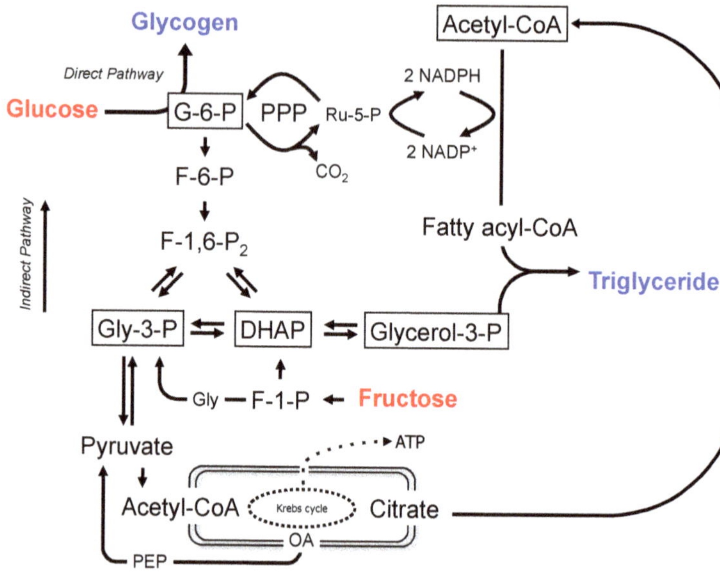

Figure 1. Metabolic model for the synthesis of glycogen and triglyceride from glucose or fructose in the liver. The model includes glucose-6-phosphate oxidation by the pentose phosphate pathway (PPP) to provide NADPH for conversion of acetyl-CoA to fatty acyl-CoA via de novo lipogenesis. The ^{13}C-enriched glucose and fructose precursors are highlighted in red and the sampled metabolites, glycogen and triglyceride, are highlighted in blue. The metabolite pools whose ^{13}C and ^{2}H enrichments are reported by the sampled metabolites, namely, glucose-6-P, triose-P (dihydroxyacetone phosphate and glyceraldehyde 3-phosphate) and lipogenic acetyl-CoA, are highlighted in boxes. Glycogen synthesis from glucose via glucose-6-P from gluconeogenic precursors, including pyruvate and triose-P sources, is also indicated (direct and indirect pathways, respectively). For simplicity, some metabolic intermediates, as well as ATP/ADP and NAD/NADH interconversions, are not shown. Abbreviations are as follows: DHAP—dihydroxyacetone phosphate; F-1-P—fructose-1-phosphate; F-6-P—fructose 6-phosphate; F-1,6-P_2—fructose-1,6-bisphosphate; G-6-P—glucose 6-phosphate; Gly—glyceraldehyde; Gly-3-P—glyceraldehyde 3-phosphate; OA—oxaloacetate; PEP—phophoenolpyruvate; Ru-5-P: ribulose-5-P.

The conversion of glucose-6-P to lipids requires the generation of NADPH. The PPP couples the oxidation of glucose-6-P to NADPH generation; hence, in principle, a portion of sugar carbons can be sacrificially oxidized such that the remainder can be converted to lipids. In the liver, NADPH can be derived from other sources [2] and, to the extent that these contribute to de novo lipogenesis (DNL) reducing equivalents, then sugar carbons are spared from PPP oxidation. The PPP is also a conduit for converting hexose sugars to pentose phosphate precursors for nucleotide biosynthesis, which is a continual requirement for hepatocyte growth and turnover.

^{13}C-Isotopomers of newly synthesized glycogen derived from [U-^{13}C]glucose and [U-^{13}C]fructose inform direct and indirect pathway fluxes [3], as well as the fraction of glucose-6-P that underwent PPP oxidation [4]. ^{13}C-isotopomers of newly synthesized triglyceride fatty acids and glycerol moieties inform the contributions of these sugars to DNL and glyceroneogenesis [5]. The main objective of this study was to integrate these measurements into a comprehensive description of hepatic glucose and fructose metabolism, starting with their initial phosphorylation to sugar phosphate intermediates and culminating with their conversion to triglycerides. Given the role of excessive sugar consumption and elevated DNL activity in the pathogenesis of non-alcoholic fatty liver disease (NAFLD) [6–8], such knowledge will improve our understanding of the role of hepatic glucose and fructose metabolic fluxes in promoting this condition.

1.2. Metabolic Model

Figure 1 shows the metabolic model for lipogenesis from glucose and fructose. Fructose is assumed to be converted to triose phosphates via the canonical fructokinase–aldolase–triokinase pathway, while glucose is converted to glucose-6-P via glucokinase. Glucose-6-P can also be synthesized from triose phosphates by gluconeogenesis (GNG). Glucose-6-P is disposed of by conversion to glycogen, by PPP oxidation, and by glycolysis. Glycerol-3-P destined for triglyceride synthesis is mostly derived from the glycolytic triose phosphate pool. The pyruvate product of glycolysis is oxidized to acetyl-CoA, which can be recruited for fatty acid synthesis via DNL. One critical aspect in interpreting the formation of glycogen and triglyceride ^{13}C-isotopomers from the ^{13}C-glucose or fructose precursors is that turnover of the product pools may not be complete over the duration of the experiment, resulting in artefactual dilutions of glycogen and lipid ^{13}C-isotopomer enrichments. To determine the fractions of glycogen and triglyceride that were synthesized while the ^{13}C-sugar precursors were present, deuterated water (^2H_2O) was administered over the same period. The ^{2}H enrichment of glycogen and triglycerides relative to body water informs these fractions [3,5,9] and, by sequential ^{2}H and ^{13}C NMR analysis, this information can be determined without interfering with the quantification of the ^{13}C-isotopomer distributions [3,5,10].

Figure 2 shows the principal ^{13}C-isotopomers of selected metabolite pools following the metabolism of [U-^{13}C]glucose. Under the experimental conditions, the ^{13}C-isotopomer distribution of newly synthesized glycogen is assumed to reflect that of glucose-6-P. The direct pathway metabolism of [U-^{13}C]glucose generates [U-^{13}C]glucose-6-P and the [U-^{13}C]glycogen isotopomer. [U-^{13}C]Glucose-6-P that undergoes PPP oxidation and recycling generates [1,2-$^{13}C_2$]glucose-6-P and other partially labeled glucose-6-P isotopomers [4]. In addition, [U-^{13}C]glucose that undergoes glycolytic–gluconeogenic recycling (either intrahepatic or via the Cori cycle) generates triose-P isotopomers, principally [1,2,3-$^{13}C_3$]- and [2,3-$^{13}C_2$]triose-P [11]. These are incorporated into glucose-6-P and glycogen via GNG, which is also historically referred to as the indirect pathway [12]. The fraction of newly synthesized glycogen derived from the indirect pathway can be estimated from the analysis of its ^{2}H enrichment from ^2H_2O [3]. Hence, the ^{13}C-isotopomer distribution of the GNG precursor pool (GNG-triose-P) can be inferred from that of glycogen after correction for the indirect pathway fraction. Glycerol-3-P for fatty acid esterification is derived from the reduction of dihydroxyacetone phosphate; hence, its ^{13}C-isotopomer distribution, read from the analysis of newly synthesized triglyceride glycerol, provides a readout of triose-P ^{13}C-isotopomers. Acetyl-CoA isotopomers that are generated from triose-P can be diluted by unlabeled non-triose substrates such as acetate before their incorporation into fatty acids. When the ^{13}C-label is provided as [U-^{13}C]fructose (Supplementary Figure S1), it generates the same set of hexose and triose-P ^{13}C-isotopomers. Note that the formation of [U-^{13}C]glucose-6-P from [U-^{13}C]fructose can occur via the condensation of [U-^{13}C]glyceraldehyde-3-P and [U-^{13}C]dihydroxyacetone-P. The probability for [U-^{13}C]glucose-6-P formation is related to the fractional enrichments of these triose-P precursors.

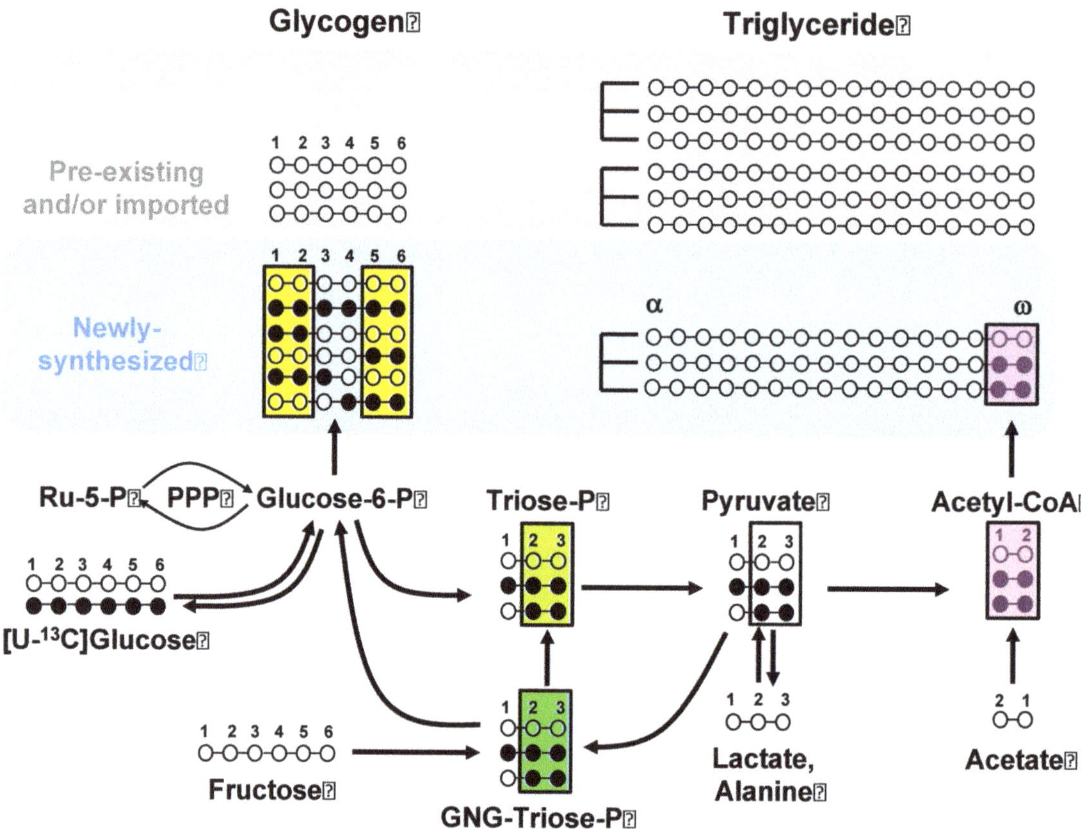

Figure 2. ^{13}C-Isotopomers of selected metabolic intermediates generated from [U-13C]glucose metabolism into lipogenic and glycogenic pathways. These include hepatic glucose-6-P—inferred from the analysis of newly synthesized glycogen; triose-P recruited for gluconeogenesis (GNG-triose-P)—inferred from the analysis of indirect pathway glycogen 13C-isotopomers; triose-P supplying glycerol-3-P for fatty acid esterification and acetyl-CoA units for de novo lipogenesis—inferred from the 13C-isotopomer analysis of newly synthesized triglyceride glycerol; and the acetyl-CoA pool supplying lipogenesis—inferred from the 13C-isotopomer analysis of newly synthesized fatty acids. For the metabolite carbon skeletons, the filled and unfilled circles represent 13C and 12C, respectively. The shading highlights those 13C-isotopomers that inform the enrichment of the lipogenic acetyl-CoA pool by [U-13C]acetyl CoA from both glycolytic precursor and fatty acid product perspectives, and the colors indicate isotopic enrichment equivalence (same color) or non-equivalence (different colors). For simplicity, in depicting the fatty acid labeling, only the 13C-isotopomers of the last two fatty acid carbons (representing the first acetyl-CoA moiety to be incorporated into de novo lipogenesis) are shown.

2. Methods

2.1. Materials

[U-^{13}C]Fructose at 99% enrichment was obtained from Omicron Biochemicals Inc., South Bend, IN, USA, and [U-^{13}C]glucose at 99% enrichment was manufactured by Cambridge Isotopes Limited, Cambridge, MA, USA, and purchased through Tracertec, Madrid, Spain. Deuterated water (^2H_2O) at 99.8% was purchased from CortecNet, Les Ulis, France.

2.2. Animal Studies

Animal studies were approved by the University of Coimbra Ethics Committee on Animal Studies (ORBEA) and the Portuguese National Authority for Animal Health (DGAV), approval code 0421/000/000/2013. A total of nine adult male C57BL/6J mice obtained from Charles River Labs, Barcelona, Spain, were housed at the University of Coimbra UC-Biotech Bioterium. They were maintained in a well-ventilated environment and a 12 h light/12 h dark cycle. Upon delivery to the Bioterium, mice were provided a two-week interval for acclimation, with free access to water and standard chow, comprising of 60% mixed carbohydrates, 16% protein, and 3% lipids. Following this period, the chow was supplemented with a 55/45 mixture of fructose and glucose present at a concentration of 30% w/v in the drinking water for a period of 12 weeks. At the beginning of the final evening, mice were administered with an intraperitoneal loading dose of 99% ^2H_2O containing 0.9 mg/mL NaCl (4 mL/100 g body weight), and the drinking water was enriched to 5% with ^2H_2O. The fructose/glucose mixture in their drinking water was replaced with mixtures of identical composition, but with 20% enriched [U-^{13}C]fructose for five mice and 20% enriched [U-^{13}C]glucose for the remaining four mice. At the end of this dark cycle, mice were deeply anesthetized with ketamine/xylazine and sacrificed by cardiac puncture. Arterial blood was immediately centrifuged, and plasma was isolated and stored at −80 °C. Livers were freeze-clamped and stored at −80 °C until further analysis.

2.3. Analysis of Glycogen and Triglyceride Isotopic Enrichments by NMR

Liver portions of ~500 mg were powdered under liquid nitrogen and extracted with methyl *tert*-butyl ether, as previously described [5]. Glycogen from the insoluble pellet was extracted, purified, and derivatized to monoacetone glucose (MAG), as previously described [3]. Triglycerides from the organic fraction were separated from other lipids, as previously described [13].

2.3.1. NMR Analysis of Glycogen ^{2}H and ^{13}C-Enrichments

Proton-decoupled ^{2}H-NMR spectra of MAG samples at 50 °C were obtained with a Bruker Avance III HD 500 spectrometer using a ^{2}H-selective 5 mm probe incorporating a ^{19}F-lock channel. Samples were resuspended in 0.5 mL 90% acetonitrile/10% ^{2}H-depleted water, to which 50 μL of hexafluorobenzene were added. ^{2}H-NMR spectra were obtained with a 90° pulse, 1.6 s of acquisition time, and a 0.1 s interpulse delay. The number of free-induction decays (f.i.d.) collected ranged from 2000 to 10,000. Positional ^{2}H enrichments were determined using the MAG methyl signals as an intramolecular standard [14]. To quantify plasma body water ^{2}H enrichments, triplicate 10 μL samples of plasma were analyzed at 25 °C by ^{2}H NMR, as previously described [15], but with 50 μL of hexafluorobenzene added to the NMR sample. Proton-decoupled ^{13}C NMR spectra at 25 °C were obtained with a Varian VNMRS 600 MHz NMR spectrometer equipped with a 3 mm broadband probe. ^{13}C NMR spectra were acquired at 25 °C using a 60° pulse, 30.5 kHz spectral width, and 4.1 s of recycling time (4.0 s of acquisition time and 0.1 s pulse delay). The number of acquisitions ranged from 2000 to 18,000. The summed f.i.d. was processed with 0.2 Hz line-broadening and zero-filled to 512 K before Fourier transform.

2.3.2. NMR Analysis of Triglyceride ^{2}H and ^{13}C Enrichments

Purified triglycerides were dissolved in ~0.5 mL $CHCl_3$. To these, 25 μL of a pyrazine standard enriched to 1% with pyrazine-d_4 and dissolved in $CHCl_3$ (0.07 g pyrazine/g $CHCl_3$), and 50 μL C_6F_6 were added. ^{1}H and ^{2}H NMR spectra were acquired with an 11.7 T Bruker Avance III HD system using a dedicated 5 mm ^{2}H probe with ^{19}F lock and ^{1}H-decoupling coil, as previously described. ^{1}H spectra at 500.1 MHz were acquired with a 90° pulse, 10 kHz spectral width, 3 s acquisition time, and 5 s pulse delay. Overall, 16 f.i.d. were collected for each spectrum. ^{2}H NMR spectra at 76.7 MHz were obtained with a 90° pulse, a 1230 Hz sweep width, an acquisition time of 0.67 s, and interpulse delay of 8 s. For ^{13}C isotopomer analysis by ^{13}C NMR, dried triglyceride samples were dissolved in 0.2 mL

99.96% enriched $CDCl_3$ (Sigma-Aldrich) and acquired using the same parameters as for the MAG samples. For each ^{13}C spectrum, 2000–4000 f.i.d. were collected.

^{13}C and ^{2}H NMR spectra were analyzed with ACD/NMR Processor Academic Edition software (ACD/Labs, Advanced Chemistry Development, Inc.).

2.4. Estimation of Substrate Contributions to Lipogenesis from Analysis of Newly Synthesized Glycogen and Triglyceride ^{13}C Isotopomers

As indicated in Figure 2, the ^{13}C-isotopomer distributions of newly synthesized glycogen informs that of glucose-6-P, while the ^{13}C-isotopomer distributions of newly synthesized triglyceride glyceryl and fatty acid moieties inform the precursor enrichments of triose-P and lipogenic acetyl-CoA pools, respectively. For each of these reporter metabolites, all ^{13}C-isotopomers that are either metabolized to form lipogenic [U-^{13}C]acetyl-CoA (i.e., glucose-6-P and triose-P) or are an immediate product (TG-fatty acid) were defined as $^{13}C_{IUA}$. These $^{13}C_{IUA}$ correspond to the shaded ^{13}C-isotopomers of glucose-6-P, triose-P, and fatty acids shown in Figure 2 and provide the basis for quantifying the isotopic dilution of the ^{13}C-enriched carbons of glucose and fructose as they are metabolized to lipids.

For the glucose-6-P precursors, [U-^{13}C]acetyl-CoA can be derived from glycolytic metabolism of [U-^{13}C]glucose-6-P, as well as from glucose-6-P isotopomers originating from recycling and/or PPP metabolism of [U-^{13}C]glucose. These include [1,2-$^{13}C_2$]-, [1,2,3-$^{13}C_3$]-, [5,6-$^{13}C_2$]-, and [4,5,6-$^{13}C_3$]glucose-6-P. Thus, as shown by equation (1), the $^{13}C_{IUA}$ for glucose-6-P can be estimated as the sum of [U-^{13}C]-, [1,2-$^{13}C_2$]-, [1,2,3-$^{13}C_3$]-, [5,6-$^{13}C_2$]-, and [4,5,6-$^{13}C_3$]glucose isotopomer enrichments of glycogen ($\Sigma_{glycogen\ isotopomers}$) multiplied by $1/f_{\text{glycogen}}$. The fraction of newly synthesized glycogen (f_{glycogen}) is estimated from the ^{2}H enrichment of position 2 relative to that of body water [3], and these data are shown in Supplementary Table S2.

$$\text{Glucose-6-P}\ ^{13}C_{IUA} = \Sigma_{glycogen\ isotopomers} \times 1/f_{\text{glycogen}} \quad (1)$$

Glucose-6-P is derived from the phosphorylation of dietary glucose and from GNG. For the [U-^{13}C]glucose tracer, enrichment of [U-^{13}C]glucose-6-P is assumed to be entirely from the direct pathway metabolism of [U-^{13}C]glucose. The direct pathway fraction (f_{direct}), which also includes sources of unlabeled glucose present in the diet, -can be estimated from the positional ^{2}H enrichment distribution of glycogen [3] (Supplementary Table S2). On this basis, $^{13}C_{IUA}$ enrichment of the dietary glucose precursor pool can be estimated as follows:

$$\text{Dietary glucose}\ ^{13}C_{IUA} = [\text{U-}^{13}\text{C}]\text{Glucose-6-P}\ ^{13}C_{IUA} \times 1/f_{direct} \quad (2)$$

Since the fraction of glucose-6-P synthesized by GNG is represented by the indirect pathway fraction of newly synthesized glycogen ($f_{indirect}$), which can be estimated from the glycogen ^{2}H enrichment distributions (see Supplementary Table S2), then $^{13}C_{IUA}$ of the GNG precursor pool can be calculated. For the [U-^{13}C]glucose tracer, [U-^{13}C]glucose-6-P needs to be excluded from $\Sigma_{glycogen\ isotopomers}$ since it is generated via the direct pathway. The glucose-6-P isotopomers formed via gluconeogenesis that can generate [1,2-$^{13}C_2$]acetyl-CoA are [1,2-$^{13}C_2$]-, [1,2,3-$^{13}C_3$]-, [5,6-$^{13}C_2$]-, and [4,5,6-$^{13}C_3$]glucose-6-P ($^{13}C_{IUA\text{-}GNG}$):

$$\text{GNG}\ ^{13}C_{IUA} = \text{Glucose-6-P}\ ^{13}C_{IUA\text{-}GNG} \times 1/f_{indirect} \quad (3a)$$

For [U-^{13}C]fructose, all glycogen isotopomers are included since they are by definition all derived via the indirect pathway:

$$\text{GNG}\ ^{13}C_{IUA} = \text{Glucose-6-P}\ ^{13}C_{IUA} \times 1/f_{indirect} \quad (3b)$$

The $^{13}C_{IUA}$ of triose-P and lipogenic acetyl-CoA are estimated by adjustment with the newly synthesized triglyceride glyceryl fraction (f_{glyceryl}) and fatty acid fractions ($f_{\text{fatty acid}}$)

estimated from the triglyceride ^{2}H enrichment distribution [5] (Supplemental Table S2), as follows:

$$\text{Triose-P}\ ^{13}C_{IUA} = \text{Triglyceride glyceryl}\ ^{13}C_{IUA} \times 1/(f_{\text{glyceryl}}) \quad (4)$$

$$\text{Acetyl-CoA}\ ^{13}C_{IUA} = \text{Triglyceride fatty acid}\ ^{13}C_{IUA} \times 1/(f_{\text{fatty acid}}) \quad (5)$$

where the measured glyceryl $^{13}C_{IUA}$ is the sum of triglyceride glyceryl isotopomers with ^{13}C in both positions 2 and 3, and the fatty acid $^{13}C_{IUA}$ is the sum of fatty acid isotopomers with ^{13}C in both ultimate (ω) and penultimate positions. The fraction of lipogenic acetyl-CoA derived from triose-P was estimated from the ratio of acetyl-CoA and triose-P $^{13}C_{IUA}$ as follows:

$$\text{Triose-P} \rightarrow \text{Acetyl-CoA} = 100 \times \text{Acetyl-CoA}\ ^{13}C_{IUA}/\text{Triose-P}\ ^{13}C_{IUA} \quad (6)$$

The fraction of acetyl-CoA derived from non-triose-P metabolites, such as acetate, was estimated as the difference:

$$\text{Non-triose-P} \rightarrow \text{Acetyl-CoA} = 100 - \text{Triose-P fraction} \quad (7)$$

For the mice provided with [U-^{13}C]glucose and unlabeled fructose, the fractional contribution of dietary glucose to triose-P was estimated from the ratio of triose-P to dietary glucose $^{13}C_{IUA}$. This fraction was adjusted for total lipogenic acetyl-CoA flux by multiplication with the fraction of Acetyl-CoA derived from triose-P (Equation (6)) and for the loss of glucose-6-P carbon 1 as CO_2 via the PPP.

$$\text{Dietary glucose} \rightarrow \text{Triose-P} = ([100 \times \text{Triose-P}\ ^{13}C_{IUA}/\text{dietary glucose}\ ^{13}C_{IUA}] \times \text{Equation(6)}) + 1/6\ \text{PPP} \quad (8)$$

The contribution of gluconeogenic precursors (GNG precursors) to triose-P was calculated as the difference between total triose-P contribution (Equation (6)) and hepatic glucose contribution (Equation (8)) and also accounted for the loss of carbon via the PPP:

$$\text{GNG precursors} \rightarrow \text{Triose-P} = (\text{Equation (6)} - \text{Equation (8)}) + 1/6\ \text{PPP} \quad (9)$$

For the mice provided with [U-^{13}C]fructose and unlabeled glucose, the contribution of triose-P to lipogenic acetyl-CoA was estimated using Equation (6). The contribution of GNG precursors to triose-P was estimated from the ratio of triose-P to gluconeogenic triose-P $^{13}C_{IUA}$, with adjustment for total lipogenic acetyl-CoA flux by multiplication by Equation (6) and the loss of glucose-6-P carbon 1 during PPP oxidation.

$$\text{GNG precursors} \rightarrow \text{Triose-P} = 100 \times (\text{Triose-P}\ ^{13}C_{IUA}/\text{GNG-Triose-P}\ ^{13}C_{IUA}) \times \text{Equation (6)} + 1/6\ \text{PPP} \quad (10)$$

The dietary glucose contribution to triose-P was calculated as the difference between total triose-P (Equation (6)) and the GNG precursor contribution (Equation (10)) and adjusted for the loss of glucose-6-P carbon 1 during PPP oxidation.

$$\text{Dietary glucose} \rightarrow \text{Triose-P} = \text{Equation (6)} - \text{Equation (10)} + 1/6\ \text{PPP} \quad (11)$$

Finally, the contributions of the 20% [U-^{13}C]glucose supplement and other unlabeled glucose sources to dietary glucose and the contribution of the 20% [U-^{13}C]fructose supplement and other unlabeled gluconeogenic precursors to GNG were calculated as follows:

$$[\text{U-}^{13}\text{C}]\text{Glucose} \rightarrow \text{dietary glucose} = [100 \times \text{dietary glucose}\ ^{13}C_{IUA}\ /20] \times \text{Equation (11)} \quad (12)$$

$$\text{Other glucose sources} \rightarrow \text{dietary glucose} = \text{Equation (11)} - \text{Equation (12)} \quad (13)$$

$$[\text{U-}^{13}\text{C}]\text{Fructose} \rightarrow \text{GNG} = [100 \times \text{GNG precursors}\ ^{13}C_{IUA}/20] \times \text{Equation (10)} \quad (14)$$

$$\text{Other GNG precursors} \rightarrow \text{GNG} = \text{Equation (10)} - \text{Equation (14)} \quad (15)$$

2.5. Estimation of the Fraction of Glucose-6-P Metabolized by the PPP

The fraction of glucose-6-P oxidized by the PPP was estimated from the ^{13}C-isotopomer distributions of glycogen, as previously described [4]. The PPP fraction was normalized to total lipogenic acetyl-CoA flux by multiplication with the product of Equation (6).

2.6. Statistical Analyses

All results are presented as means ± standard deviations. All datasets were submitted to a Shapiro–Wilk normality test and homoscedasticity test (F test of equality of variances). If both groups presented a normal distribution, then an unpaired Student's *t*-test was applied (Welch-corrected if variances were unequal). Otherwise, the Mann–Whitney U-test was employed.

3. Results

3.1. Enrichment of Hepatic Metabolic Pools from [U-^{13}C]Glucose and [U-^{13}C]Fructose

The ^{13}C-isotopomer distributions in the glucose-6-P and triose-P pools were almost all accounted for by $^{13}C_{IUA}$ species (Supplementary Table S1). For the mice provided with [U-^{13}C]glucose, the glucose-6-P pool had the highest $^{13}C_{IUA}$ abundance, with the principal isotopomer being [U-^{13}C]glucose-6-P. From glucose-6-P to glycerol-3-P and acetyl-CoA, there was a stepwise dilution in $^{13}C_{IUA}$ consistent with an inflow of unlabeled triose-P and acetyl-CoA carbons, respectively (Table 1). The enrichment of the gluconeogenic triose-P pool via indirect pathway metabolism or Cori cycling was relatively low, with the principal contribution coming from PPP activity, as seen by the dominance of [1,2-$^{13}C_2$]glucose-6-P over that of [5,6-$^{13}C_2$]glucose-6-P (Supplementary Table S1) [16]. Following its ingestion and subsequent absorption, the [U-^{13}C]glucose supplement was diluted almost four-fold by other unlabeled glucose sources by the time it reached the liver (Table 1).

Table 1. Fractional enrichments (%) of ^{13}C-isotopomers that generate or are associated with lipogenic [U-^{13}C]acetyl-CoA ($^{13}C_{IUA}$) for selected hepatic metabolite pools for a group of four mice provided with ^2H_2O and [U-^{13}C]glucose tracers, and a group of five mice provided with ^2H_2O and [U-^{13}C]fructose. Values are reported as means ± SE. N.D. not determined.

	$^{13}C_{IUA}$ (Equation)				
Experiment	**Hepatic Glucose-6-P (Equation (1))**	**Dietary Glucose (Equation (2))**	**GNG precursors (Equation (3a,b))**	**Triose-P (Equation (4))**	**Acetyl-CoA (Equation (5))**
[U-^{13}C]Glucose and unlabeled fructose (n = 4)	3.78 ± 0.74	5.47 ± 1.25	1.30 ± 0.15	2.50 ± 0.66	1.53 ± 0.43
Unlabeled glucose and [U-^{13}C]fructose (n = 5)	5.06 ± 0.34	N.D.	14.99 ± 1.24	8.99 ± 1.05	4.32 ± 0.50

For mice provided with [U-^{13}C]fructose, the highest $^{13}C_{IUA}$ abundances were found in the GNG precursor and triose phosphate pools with dilution at both glucose-6-P and acetyl-CoA pools (Table 1). This enrichment distribution indicates that, under our experimental conditions, fructose was mostly metabolized to triose-P by the liver, followed by carbon flows into both glycogenic and lipogenic pathways. Had the fructose been fully metabolized to glucose in the intestine prior to reaching the liver [17], this would have resulted in a $^{13}C_{IUA}$ distribution resembling that observed with [U-^{13}C]glucose, i.e., highest for glucose-6-P, then progressive dilution at triose-P and acetyl-CoA pools. Finally, in contrast to [U-^{13}C]glucose, the dietary [U-^{13}C]fructose supplement underwent relatively minor dilution (~1.3-fold) from competing gluconeogenic precursors at its point of entry into the GNG pool.

3.2. Sourcing of Lipogenic Acetyl-CoA Carbons Reported by [U-^{13}C]Glucose and [U-^{13}C]Fructose and PPP Activity

A comparison of the contributions of different sources to lipogenic acetyl-CoA estimated from [U-^{13}C]glucose and [U-^{13}C]fructose tracers is shown in Table 2. Both tracers report a substantial contribution (40–50%) of non-sugar substrates such as acetate to the lipogenic acetyl-CoA pool, even with chronic high-sugar feeding. Under our study conditions, the bulk of triose-P destined for lipogenesis was derived from either dietary glucose or fructose, with only minor contributions from other gluconeogenic precursors. For the four common component fluxes reported by both tracers, the biggest divergence was found for the triose-P and non-triose-P acetyl-CoA sources, while estimates for the contributions of dietary glucose and GNG precursors to the lipogenic triose-P were in better agreement. Figure 3 shows the values of these fluxes obtained by combining and averaging the data derived from the [U-^{13}C]glucose and [U-^{13}C]fructose measurements. This includes the overall PPP flux, which represents the sum of PPP fluxes attributed to glucose-6-P derived from dietary glucose (i.e., direct pathway) and glucose-6-P derived from GNG sources (indirect pathway) reported by [U-^{13}C]glucose and [U-^{13}C]fructose, respectively. Our data indicate that about 11% of glucose-6-P had undergone PPP oxidation. While our previous measurement of fractional PPP utilization of glucose-6-P in these livers showed modest but significant differences between [U-^{13}C]glucose and [U-^{13}C]fructose tracers [4], the significance was lost after the values were normalized to that of lipogenic acetyl-CoA flux (Table 2).

Table 2. Estimates of substrate fluxes contributing to lipogenic acetyl-CoA expressed as a fraction of total lipogenic acetyl-CoA flux into fatty acid synthase from ^{2}H enrichment and ^{13}C-isotopomer analysis of a group of mice provided with $^{2}H_2O$ and [U-^{13}C]glucose tracers (n = 4), and a group provided with $^{2}H_2O$ and [U-^{13}C]fructose (n = 5). The estimated pentose phosphate pathway (PPP) fluxes involved in glucose-6-P oxidation and carbon recycling to regenerate glucose-6-P (Glucose-6-P → PPP → Glucose-6-P) are also shown.

Pathway Component		[U-^{13}C]Glucose	[U-^{13}C]Fructose	*p* Value
Acetyl-CoA → Fatty acids		100	100	N.D.
Non-Triose-P → Acetyl-CoA	Equation (7)	40 ± 4	51 ± 8	0.08
Triose-P → Acetyl-CoA	Equation (6)	60 ± 4	49 ± 8	0.08
Dietary glucose → Triose-P	Equations (8) and (11)	28 ± 4	21 ± 9	0.32
[U-^{13}C]glucose → Dietary glucose	Equation (12)	8 ± 4	N.D.	N.D.
Other dietary glucose sources → Dietary glucose	Equation (13)	19 ± 4	N.D.	N.D.
GNG precursors → Triose-P	Equations (9) and (10)	34 ± 8	29 ± 4	0.38
[U-^{13}C]fructose → GNG	Equation (14)	N.D.	22 ± 6	N.D.
Other precursors → GNG	Equation (15)	N.D.	7 ± 4	N.D.
Glucose-6-P → PPP → Glucose-6-P		7 ± 1	5 ± 1	0.13

Values are reported as means ± SD. N.D. not determined.

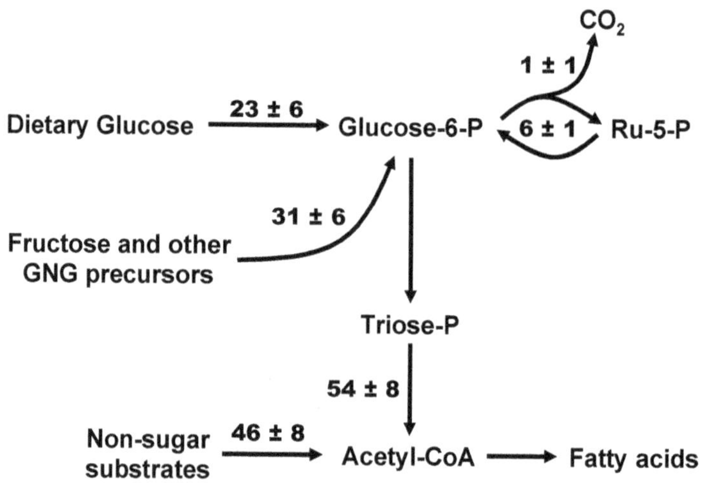

Figure 3. Fractional contributions of sugar and non-sugar sources to lipogenic acetyl-CoA estimated by combining the data of both [U-13C]glucose (n = 4) and [U-13C]fructose (n = 5) analyses. Fractional values were adjusted to that of acetyl-CoA conversion to fatty acids (arbitrarily set to 100) and the standard deviations are shown alongside the means.

4. Discussion

4.1. General Overview

We developed a method for quantifying the major fluxes associated with hepatic sugar metabolism that can be easily applied to mice and other small animal models. We demonstrated that this approach can utilize ^{13}C-isotopomer information from either [U-^{13}C]glucose or [U-^{13}C]fructose. In principle, it could also function with other ^{13}C-sugar tracers that have been used as probes of hepatic carbohydrate metabolism such as galactose [18,19] or glycerol [16,20,21]. Alongside the $^{2}H_2O$ tracer, these can be formulated into the animal's food or drinking water, allowing hepatic metabolic activity to be measured in unperturbed ad libitum feeding conditions. Although dietary glucose is metabolized by most, if not all, tissues, we can nevertheless identify that which is metabolized first-pass by the liver as intact [U-^{13}C]glucose. Paradoxically, although fructose metabolism is more strongly associated with the liver compared to glucose, our metabolic analysis does not provide direct information on hepatic [U-^{13}C]fructose prior to it being metabolized to sugar phosphates. This means that, unlike the first-pass hepatic metabolism of [U-^{13}C]glucose, we cannot be certain that the observed labeling of hepatic glucose-6-P and triose-P from [U-^{13}C]fructose was entirely the result of hepatic [U-^{13}C]fructose metabolism.

4.2. Hepatic Versus Extrahepatic Fructose Metabolism

The liver was long believed to be the principal site for fructose metabolism, but this has been recently challenged with evidence of other tissues, notably the intestine, with the capacity of enterocytes for fructose phosphorylation and incorporation into glycolytic and gluconeogenic fluxes [17]. Moreover, and perhaps not surprisingly, any fructose that is not immediately absorbed can also be avidly metabolized by the intestinal microbiome [22,23], with products such as acetate being subsequently absorbed and recruited as lipogenic substrates by the liver [23]. As proposed by Jang et al., [17], the extent of intestinal versus hepatic fructose metabolism may be related to the total amount of sugar ingested, with low intakes being accommodated entirely by the intestine, and the liver metabolizing any surplus above and beyond the intestinal capacity for fructose disposal. Our mice were kept for 18 weeks on standard chow that was accompanied by drinking water containing 30 g/100 mL of a 55/45 fructose/glucose mixture. There was no other source of drinking water provided. Assuming a daily water intake of ~7 mL water per mouse [24], this would require ingestion of ~10 mL of the mixture, resulting in about 2.5 g of ingested sugar (1.38 g fructose and 1.12 g glucose). Given the average mouse mass of 35 grams, this translates to 39 g of fructose and 32 g of glucose per kg body mass over 24 h, or an average of ~1.6 g kg^{-1} fructose and ~1.3 g kg^{-1} of glucose per hour. If we compare these quantities to the criteria of low and high-dose sugar intake established by Jang et al. based on single gavages of 0.5 g kg^{-1} and 2 g kg^{-1} of a 1:1 fructose/glucose mixture, respectively [17], then our mice had a sugar intake that was well beyond the high dose defined by Jang et al. Under our study conditions, much, if not most, of the fructose would be expected to be metabolized by the liver, which is consistent with our observed hepatic metabolite ^{13}C enrichment patterns from [U-^{13}C]fructose.

4.3. PPP Flux in Relation to De Novo Lipogenesis

The fraction of glucose-6-P that was oxidized by the PPP was estimated to be 11%. The incorporation of n equivalents of acetyl-CoA into the fatty acid polymer requires 2n-2 equivalents of NADPH; hence, the synthesis of palmitate from 8 acetyl-CoA consumes a total of 14 NADPH. Since two NADPH are generated for each glucose-6-P carbon oxidized to CO_2 via the PPP, a total of 1.17 glucose-6-P equivalents are required to generate the necessary number of NADPH for the synthesis of each palmitate as follows:

$$4 \text{ Glucose-6-P} \rightarrow 8 \text{ Acetyl-CoA} \rightarrow 1 \text{ Palmitate}$$

$$1.17 \text{ Glucose-6-P} \rightarrow 14 \text{ NADPH} \rightarrow 1 \text{ Palmitate}$$

Therefore, if glucose-6-P is the sole contributor of lipogenic acetyl-CoA and if the PPP is the sole source of NADPH, then the fraction of glucose-6-P that is utilized by the PPP relative to the total used for lipogenesis (i.e., PPP oxidation plus acetyl-CoA generation) is $1.17/(4 + 1.17) = 23\%$ (this relationship also approximates for C18 fatty acids: 22.9% versus 22.6% for C16). In adipose tissues, glucose-6-P is considered to be the main precursor of acetyl-CoA [25], with the PPP considered to be the principal source of NADPH [26]. An in situ measurement of PPP flux in human adipose tissue via a microdialysis method yielded a PPP fraction of 17–22%, approaching the theoretical value for quantitative glucose-6-P conversion to fatty acids [27]. In the liver, lipogenic acetyl-CoA is derived from sources other than glucose-6-P, notably acetate. Therefore, under these conditions, if the PPP was the sole source of NADPH, then a higher fractional PPP flux per equivalent of glucose-6-P converted to acetyl-CoA would be required. For example, if acetate and glucose-6-P each contribute 50% of acetyl-CoA for palmitate synthesis as follows:

$$4 \text{ Acetate} \rightarrow 4 \text{ Acetyl-CoA}$$

$$2 \text{ Glucose-6-P} \rightarrow 4 \text{ Acetyl-CoA}$$

$$1.17 \text{ Glucose-6-P} \rightarrow 14 \text{ NADPH}$$

then, to provide the theoretical amount of NADPH, the fraction of glucose-6-P that undergoes PPP oxidation would need to increase to $1.17/(2 + 1.17) = 37\%$. Our data indicate that glucose-6-P accounted at most for about half of lipogenic acetyl-CoA, but only 11% was oxidized by the PPP. This suggests that the PPP accounted, at the most, for only about 11/37, or about 30%, of the total NADPH demand for DNL under these conditions (If NADPH derived from PPP oxidation was also consumed by other processes, such as the reduction of oxidized glutathione, then its fractional contribution to DNL would be even less than 30%). Other possible sources of cytosolic NADPH include cytosolic NADP-malic enzyme 1 and NADP-isocitrate dehydrogenase 1 [2] and folate-mediated serine catabolism [28].

4.4. Limitations of the Approach

There are several important limitations of our approach that must be taken into account when interpreting the results. As previously discussed, our mouse model involved a very high intake of sugar that ensured that the fructose component was predominantly metabolized by the liver. If the amount of sugar was reduced, then it is likely that a much higher proportion of the [U-^{13}C]fructose would be metabolized by the intestine to form ^{13}C-isotopomers of glucose, lactate, and other metabolites [17], and these would be the principal products seen by the liver rather than [U-^{13}C]fructose. Nevertheless, aside from the uncertainty in determining the contribution of fructose to the hepatic gluconeogenic triose-P pool, the ^{13}C-isotopomer distributions of glycogen and triglycerides would still provide valid information on PPP fluxes, glyceroneogenesis, and the contribution of glucose-6-P and non-glucose-6-P sources to DNL. Under high sugar intake conditions, Jang et al. reported a substantial amount of fructose metabolism by the intestinal microbiota [17], with acetate being a principal product [23]. The microbial fermentation of [U-^{13}C]fructose results in the formation of [U-^{13}C]acetate, whose incorporation into DNL is indistinguishable from that of [U-^{13}C]acetyl-CoA derived from hepatic [U-^{13}C]fructose metabolism. To the extent that the fermentative metabolism of [U-^{13}C]fructose contributes to the fatty acid ^{13}C-isotopomer enrichment, then the fraction of acetyl-CoA derived from non-glucose-6-P sources would be expected to be underestimated, and, accordingly, the contribution of glucose-6-P to DNL overestimated. However, when these parameters obtained from [U-^{13}C]fructose are compared with those derived from [U-^{13}C]glucose (Table 2), they show a strong tendency to report higher non-glucose-6-P and lower glucose-6-P fractions. One possibility is that, given the very high sugar intake, there may have also been extensive microbial metabolism of [U-^{13}C]glucose. Glucose is normally efficiently absorbed in the small intestine, but small intestinal bacterial overgrowth [29,30], possibly induced by high sugar diets [31], can result in a portion of the glucose being fermented instead. Finally, the PPP flux is based on the sugar phosphates that are recycled back to fructose-6-P and glucose-6-P and does not take into account those pentose-P equivalents that were recruited for nucleotide biosynthesis. Thus, the PPP estimate represents a lower limit of the real oxidative glucose-6-P flux.

4.5. Conclusions

Hepatic metabolism and assimilation of dietary sugar involves the co-ordination of gluconeogenic, glycogenic, PPP, glycolytic, and lipogenic fluxes. While there are longstanding methodologies for measuring these fluxes individually, until now there has been no approach for quantifying fluxes through the entire ensemble. We demonstrate that, with a combination of ^2H_2O and a [U-^{13}C]hexose sugar that can be either glucose or fructose, these fluxes can be quantified in mice under natural feeding conditions by analysis of liver glycogen and triglyceride ^{13}C-isotopomers. In addition to confirming a previous study that a substantial fraction of lipogenic acetyl-CoA is derived from sources other than glucose-6-P, even during high sugar feeding [5], our analysis also reveals that the PPP was not the main supplier of NADPH for DNL, at least under our study conditions. Such information could be valuable in improving our understanding of hepatic sugar metabolism under different physiological and pathophysiological states.

Supplementary Materials: The following supporting information can be downloaded at: https://www.mdpi.com/article/10.3390/metabo12111142/s1, Figure S1: ^{13}C-Isotopomers of selected metabolic intermediates generated from [U-^{13}C]fructose metabolism into lipogenic and glycogenic pathways. These include hepatic glucose-6-P—inferred from the analysis of newly-synthesized glycogen; triose-P recruited for gluconeogenesis (GNG-triose-P)—inferred from the analysis of indirect pathway glycogen ^{13}C-isotopomers; triose-P supplying glycerol-3-P for fatty acid esterification and acetyl-CoA units for de novo lipogenesis—inferred from the ^{13}C-isotopomer analysis of newly-synthesized triglyceride glycerol, and the acetyl-CoA pool supplying lipogenesis—inferred from the ^{13}C-isotopomer analysis of newly-synthesized fatty acids. For the metabolite carbon skeletons, the red filled and unfilled circles represent ^{13}C and ^{12}C, respectively. The shading highlights those isotopomers that form [U-^{13}C]acetyl CoA and the colors indicate isotopic equivalence (same color) or non-equivalence (different colors). For simplicity, in depicting the fatty acid labeling, only the ^{13}C-isotopomers of the last two fatty acid carbons (representing the first acetyl-CoA moiety to be incorporated into DNL) are shown.; Table S1: Liver glycogen ^{13}C-isotopomer enrichments from mice provided with [U-^{13}C]glucose (n = 4) and [U-^{13}C]fructose (n = 5). The glycogen ^{13}C-isotopomers shown in bold text are metabolized to [U-^{13}C]acetyl-CoA. Table S2: Newly synthesized glycogen fraction ($f_{glycogen}$) with direct and indirect pathway contributions to the newly synthesized glycogen (f_{direct} and $f_{indirect}$), and newly synthesized triglyceride glyceryl and fatty acid fractions ($f_{glyceryl}$ and $f_{fatty\ acid}$) from ^{2}H-enrichment data of liver glycogen and triglyceride, respectively.

Author Contributions: J.G.J., C.B., L.C.T. and G.D.B. designed the experiments. I.V., G.D.N., G.D.B., J.G.S., L.P. and L.C.T. conducted the experiments. J.G.J. and I.V. provided facilities to perform the experiments and provided material and instrumentation to perform the experiments and analyze samples. Results were discussed and analyzed by J.G.J., I.V., G.D.N., G.D.B., J.G.S., L.P., C.B., A.N.T. and L.C.T. The manuscript was written by J.G.J. and I.V. All authors reviewed the manuscript. All authors have read and agreed to the published version of the manuscript.

Funding: The authors acknowledge financial support from the Portuguese Foundation for Science and Technology (research grant FCT-FEDER-02/SAICT/2017/028147). Structural funding for the Center for Neurosciences and Cell Biology and the UC-NMR facility is supported in part by FEDER—European Regional Development Fund through the COMPETE Programme, Centro 2020 Regional Operational Programme, and the Portuguese Foundation for Science and Technology through grants UIDB/04539/2020; UIDP/04539/2020, LA/P/0058/2020, POCI-01-0145-FEDER-007440; REEQ/481/QUI/2006, RECI/QEQ-QFI/0168/2012, CENTRO-07-CT62-FEDER-002012, and Rede Nacional de Ressonancia Magnética Nuclear. The National Mass Spectrometry Network (RNEM) provided funding under the contract POCI-01-0145-FEDER-402-022125 (ref.: ROTEIRO/0028/2013). GDB was supported by the European Union's Horizon 2020 Research and Innovation programme under the Marie Skłodowska-Curie Grant Agreement No. 722619 (Project FOIE GRAS).

Institutional Review Board Statement: The study was conducted in accordance with the University of Coimbra Ethics Committee on Animal Studies (ORBEA) and the Portuguese National Authority for Animal Health (DGAV), approval code 0421/000/000/2013.

Informed Consent Statement: Not applicable.

Data Availability Statement: The datasets generated during and/or analyzed during the current study are available from the corresponding author on reasonable request. The data are not publicly available due to privacy.

Conflicts of Interest: The authors declare no competing interest.

References

1. Sclafani, A. Gut-brain nutrient signaling. Appetition vs. satiation. *Appetite* **2013**, *71*, 454–458. [CrossRef] [PubMed]
2. Xiao, W.S.; Wang, R.S.; Handy, D.E.; Loscalzo, J. NAD(H) and NADP(H) Redox Couples and Cellular Energy Metabolism. *Antioxid. Redox Signal.* **2018**, *28*, 251–272. [CrossRef]
3. DiNunzio, G.; Belew, G.D.; Torres, A.N.; Silva, J.G.; Silva, L.P.; Barosa, C.; Tavares, L.; Jones, J.G. Determining the contribution of a high-fructose corn syrup formulation to hepatic glycogen synthesis during ad-libitum feeding in mice. *Sci. Rep.* **2020**, *10*, 12852. [CrossRef] [PubMed]
4. Belew, G.D.; Di Nunzio, G.; Tavares, L.; Silva, J.G.; Torres, A.N.; Jones, J.G. Estimating pentose phosphate pathway activity from the analysis of hepatic glycogen C-13-isotopomers derived from U-C-13 fructose and U-C-13 glucose. *Magn. Reson. Med.* **2020**, *84*, 2765–2771. [CrossRef] [PubMed]
5. Silva, J.C.P.; Marques, C.; Martins, F.O.; Viegas, I.; Tavares, L.; Macedo, M.P.; Jones, J.G. Determining contributions of exogenous glucose and fructose to de novo fatty acid and glycerol synthesis in liver and adipose tissue. *Metab. Eng.* **2019**, *56*, 69–76. [CrossRef] [PubMed]
6. Jensen, T.; Abdelmalek, M.F.; Sullivan, S.; Nadeau, K.J.; Green, M.; Roncal, C.; Nakagawa, T.; Kuwabara, M.; Sato, Y.; Kang, D.-H.; et al. Fructose and sugar: A major mediator of non-alcoholic fatty liver disease. *J. Hepatol.* **2018**, *68*, 1063–1075. [CrossRef] [PubMed]
7. Chiu, S.; Mulligan, K.; Schwarz, J.-M. Dietary carbohydrates and fatty liver disease: De novo lipogenesis. *Curr. Opin. Clin. Nutr. Metab. Care* **2018**, *21*, 277–282. [CrossRef]
8. Hodson, L.; Gunn, P.J. The regulation of hepatic fatty acid synthesis and partitioning: The effect of nutritional state. *Nat. Rev. Endocrinol.* **2019**, *15*, 689–700. [CrossRef] [PubMed]

9. Jarak, I.; Barosa, C.; Martins, F.O.; Silva, J.C.P.; Santos, C.; Belew, G.D.; Rito, J.; Viegas, I.; Teixeira, J.; Oliveira, P.J.; et al. Sources of hepatic glycogen synthesis in mice fed with glucose or fructose as the sole dietary carbohydrate. *Magn. Reson. Med.* **2019**, *81*, 639–644. [CrossRef] [PubMed]
10. Soares, A.F.; Carvalho, R.A.; Veiga, F.J.; Alves, M.G.; Martins, F.O.; Viegas, I.; Gonzalez, J.D.; Meton, I.; Baanante, I.V.; Jones, J.G. Restoration of direct pathway glycogen synthesis flux in the STZ-diabetes rat model by insulin administration. *Am. J. Physiol.-Endocrinol. Metab.* **2012**, *303*, E875–E885. [CrossRef]
11. Perdigoto, R.; Rodrigues, T.B.; Furtado, A.L.; Porto, A.; Geraldes, C.; Jones, J.G. Integration of U-C-13 glucose and (H_2O)-H-2 for quantification of hepatic glucose production and gluconeogenesis. *Nmr Biomed.* **2003**, *16*, 189–198. [CrossRef] [PubMed]
12. Newgard, C.B.; Hirsch, L.J.; Foster, D.W.; McGarry, J.D. Studies on the Mechanism by Which Exogenous Glucose Is Converted into Liver-Glycogen in the Rat—A Direct or an Indirect Pathway. *J. Biol. Chem.* **1983**, *258*, 8046–8052. [CrossRef]
13. Belew, G.D.; Silva, J.; Rito, J.; Tavares, L.; Viegas, I.; Teixeira, J.; Oliveira, P.J.; Macedo, M.P.; Jones, J.G. Transfer of glucose hydrogens via acetyl-CoA, malonyl-CoA, and NADPH to fatty acids during de novo lipogenesis S. *J. Lipid Res.* **2019**, *60*, 2050–2056. [CrossRef] [PubMed]
14. Jones, J.G.; Barosa, C.; Gomes, F.; Mendes, A.C.; Delgado, T.C.; Diogo, L.; Garcia, P.; Bastos, M.; Barros, L.; Fagulha, A.; et al. NMR derivatives for quantification of H-2 and C-13-enrichment of human glucuronide from metabolic tracers. *J. Carbohydr. Chem.* **2006**, *25*, 203–217. [CrossRef]
15. Jones, J.G.; Merritt, M.; Malloy, C. Quantifying tracer levels of 2H2O enrichment from microliter amounts of plasma and urine by 2H NMR. *Magn. Res. Med.* **2001**, *45*, 156–158. [CrossRef]
16. Jin, E.S.; Sherry, A.D.; Malloy, C.R. Interaction between the Pentose Phosphate Pathway and Gluconeogenesis from Glycerol in the Liver. *J. Biol. Chem.* **2014**, *289*, 32593–32603. [CrossRef] [PubMed]
17. Jang, C.; Hui, S.; Lu, W.; Cowan, A.J.; Morscher, R.J.; Lee, G.; Liu, W.; Tesz, G.J.; Birnbaum, M.J.; Rabinowitz, J.D. The Small Intestine Converts Dietary Fructose into Glucose and Organic Acids. *Cell Metab.* **2018**, *27*, 351–361. [CrossRef] [PubMed]
18. Soares, A.F.; Carvalho, R.A.; Veiga, F.J.; Jones, J.G. Effects of galactose on direct and indirect pathway estimates of hepatic glycogen synthesis. *Metab. Eng.* **2010**, *12*, 552–560. [CrossRef] [PubMed]
19. Rother, K.I.; Schwenk, W.F. Glucose-Production in Glycogen-Storage-Disease-I Is Not Associated with Increased Cycling through Hepatic Glycogen. *Am. J. Physiol.-Endocrinol. Metab.* **1995**, *32*, E774–E778. [CrossRef] [PubMed]
20. Jin, E.S.; Sherry, A.D.; Malloy, C.R. An Oral Load of C-13(3) Glycerol and Blood NMR Analysis Detect Fatty Acid Esterification, Pentose Phosphate Pathway, and Glycerol Metabolism through the Tricarboxylic Acid Cycle in Human Liver. *J. Biol. Chem.* **2016**, *291*, 19031–19041. [CrossRef]
21. Jones, J.G.; Garcia, P.; Barosa, C.; Delgado, T.C.; Diogo, L. Hepatic anaplerotic outflow fluxes are redirected from gluconeogenesis to lactate synthesis in patients with Type 1a glycogen storage disease. *Metab. Eng.* **2009**, *11*, 155–162. [CrossRef] [PubMed]
22. Zhang, X.F.; Grosfeld, A.; Williams, E.; Vasiliauskas, D.; Barretto, S.; Smith, L.; Mariadassou, M.; Philippe, C.; Devime, F.; Melchior, C.; et al. Fructose malabsorption induces cholecystokinin expression in the ileum and cecum by changing microbiota composition and metabolism. *Faseb J.* **2019**, *33*, 7126–7142. [CrossRef] [PubMed]
23. Zhao, S.; Jang, C.; Liu, J.; Uehara, K.; Gilbert, M.; Izzo, L.; Zeng, X.; Trefely, S.; Fernandez, S.; Carrer, A.; et al. Dietary fructose feeds hepatic lipogenesis via microbiota-derived acetate. *Nature* **2020**, *579*, 586–591. [CrossRef] [PubMed]
24. Bachmanov, A.A.; Reed, D.R.; Beauchamp, G.D.; Tordoff, M.G. Food intake, water intake, and drinking spout side preference of 28 mouse strains. *Behav. Genet.* **2002**, *32*, 435–443. [CrossRef] [PubMed]
25. Yoo, H.; Stephanopoulos, G.; Kelleher, J.K. Quantifying carbon sources for de novo lipogenesis in wild-type and IRS-1 knockout brown adipocytes. *J. Lipid Res.* **2004**, *45*, 1324–1332. [CrossRef] [PubMed]
26. Liu, L.; Shah, S.; Fan, J.; Park, J.O.; Wellen, K.E.; Rabinowitz, J.D. Malic enzyme tracers reveal hypoxia-induced switch in adipocyte NADPH pathway usage. *Nat. Chem. Biol.* **2016**, *12*, 345–352. [CrossRef]
27. Minehira, K.; Bettschart, V.; Vidal, H.; Vega, N.; Di Vetta, V.; Rey, V.; Schneiter, P.; Tappy, L. Effect of carbohydrate overfeeding on whole body and adipose tissue metabolism in humans. *Obes. Res.* **2003**, *11*, 1096–1103. [CrossRef]
28. Zhang, Z.; TeSlaa, T.; Xu, X.; Zeng, X.; Yang, L.; Xing, G.; Tesz, G.J.; Clasquin, M.F.; Rabinowitz, J.D. Serine catabolism generates liver NADPH and supports hepatic lipogenesis. *Nat. Metab.* **2021**, *3*, 1608–1620. [CrossRef]
29. Wu, W.C.; Zhao, W.; Li, S. Small intestinal bacteria overgrowth decreases small intestinal motility in the NASH rats. *World J. Gastroenterol.* **2008**, *14*, 313–317. [CrossRef]
30. Batt, R.M.; Hall, E.J.; McLean, L.; Simpson, K.W. Small intestinal bacterial overgrowth and enhanced intestinal permeability in healthy beagles. *Am. J. Vet. Res.* **1992**, *53*, 1935–1940.
31. O'Brien, P.; Han, G.; Ganpathy, P.; Pitre, S.; Zhang, Y.; Ryan, J.; Sim, P.Y.; Harding, S.V.; Gray, R.; Preedy, V.R.; et al. Chronic Effects of a High Sucrose Diet on Murine Gastrointestinal Nutrient Sensor Gene and Protein Expression Levels and Lipid Metabolism. *Int. J. Mol. Sci.* **2021**, *22*, 137. [CrossRef] [PubMed]

MDPI
St. Alban-Anlage 66
4052 Basel
Switzerland
Tel. +41 61 683 77 34
Fax +41 61 302 89 18
www.mdpi.com

Metabolites Editorial Office
E-mail: metabolites@mdpi.com
www.mdpi.com/journal/metabolites

www.ingramcontent.com/pod-product-compliance
Lightning Source LLC
Chambersburg PA
CBHW061600210926
51850CB00009B/2165
* 9 7 8 3 0 3 6 5 7 4 2 3 3 *